MADE MODERN

MADE MODERN
Science and Technology in Canadian History

Edited by Edward Jones-Imhotep
and Tina Adcock

UBCPress·Vancouver·Toronto

© UBC Press 2018

All rights reserved. No part of this publication may be reproduced, stored in a retrieval system, or transmitted, in any form or by any means, without prior written permission of the publisher, or, in Canada, in the case of photocopying or other reprographic copying, a licence from Access Copyright, www.accesscopyright.ca.

27 26 25 24 23 22 21 20 19 18   5 4 3 2 1

Printed in Canada on FSC-certified ancient-forest-free paper (100% post-consumer recycled) that is processed chlorine- and acid-free.

Library and Archives Canada Cataloguing in Publication

Made modern : science and technology in Canadian history / edited by Edward Jones-Imhotep and Tina Adcock.

Includes bibliographical references and index.
Issued in print and electronic formats.
ISBN 978-0-7748-3723-1 (hardcover). – ISBN 978-0-7748-3724-8 (pbk). –
ISBN 978-0-7748-3725-5 (PDF). – ISBN 978-0-7748-3726-2 (EPUB). –
ISBN 978-0-7748-3727-9 (Kindle)

1. Science – Canada – History. 2. Science – Social aspects – Canada – History. 3. Technology – Canada – History. 4. Technology – Social aspects – Canada – History. 5. Science and civilization. 6. Technology and civilization. 7. Technological innovations – Canada – History. I. Jones-Imhotep, Edward, editor II. Adcock, Tina, editor

Q127.C2M33 2018   509.71   C2018-904321-0
                                          C2018-904322-9

Canadä

UBC Press gratefully acknowledges the financial support for our publishing program of the Government of Canada (through the Canada Book Fund), the Canada Council for the Arts, and the British Columbia Arts Council.

This book has been published with the help of grants from the Canadian Federation for the Humanities and Social Sciences, through the Awards to Scholarly Publications Program, using funds provided by the Social Sciences and Humanities Research Council of Canada, and from Simon Fraser University's Publications Fund.

Printed and bound in Canada by Friesens
Set in Futura and Warnock by Artegraphica Design Co. Ltd.
Copy editor: Dallas Harrison
Indexer: Sergey Lobachev

UBC Press
The University of British Columbia
2029 West Mall
Vancouver, BC V6T 1Z2
www.ubcpress.ca

For Rich Jarrell

Contents

List of Figures and Tables / ix

Acknowledgments / xi

Introduction: Science, Technology, and the Modern in Canada / 3
EDWARD JONES-IMHOTEP and TINA ADCOCK

Part 1: Bodies

1 Civilizing the Natives: Richard King and His Ethnographic Writings on Indigenous Northerners / 39
EFRAM SERA-SHRIAR

2 Scientist Tourist Sportsman Spy: Boundary-Work and the Putnam Eastern Arctic Expeditions / 60
TINA ADCOCK

3 Nature's Tonic: Electric Medicine in Urban Canada, 1880–1920 / 84
DOROTEA GUCCIARDO

4 Cosmic Moderns: Re-Enchanting the Body in Canada's Atomic Age, 1931–51 / 104
BETH A. ROBERTSON

Part 2: Technologies

5 The Second Industrial Revolution in Canadian History / 125
JAMES HULL

6 Mysteries of the New Phone Explained: Introducing Dial Telephones and Automatic Service to Bell Canada Subscribers in the 1920s / 143
JAN HADLAW

7 Small Science: Trained Acquaintance and the One-Man Research Team / 166
DAVID THEODORE

8 Paris–Montreal–Babylon: The Modernist Genealogies of Gerald Bull / 185
EDWARD JONES-IMHOTEP

9 Percy Schmeiser, Roundup Ready® Canola, and Canadian Agricultural Modernity / 216
EDA KRANAKIS

Part 3: Environments

10 Landscapes of Science in Canada: Modernity and Disruption / 251
STEPHEN BOCKING

11 "For Canada and for Science": Transnational Modernity and the *Report of the Canadian Arctic Expedition 1913–1918* / 279
ANDREW STUHL

12 North Stars and Sun Destinations: Time, Space, and Nation at Trans Canada Air Lines/Air Canada, 1947–70 / 305
BLAIR STEIN

13 Negotiating High Modernism: The St. Lawrence Seaway and Power Project / 326
DANIEL MACFARLANE

Epilogue: Canadian Modernity as an Icon of the Anthropocene / 348
DOLLY JØRGENSEN

Contributors / 358

Index / 362

Figures and Tables

FIGURES

2.1 Putnam Baffin Island Expedition commercial tie-in: Daisy air rifle advertisement, 1928 / 73

6.1 Early Bell advertising portraying the telephone as dependable employee and helpful servant, 1911 / 149

6.2 Cartoon making fun of telephone subscribers, 1923 / 150

6.3 Bell instructional blotter explaining how to use a dial telephone, 1929 / 156

6.4 Bell dial telephone window display in the G.F. Glassco Fur Store, Hamilton, 1928 / 157

6.5 How-to-dial demonstration in the Bell office in Montreal, 1925 / 159

6.6 A Bell employee demonstrating the operation of the new dial system, 1928 / 160

7.1 Digital Equipment Corporation PDP-12 minicomputer at the Montreal Neurological Institute, c. 1970 / 167

7.2 Display on a Tektronix 4002 terminal during an operation at the Montreal Neurological Institute / 168

7.3 A contemporary view of the 1934 Montreal Neurological Institute / 175

7.4 Polaroid image of computer display extracting features from electroencephalography / 176

8.1 Dorothy Fisk's portrayal of the upper atmosphere, 1934 / 190

8.2 Flight envelopes for Bull's gun-launched Martlet "missiles" / 193

8.3 Atlantic Missile Range / 195

8.4 Layout of inter-island and down range instrumentation and communications via radio link / 197

8.5 Map of Barbados showing nominal geographic location of major installations / 198

8.6 Bull's vision of launching satellites into orbit from the surface of the Earth / 202

8.7 Depiction of Bull's supergun, 1964 / 204

8.8 CIA reconstruction of Bull's supergun for Project Babylon / 205

11.1 The title page of the *Report of the Canadian Arctic Expedition 1913–1918*, volume 8, part H / 298

12.1 1949 Trans-Canada Air Lines advertisement designed to foster "all-weather" confidence in air travel / 316

12.2 1951 Trans-Canada Air Lines advertisement / 316

12.3 1960s Air Canada advertisement touting the appeal of a "sun destination" winter / 319

13.1 Map of the St. Lawrence Seaway / 330

13.2 Map of Lake St. Lawrence and Lost Villages / 330

13.3 Ontario Hydro schematic outline of Lake St. Lawrence and rehabilitated area, including Lost Villages / 332–33

13.4 Ontario Hydro blueprint of plans for New Town No. 1 (Ingleside) / 334

13.5 Ontario Hydro model in warehouse at Islington, Ontario / 338

13.6 Moses-Saunders power dam under construction / 339

14.1 Canadian $5 bill issued in 1935 / 350

TABLES

11.1 Geographical distribution of scientists recruited to analyze Canadian Arctic Expedition collections / 284

11.2 Institutional affiliation of scientists hired to analyze Canadian Arctic Expedition collections / 284

Acknowledgments

This volume had its origins in the conference Science, Technology, and the Modern in Canada, held in April 2015 at York University. It was organized in memory of Richard Jarrell, who passed away suddenly in December 2013 at the height of a forty-three-year career at York. Rich helped establish the field of history of Canadian science and technology, cofounding its premier scholarly organization, helping to inaugurate its annual conference, and founding its journal. He was also a dear friend and colleague. With great pleasure, we dedicate this volume to him.

For their generous financial support of this conference, we would like to thank a number of individuals and bodies at York: the Institute for Science and Technology Studies (particularly Bernie Lightman, Natasha Myers, and Bernie Michael Anderson), the Faculty of Science, the Lassonde School of Engineering, the Office of the Vice-President of Research and Innovation, and the Robarts Centre for Canadian Studies. We would also like to thank the Social Sciences and Humanities Research Council Situating Science Research Cluster. For their invaluable help in organizing and running the conference, we thank York University graduate students (and now newly minted PhDs) Jordan Bimm and Yana Boeva.

We are grateful to the estate of Gerald Bull for allowing us to reproduce images in Chapter 8. Frederic Witt and Stefanie Valjeur at Tamm Media helped track down and provide image permissions. At UBC Press, Darcy Cullen, our acquisitions editor, guided the book through its early phases, and Katrina Petrik ably saw it through its final production. It has been a pleasure working with them.

An edited volume is a complex undertaking. We have been truly fortunate in working with a group of superb scholars who have also been conscientious collaborators, meeting deadlines and following guidelines with diligence and good humour. Beyond the volume, we would like to thank not only our families and friends, but also our colleagues who took an early interest in the project and who provided support while it developed: Katey Anderson, Tina Choi, Yves Gingras, Chris Green, Nicolas Kenny, Sean Kheraj, Bernie Lightman, Eric Mills, David Pantalony, Mike Pettit, and Asif Siddiqi.

Finally, the editors would like to pay tribute to each other. Tina would like to thank Edward for inviting her to coedit this volume; she could not have asked for a better partner. His brilliance and warmth have enriched our collaboration and this project from beginning to end. Edward, in turn, would like to thank Tina. It has been a true delight to talk, think, and write together with her since the start of this project. Every element of the book is better because of her intelligence, scholarship, and dedication.

MADE MODERN

Introduction
Science, Technology, and the Modern in Canada

EDWARD JONES-IMHOTEP and TINA ADCOCK

For much of the twentieth century, scholars cast science and technology as great engines of modernity. In a world where revolutions provided the machinery of historical development and Western Europe was its focus, systematic knowledge and radical invention were prime movers.[1] The revolutions of science and industry, that story held, contributed as much to the origins of the modern world as did the political upheavals that shook the eighteenth and nineteenth centuries. Originating in Western Europe, those developments spread over the globe through conquest and commerce. Contemporaries expressed the experiences of those changes in the powerful language of geology and celestial physics – abysses, earthquakes, eruptions, crushing gravitational forces – reaching for a vocabulary to capture magnitudes beyond the scale of the human.[2] Their historians cast knowledge and machines in that disembodied role. With their rationalism and empiricism, their novelty and utility, and their intellectual and physical mastery over the natural world, science and technology provided a model of profound rupture with the past, helping to usher in the rationalized, industrial, democratic, secular societies of the modern age. They furnished "the real origin both of the modern world and [of] the modern mentality" as well as the fountains of its disenchantment.[3]

We have since become wary of these kinds of sweeping claims. The huge variety of historical transformations over the past four centuries makes the neat clockwork narrative of modernity ring false. Revolutions were never the singular ruptures we were taught. The Scientific Revolution, mirroring Voltaire's quip about the Holy Roman Empire, was neither singular nor "scientific" nor a "revolution."[4] The transformations in production that go under the label of the Industrial Revolution coexisted alongside traditional and adaptive modes of working life and labour.[5] And, rather than strip the world of its charms, modern observers have long attempted to conjure up and preserve enchantment in both natural knowledge and material inventions.[6] The origin story of European modernity has also been challenged by scholars aiming for a more global history of our times.[7] Yet, rather than undermine the relationship between science, technology, and the modern, those studies have made that relationship stronger, more vibrant, and more complex. Knowledge and machines were as much products of their societies as they were progenitors of them. Solutions to the problems of science and technology – how knowledge should be produced, how machines should be designed and built – were simultaneously solutions to problems of social order.[8] Far from being external and autonomous forces, scientific knowledge and technical artifacts formed the very fabric of modernity. They became implicated not only in factory machinery, economic empires, and industrial landscapes, but also in the identities, anxieties, and understandings of people living through those revolutionary yet all too recognizable times.

This volume contributes to this rich and varied scholarship. It explores the utility of thinking about the history of Canadian science and technology through the lens of the "modern." Rather than telling a story of diffusion from European origins, it asks how the relations among scientific knowledge, material artifacts, and the modern were made, mobilized, and challenged in a nation at once deeply embedded in European political, social, and cultural norms and profoundly shaped by colonial anxieties. The chapters tease out the ambiguities, contradictions, and instabilities in Canadian scientific and technical activities throughout the nineteenth and twentieth centuries to engage diverse iterations of the modern. They unsettle historical and contemporary assumptions about the meanings and experiences of modernity, and they make important interventions in national and international historiographical debates. Most of all, they seek to explore how science and technology have formed the sites for Canadians to imagine, renounce, and reshape themselves as modern.

As a category of historical analysis, the "modern" in its various forms – modernity, modernism, modernization – has come under intense criticism in recent decades.[9] As a term of art, the root "modern" traces its improbable origins back to late fifth-century Latin, where it denoted something "of the present time." By the seventeenth-century battle between the Ancients and the Moderns, it was already linked to the idea of linear time, tying it to a view of the past as both irrevocably lost and somehow flawed, and to ideological programs that saw the present as the first step toward a more perfect future.[10] For nineteenth-century observers such as Gustave Flaubert, it meant newness and the fleeting present, and it was expressed in a series of "modernisms" – literary, artistic, scientific, economic, political – that laid out what it meant to be "of the age." Of all its protean forms, though, the most mist-shrouded and contentious has been its status as a historical condition – "modernity."[11] For scholars who assert its analytical force, modernity signals shifts so rapid that they represent a step-change in human social organization: the rise of the centralized nation-state, of the public sphere, of industrial capitalism, of global commerce, of large-scale bureaucracy, of mass urbanization, and so on.[12] It also signals a set of experiences, often in tension with those broader processes, that call for critical examination – the creation of a society of strangers; a sense of constant change, rupture, and upheaval; a valuation of newness; an attendant narrative of loss; an appeal to overarching rational schemes; and, most consistently, a recurrent set of contradictions, paradoxes, and tensions: hope and fear, despair and exhilaration, degeneration and new birth.[13] For its critics, modernity complacently assumes a linear understanding of historical development, a sweeping narrative of progress and change, implicit judgments of Western European superiority, and a sanctioning of the darker histories of science and technology. As such, it not only normalizes the European experience, but also casts the rest of the world as peoples waiting to become modern – consigned, as Dipesh Chakrabarty has observed, to the "waiting room of history."[14]

As proxies for modernity, Western science and technology are deeply implicated in that history and its problems. As Immanuel Wallerstein has noted, those two enterprises stand for the supposed triumph of humankind over the natural world.[15] Their origin stories are tied to the very idea of the modern. The term "science" as denoting a disciplined inquiry into the phenomena and order of nature, for example, had no coherent counterpart in the sixteenth and seventeenth centuries. The very category of science as we understand it was a creation of the modern period, deeply entwined with its ideologies of progress and perfection, but also of doom, foreboding, and danger. Similarly, the idea of technology as an autonomous historical force took shape amid the public spectacle of production, transportation, and innovation in the late nineteenth century and early twentieth century.[16] As such, science and technology (as we understand them) carry the contradictions of the modern age. They represent both the abstract, modern ideals of betterment and the concrete agents of change and destruction – of tradition, of values, of the natural world – that have sparked the fury and frustration of antimodernists, saboteurs, luddites, and environmentalists alike. They have been invoked as instruments of freedom, ones that smoothed the transoceanic and transcontinental passages of people, goods, and ideas through networks of exploration and trade. Yet they were also deployed to classify, enslave, and control non-Western peoples, reifying ideas of "purity" that set modern objects against the hybridity of premodern cultures and that ranked premodern peoples beneath the "enlightened" nations of the world.[17] Those contradictions were inscribed in the very fabric of scientific theories and material technologies that, in turn, furnished the central metaphors of utopian social and political ideologies such as Taylorism, Bolshevism, and fascism; that cast human relations in increasingly abstract and quantitative terms; and that inspired the aesthetic modernism of the Futurists, the Dadaists, jazz, and the Precisionist movement of the early twentieth century.

One response to these criticisms has been to expand what counts as modernity – to explore multiple modernities and multiple ways of being modern.[18] These critical responses have the great merit of turning the modern into a global phenomenon, and therefore into the subject of global history.[19] They recognize the wide array of possible actions, reactions, and responses as people grappled with larger historical processes and struggled to feel part of them. At stake, in this view, is the right to historical representation and agency – the ability of the world's peoples to be modern on their own terms, and therefore to appear as full participants in the history of modernity.[20] Others see these attempts to democratize modernity as representing their own presentist judgments, as obscuring and diluting the concept, robbing it of its analytical force.[21] From this view, the modern is a useful category only if it denotes a singular (though possibly contested) process or condition, with commonalities stretching across continents and centuries, even if different societies experience them differently in different places and times. As Carol Gluck writes, modernity is indispensable for historians precisely because societies could not choose another historical condition for themselves; the similarities of their experiences make truly alternative modernities impossible.[22]

As a simple story of origins and superiority, then, modernity holds little promise. Its great value as a category of historical analysis lies in its enormous appeal to people around the globe – to people who helped to coproduce it, particularly in nations that often saw themselves as marginal or diminished.[23] Those people and nations helped to make modernity a global phenomenon rather than "a virus that spreads from one place to another," in Sanjay Subrahmanyam's words.[24] Ignoring this aspirational modernity and its enormous appeal outside Europe simply reinforces the Eurocentrism that threatened to ignore non-European hopes and contributions in the first place.[25]

This book aims to explore the usefulness of the modern, as both historical condition and historical aspiration, for thinking and writing about science and technology in Canada. It begins from the premise that modernity represents a specific and concrete historical condition that shaped Canadian society and Canadians' experiences of themselves and their place in the world. For people around the globe, modernity provided a powerful attraction as a form of life and a way of being.[26] In Canada, its intersections with science and technology formed sites through which contemporaries channelled and expressed their own conscious modernity, sought to bring their modernist visions into being, or reacted to modern anxieties and excesses. Science and technology also gave rise to occasions when Canadians internalized modern values, presumptions, and attitudes that shaped their choices and actions as much as any conscious criteria.[27] As for so many others, for these people, positioning themselves on the map of modernity mattered crucially for their own identities.[28] By studying the historical uses of the idea of the modern and its links to science and technology, scholars can better understand historical experience in general and Canadian experience in particular. Chapters in this volume use midlevel concepts and specific objects – patents, the occult, exploration, scientific rationality, and infrastructure, to name a few – to illuminate the contours of Canadian engagements with science and technology and to situate them within larger national and transnational developments characteristic of the modern world.[29]

Canada Has Always Been Modern

Born in the late nineteenth century, the settler-colonial nation of Canada has always been modern.[30] It is no surprise, then, that historians of English and French Canada – the majority of whom, like those working in other national schools of history, still tend to use the nation or nation-state as their referent – have been keen to interrogate Canadians' multifaceted experiences of modernity in different times, places, and polities.[31] A handful of historians, including Keith Walden, Christopher Dummitt, Jarrett Rudy, Nicolas Kenny, and Jane Nicholas, have studied iterations of modernity in Canada directly. Their studies range usefully from the late nineteenth century to the late twentieth century, and from Vancouver to Montreal.[32] Most recent Canadian historical writing on the modern, however, has approached the subject more obliquely, through one of three lenses: modernization, as connected most immediately to urbanization and industrialization; antimodernism; or high modernism.[33] A brief look backward through these lenses will help readers to sight similar themes in the chapters that follow.

With the social historical turn well embedded in mainstream Canadian historiography, many scholars have attended to the greatly variable pressures of modernization in this nation. These have left uneven and often inequitable imprints on gendered, racialized, classed, aged and aging, and regionalized bodies. We know more than we did a generation ago about the effects of modernization on men in their private and public lives, including how both managers and the managed fared in newly industrialized and rationalized sites of work.[34] We know much more about women's experiences of modernization at different points in their lives and labours, including those of young women, working women, and mothers, and about the ways in which children and adolescents were prepared in modernizing homes and schools to enter modernizing workplaces.[35] We continue to learn about Indigenous peoples' encounters with modernization and about those of people living in parts of the country coded as "remote" or "un(der)-developed" by metropolitan gazes both historical and contemporary.[36]

These studies collectively demonstrate that, in differing fashions throughout late nineteenth-century and twentieth-century Canada, the bodies of women, men, and children all came, by turns, to signify one or more of the perils of modernization and hence modernity: physical enervation and degeneration through overcivilization, social licentiousness, sexual immorality, and the like. As the subtitle of Carolyn Strange's classic monograph suggests, these perils were all the more urgent since they derived from, or were often experienced most immediately as, distinctly modern pleasures – the newfound freedom that the city offered to young women, or the delight that teenagers of all genders drew from commercialized leisure.[37] These pleasures threatened the integrity not only of individual bodies, but also of the body politic. Given its relative youth compared with other Western countries, Canada was often discursively figured as an immature or adolescent nation throughout the modern era.[38] It was therefore perceived as all the more important that the forces of modernization be bent toward developing Canada's nascent social and economic potential, and that they produce a fully-fledged nation that used its flourishing human and nonhuman resources *ad majorem patriae gloriam.*

But important to whom, precisely? The will to modernize flowed along sometimes well-worn channels in Canadian society. Power pooled at the feet of particular groups: politicians and civil servants; manufacturers and purveyors of commercial goods; moral reformers and their descendants, social workers and public health officials; medical, educational, and psychiatric professionals; urban planners; and, of greatest interest to this volume, scientists, technicians, and engineers. Although projects undertaken by such modernizers often came cloaked in sleek, modern raiment, they tended to buttress well-established premodern structures of power, as Cynthia Comacchio has noted of early twentieth-century modernization schemes aimed at Ontario's mothers and children.[39] These top-down efforts often had substantial, long-lasting, and sometimes unintended effects on people's lives. Yet Canadian historians have also concluded that the results were rarely as comprehensive or as permanent as modernizers would have liked.

Canadians responded individually as well as collectively to the perils of modernization. Studies of paradigmatic sites in which Canadians were taught, and taught themselves, how to be modern subjects – the department store, the summer camp, the industrial exhibition – emphasize that learning to be modern could be difficult, stressful, and even frightening.[40] Even as they sought the pleasures of modern life, many Canadians also sought to escape its perils. They found refuge in places and practices that historians today term "antimodern," or brimming with intense, authentic, or risk-laden experiences not easily found amid the supposed banality and homogeneity of modern life.[41] In Canada, as elsewhere, antimodernism was often infused with notions of the martial, primitive, folksy, rustic, or wild, and not infrequently some blend thereof. Scholars have demonstrated how both Nova Scotia and the North were constructed as antimodern spaces par excellence.[42] They have also charted the plethora of activities that brought antimodernists physical and psychological relief from the exhausting or otherwise undesirable features of modern life. These activities included, but were not limited to, hunting and fishing, camping, climbing mountains, painting *en plein air*, collecting folklore and handicrafts, writing and reading sentimental poetry and Mountie adventure stories, engaging in wilderness tourism, and participating in organized sports and drill.[43]

Despite their best and, one assumes, sincerest efforts, Canada's antimodernists did not so much jettison the modern as force it underground. Not only were their actions and perceptions ineluctably framed by the circumstances and concerns arising from modernization, but they also often carried with them into the bush or the village the very modern habits and values that they ostensibly sought to leave behind. Indeed, some of these outcroppings of antimodernism, particularly those associated with the North and the wilderness, have greatly influenced modern imaginings of the Canadian nation. The (in)famous cultural-mythical-nationalist complex constructed around the Group of Seven and their hinterland art over the last near-century best exemplifies how individual Canadian reactions to modernity could be writ large on a national canvas, shaping others' actions in turn.[44]

Nowhere have the perils of modernity been delineated more acutely in recent years, perhaps, than in the growing scholarly literature on high modernism in Canada. Historians have concentrated especially on hydroelectric dams and power projects, the signature charismatic megastructures of the postwar era. Even before the Second World War, modernization projects were not confined to urban centres, of course. As Stephen Bocking relates in this volume, and as Tina Loo and John Sandlos, among others, have demonstrated elsewhere, state and nonstate conservationists lobbied for and enacted various policies and programs to manage the hinterland's nonhuman (and, indirectly, human) populations and resources in what they considered to be an efficient and rational manner.[45] But the physical, financial, and administrative scales of megaprojects after 1945 dwarfed all previous efforts, sometimes alarmingly so, to those Canadians who stood literally and figuratively in their paths.

Even as Canadian historians have fit James C. Scott's seminal work on high modernism to homegrown contexts, they have inflected it with considerably more nuance. To Scott, "seeing like a state" meant the inability or unwillingness to leaven the characteristic top-down, synoptic gaze of high modernity with "mētis," or local knowledge about the social and environmental particularities of sites slated for development.[46] Yet Tina Loo and Meg Stanley have found just such a downward slide in scale in the preparations to construct hydroelectric dams on the Peace and Columbia Rivers. Engineers and geologists spent years studying sites at first hand and drawing up detailed maps and surveys that contained what Loo and Stanley term "high modernist local knowledge," or "the knowledge that provided pixels for the larger synoptic picture."[47]

Nor did Canadians experience the "strong" version of high modernism that Scott originally laid out in his case studies of mid-twentieth-century authoritarian states. As a liberal democratic nation, Canada played host to what Daniel Macfarlane has named "negotiated high modernism."[48] Civil servants, engineers, planners, and corporate representatives held community consultations and conducted public relations campaigns to persuade people in the way of these projects to cede their lands and uproot their lives in the name of progress and prosperity. Many citizens took the opportunity to speak back to power and to present alternative visions of modernity that privileged different modes of social and economic life in common.[49] Ultimately, however, their visions did not hold water; hydroelectric dams did. Political and economic powers manufactured and extracted consent with a characteristic high modernist cudgel in hand – the technological power to flood entire landscapes and homelands – that local residents found almost impossible to resist. They paid a heavy price. Abruptly sundered from the natural and social spaces to which they had long been accustomed, they had to deal with the ensuing trauma even as they painstakingly forged new relationships with environments new to them, and often new under the sun, remade as they were through concrete and steel and water.[50]

In our professional duty as critical analysts of the past, and perhaps in our own *fin de siècle* disenchantment with late modernity (beautifully articulated in Eda Kranakis's chapter), historians after the social turn have tended to emphasize the weak, failed, and destructive aspects of modernization and high modernism in Canada. Science and technology loom large in these dark-edged narratives. Motherhood never became as scientific as doctors and child welfare professionals had prophesied; bodies proved to be flawed organic machines in factories and other workplaces; the failures of other machines, as much as their triumphs, defined natural worlds and national identities.[51] Meanwhile, engineers revivified the power, if not the judgment, of God of the Old Testament, flooding plains and valleys and drowning all in their wake. The optimism that many nineteenth-century and twentieth-century Canadians felt about science, technology, and modernity, on display in many of the chapters herein, can strike historians as foreign, stale, or sadly misplaced in the light of subsequent events.

Yet the affective pendulum may be starting to swing in the other direction once more. Historians such as Tina Loo and Joy Parr have drawn lessons of redemption and hope from Canadians' sometimes misguided experiments with modernity. In a recent review of the Canadian historiography on high modernism, Loo emphasizes megaprojects' creative and transformative as well as destructive powers and demonstrates how a synoptic view can enhance our ability to steward the nonhuman world wisely. Parr concludes her monograph on postwar megaprojects by focusing on residents' resilience in the face of worlds remade.[52] These scholars' turn toward the positive may have stemmed from their familiarity with environmental history and, in particular, that field's struggles with declensionism, or the propensity to tell narratives of unrelenting decline that leave readers with a sense of futility rather than agency in the face of continued large-scale environmental issues.[53] Canadians still live in a modern era, and the problems of late modernity are, alas, not limited to anthropogenic climate change. Historians of modernity in Canada need not don antique pairs of rose-tinted glasses when casting an analytical gaze backward. But they might consider the value of finding and telling stories that showcase the pragmatic, recuperative, hopeful – and yes, even optimistic – ways that Canadians faced and dealt with the challenges and setbacks of modernity in this country, not least because their descendants might find some grains of wisdom or inspiration therein.

Canadian Science and Technology Have Never Been (Just) Canadian

Histories of Canadian science and technology have not traditionally held centre stage in either the national or the thematic historiographies to which they contribute. However, as Beth Robertson reminds us in her chapter, views from the fringe can productively challenge received narratives not only about science and technology but also, indeed, about Canada itself. Take, for example, the multiple perspectives from the country's geographical fringe, the North, on display in this volume. Andrew Stuhl and Tina Adcock, respectively, demonstrate how northern scientific reports and exploratory permits simultaneously enhanced and undermined Canada's sovereign and epistemic power on the North Atlantic stage. Their chapters, along with Blair Stein's, also demonstrate the often tenuous nature of Canada's political and cultural claims to nordicity. Could eschewing their country's long, cold winters in favour of "sun destinations" make Canadians less Canadian? Is snowbirding a national betrayal? Scholars of Canada have regularly presented both technoscience and the North as enrolled in the service of the state, supporting top-down expressions of soft and hard power. But the fidelity of these instruments was not assured. Even when deployed ostensibly to shore up state authority, they could just as easily (and sometimes unintentionally) draw attention to cracks in the nation's façade.

If we look at another set of the nation's liminal spaces, its borders, writing the history of science and technology there reveals not only the instability but also the porosity of the modern nation. A steady procession of humans, nonhumans, and material artifacts circulated freely across its thresholds in the pursuit of these practices. As the volume's first two chapters illustrate, English ethnologists such as Richard King and American explorers such as George Palmer Putnam came to study the human and natural populations of northern Canada. State agents sometimes scrutinized these foreigners suspiciously in turn. Meanwhile, Canadian specimens, scientists, and technologies ranged far beyond Canada's borders, sketching out both familiar and surprising trajectories (as discussed below). From this brief borderland reconnaissance, we can draw two provisional conclusions, which are supported empirically by many of the chapters that follow. First, science and technology in Canada have never been unambiguously Canadian. Second, the modern history of Canadian science and technology does not stop at the border; it is as much a transnational as a national story.

Neither of these points is novel, historiographically speaking. Once-dominant Western narratives of development presented the progress of colonial and settler-colonial states such as Canada toward modernity as the product of international diffusion and transfer. Historians traced the movement of constitutional, scientific, and technological models from imperial centres to territories characterized as peripheral or "virgin," where they were planted atop those of Indigenous cultures and societies. They demonstrated how colonial politicians and scientists drew their countries along neat, unidirectional pathways, past uniform and universal milestones of achievement. By journey's end, they argued, these colonies had not only become nations but had also developed uniquely national styles of technoscientific endeavour. Although universalist, positivist teleologies of scientific diffusionism and technological transfer have now almost completely fallen out of scholarly fashion, they formed part of the intellectual backdrop against which the field of Canadian science and technology emerged in the mid-1970s and helped to shape the first generation of inquiries into these subjects.[54]

Those initial efforts focused heavily, if implicitly, on central features of technoscientific, Eurocentric modernity – institutions and education, disciplines and professionalization, marginalization and identity, nation building and market economies, technocracies and political power.[55] Reflecting both the historiographical currents traced above and contemporary concerns about Canadian sovereignty and identity beyond the academy, early scholarship in this field sought to locate the distinctive role of science and technology in Canadian culture and history, even proposing a Canadian "style" of technoscientific development.[56] Here, the work of Richard Jarrell merits particular attention, both for inspiring the conference that gave rise to this volume and for providing the Canadian field with much of its foundation. Trained in both astronomy and the history and philosophy of science at Indiana University, but increasingly disillusioned by political developments in the United States, Jarrell represents a double displacement – an American choosing Canada as both adoptive country and intellectual focus. Across a wide array of topics, regions, and periods, he returned again and again to the central place of science and technology in the social and cultural development of Canada. Like so many scholars in this nascent field, Jarrell was concerned to clarify what distinguished Canadian technoscience from its European and American counterparts.[57] His research explored the ideas and processes of Canadian nationhood; the geographical, political, and environmental factors that gave Canadian science its utilitarian focus; the relationship between seemingly radical European metropoles and conservative Canadian peripheries; and the historical conditions on the ground – population density, political control, economic resources – that enabled a shift to what he (and others) considered a fully national iteration of science and technology in Canada.[58]

At stake in those investigations was an urgency to articulate Canada's place and significance in a world at once defined and transformed by the practices, priorities, and products of modern science and technology.[59] Jarrell, along with Carl Berger, Robert Bothwell, Yves Gingras, Trevor Levere, Suzanne Zeller, and others, took the nation as the foundational unit of analysis.[60] Subsequent scholarship by Canadian historians of science and technology has increasingly probed that assumption, problematizing the nation itself as a historical actor. Few areas better reflect this shift than the emerging body of work linking science and technology to commerce and political economy. In his recent book on the Hudson's Bay Company's scientific networks, Ted Binnema demonstrates how eighteenth-century and early nineteenth-century corporate patronage provided international material and symbolic networks that predated the Canadian nation, while modelling symbiotic relations between science, political status, and commercial profit for the future Canadian state.[61] Similarly, social histories of commercial technology in Canada can reveal things that the view from the nation-state conceals. Building upon the work of business and economic historians, Dorotea

Gucciardo's study of the late nineteenth-century and early twentieth-century electrification of Canada moves away from the view of electricity as an object of political governance. It presents electrification as an international social phenomenon, one profoundly linked to the gendered histories of consumption, domesticity, and labour in Canada.[62] A focus on corporations, consumers, commercial infrastructures, and networks provides an alternative perspective on "national" science and technology, one that points us to the scales, systems, and processes that underlie and supersede the modern nation-state. It also transforms our understanding of the kind of nation Canada has been and has been understood to be, both inside and outside its borders. Pierre Bélanger's recent engagement with the subject of mining, a topic that has long preoccupied Canadian historians, does more than illustrate how modern Canadian life came to be mediated through mineral extraction. By placing Canadian mining activities within the context of global economic infrastructures, Bélanger also charts the nation's surprising course from *colony* to *empire*: from a resource-producing colony of Great Britain to an extractive empire in its own right. What would it mean for Canadians to think of themselves, as Bélanger argues they should, as the planet's pre-eminent "extraction nation" – as not just a giver but also a taker of natural wealth the world over?[63]

As these examples indicate, scholars have come to favour scales both smaller and larger than the nation. Historians of new imperial, postcolonial, and global science and technology encourage us to focus upon the dynamic tension between the local, contextual features that play an important role in determining how science and technology are practised and experienced in a given place (and time) and the transnational flows of people, goods, and ideas that connect disparate sites of technoscience and bring new and transformative influences to bear on modes of scientific and technological conduct there.[64] Such methods draw from and are further developed in tandem with those now used in the broader fields of postcolonial, transnational, and global history. Their findings are often of interest to scholars working in those traditions as well, including those whose work nourishes a small but growing body of transboundary, transatlantic, and transnational histories of Canada.[65]

Even as it contributes to the wider transnational turn in Canadian history, then, this volume initiates a similar turn in the history of Canadian science and technology. Taken together, its case studies move between micro- and macroscales, revealing distinctive local elements of scientific and technological cultures, practices, and artifacts across Canada and situating them within larger historical circuits and historiographical conversations. These chapters begin to locate the place of knowledge in Canada. They demonstrate how the particularities of specific sites and localities were integral to the fashioning of Canadian science and technology.[66] By establishing a handful of discrete data points, the authors contribute to the ongoing project of assembling a more coherent, if inevitably pointillist, history of these activities in modern Canada.[67] Even if we discard notions of some kind of inevitable or hegemonic Canadian style of science and technology, we still know far too little, as David Theodore notes in his chapter, about the precise contours of modern science and technology in Canada. Is the figure of Gerald Bull, the aeronautical engineer turned arms dealer, as atypical as Edward Jones-Imhotep suggests? (In all fairness, we suspect so.) In piecing together the larger technoscientific history of modern Canada from smaller actor- and site-specific narratives, we aim to help future historians better distinguish between the normal and the exceptional, the usual and the unusual, and the mean and the deviations, and to adjust their hypotheses and methods accordingly.

While eschewing an exceptionalist approach to the modern history of science and technology in Canada, we build upon scholarship that tries to identify signature influences, themes, and patterns in Canadian scientific and technological practices, whether at the local, regional, or national scale. Both historians of science and technology and environmental historians have weighed in, pinpointing features such as the central role of technology in Canadian history, the diversity and extremity of Canadian ecosystems and climates, the country's abundance of renewable and nonrenewable natural resources, and the prominence of the state in scientific and technological endeavours.[68] The authors herein add their own suggestions, most notably David Theodore, who asserts that "normal science in Canada is small science." In light of the considerable archival and analytical spadework that lies ahead for historians of Canadian science and technology, we offer such statements more as hypotheses that require further testing than as definitive assertions about the nature of science, technology, and the modern in this country.

The authors also position these local and localist histories of science and technology as nodes within larger transnational networks of circulation and exchange. They inquire into how the movements of scientific and technical personnel, objects, and knowledge across borders connected Canadian manifestations of science and technology to those practised elsewhere, and they suggest how Canadian science and technology transformed, and were transformed by, these bodies and ideas in motion.[69] Many of the human and nonhuman itineraries charted in these pages will be at least somewhat familiar to readers. They follow paths similar to those laid out in older continental and imperial histories of Canada, as well as in newer borderland and transnational ones.[70] The chapters reveal how different streams of medical, technological, and engineering knowledge, actors, and practices flowed north from the United States and west across the North Atlantic from Europe. For its part, Canada offered the world both raw and processed versions of technoscientific personnel, knowledge, techniques, and objects – the most delicious of which was likely vanillin, as James Hull details in his chapter.

This volume also bears witness to transnational trajectories and circuits less often discussed in the pages of globally minded Canadian histories. All-weather, jet-age airplanes helped mid-twentieth-century Canadian vacationers to engage, albeit a little uneasily, with the sun-soaked environments of Florida and the Caribbean. Similar landscapes served as sites of technologically mediated work as well as leisure, as we see in the case of McGill's High Altitude Research Project on Barbados.[71] Canada has entered the scholarly literature on Cold War geographies of hostile or "intemperate" environments – arctic, desert, tropic, alpine – chiefly by way of the Arctic.[72] Edward Jones-Imhotep's chapter reveals that Canadian scientists contributed to research on the Tropics, too. Following Gerald Bull's peripatetic career reveals what is perhaps the most surprising extra-Canadian connection in this volume – Saddam Hussein's Iraq. Bull repackaged ballistics technology originally intended to help a middle-power nation such as Canada compete with better-funded nations in the space race and resold it to the Iraqi government as a means by which to modernize Iraq in the late twentieth century. The example of Bull underscores the potential value of individual "life geographies" in delineating the scope and trajectories of Canadian histories of science and technology outside Canada. Such travelling actors can double as ad hoc dye injections, their movements illuminating pathways previously hidden from the historian's gaze.[73]

By telling stories about modern Canadian science and technology that reach well beyond the nation's borders, this volume joins other transnational histories of Canada in articulating and analyzing Canada's contributions to global social and economic networks, including the defining structures of industrial capitalism. The chapters by Jan Hadlaw, James Hull, and Eda Kranakis reveal Canadian reactions and contributions to sweeping historico-economic developments: how people were taught to be customers and consumers; how industrial production was reimagined as a scientific process; how natural objects were recast as intellectual property. Their accounts reshape global histories of capitalism by centring perspectives often considered to be peripheral. As Adele Perry notes, there is "power in studying empire from its ragged margins."[74] We submit that the same holds true for studying modern Canadian history from the perspective of science and technology, which have often been relegated to the margins of Canadian historiography. The stories told from the fringes in these pages subtly upend well-known narratives of nation. Listening to them, one begins to sense that modern Canada is less northern (Adcock, Stuhl, Stein), less peaceful (Jones-Imhotep, Kranakis), and smaller (Theodore) than we have been led to believe.

Centres and peripheries are relative, mutable spaces, however. They attain such status by dint of their social, economic, intellectual, and institutional capital (or lack thereof) as much as by their physical placement.[75] Are science and technology really marginal in modern Canadian history, then? Or have they simply enjoyed the attention of fewer historians for a briefer period of time? With this in mind, we further submit, in a twist on Percy Bysshe Shelley's phrase, that science and technology have been the unacknowledged legislators of modern Canada.[76] They have been essential to Canada's (high) modernization, for better *and* for worse, and they richly embody the hopes and contradictions of modern life in this country, as the chapters that follow ably demonstrate.

Bodies, Technologies, Environments

This volume is organized around three key themes – bodies, technologies, and environments – that reflect traditional and contemporary strengths among historians of Canadian science and technology, including historians of medicine, architecture, and the environment. The three are interlinked in modern Canada, as Dolly Jørgensen's epilogue shows. Guided by these themes, the authors train their analytical gaze on exemplary discursive and material sites, asking and answering a series of linked questions. How do bodies help us to understand the confluence of science, technology, and the modern in Canada? How do technologies? How do environments?

The chapters corresponding to the first theme join a growing Canadian literature on sensuous and corporeal histories, often grounded in experiences of modernity.[77] They also join a rich literature on the conceptual, material, and social intersections of medicine with science, technology, health, and the body in Canada.[78] Their particular interest, however, lies in studying how Canadian bodies were remade and reimagined through new scientific categories and theories and technological therapies, and how they were taught to be modern through the mastery of new practices and disciplines. In different ways, they focus on bodies or activities on the fringes, whether geographical, intellectual, or social. These fresh perspectives on the modern invite readers to reconsider the relationship between Canadian bodies and the Canadian nation, a matter of interest in other recent scholarship.[79]

We begin at the temporal edge of the modern and at the spatial edge of what would become Canada with Efram Sera-Shriar's chapter on the British physician Richard King and his early nineteenth-century expedition to what are today Nunavut and the Northwest Territories. Sera-Shriar places the ethnographic view from nowhere in northern Canada, illustrating how King's experiences there led directly to the reform of British race science.[80] Even as King sought to modernize ethnographic witnessing and reporting, providing both data for armchair ethnologists and methods for fieldworkers to follow, he was unexpectedly touched by the hardships that northern Indigenous peoples faced, which he believed the Hudson's Bay Company's policies had exacerbated. For King, modern ethnography was inherently humanitarian. Its duty was to civilize the people whom it studied and so to lead them into the modern world, where they could interact with other moderns on common ground.

Remaining on northern soil, Tina Adcock examines two expeditions of more recent vintage, those led by the American publisher George Palmer Putnam to the eastern Canadian Arctic in the mid-1920s. Adcock uses these expeditions and the diplomatic furor they caused as a vehicle to analyze the boundary-work surrounding the state's regulation of foreign scientists and explorers wishing to work in the Northwest Territories. The newly minted Scientists and Explorers Ordinance gave civil servants the ability to compel fieldworkers to abide by certain rules and regulations, but it also produced an unexpected, if rather modern, problem of classification (and, by extension, professionalization). What, exactly, constituted science or exploration, as opposed to sport hunting or tourism? Putnam's expeditions appeared legitimate enough on paper, but their use of migratory birds for target practice and consumption transgressed the epistemic and legal boundaries of northern field science. Although American diplomats rushed to mend the damage, the incident exposed the limits of Canadian scientific as well as political sovereignty in the Arctic Archipelago.

Both Adcock's chapter and the one that follows, by Dorotea Gucciardo, reflect different therapeutic solutions to the enervating hustle of late nineteenth-century and early twentieth-century urban life. While Putnam and his son David turned to the archetypal antimodernist cures of hunting and fishing in "wild" surrounds, other city dwellers sought technological cures for modern problems. Gucciardo charts the rise and fall of electrotherapy in urban Canada between 1880 and 1920. Electricity was a double-edged sword, she argues: it raised modern life to the fevered pitch that caused neurasthenia and hysteria, but it could also supposedly cure those same ills. Like Adcock, Gucciardo probes the boundary between licit and illicit practices of modern science. Electrotherapeutists rarely had medical training, but the presence of such technologies in physicians' offices helped to legitimate their wider use. In pulling back the curtains shielding these spaces from view, Gucciardo reveals the intimate and gendered nature of modern encounters with medical technology. Both rhetorically and materially, electricity penetrated and modernized Canadian bodies for a time. But the changing social and intellectual contours of medicine eventually disconnected electrotherapy from mainstream currents of thought, leading to its interwar demise.

Beth Robertson's chapter evinces a similarly intimate gaze with respect to bodies and modernity. It examines another set of fringe discourses and practices, this time pertaining to the occult. Focusing on a group of spiritualists based in Kitchener-Waterloo and led by the medium Thomas Lacey, Robertson charts their engagements with new theories of atomic energy from the 1930s to the 1950s. These spiritualists not only integrated knowledge produced by modern physicists into their discourses on the nature of bodily and psychical energies but also questioned the very boundary between spiritualism and science, as captured in the phrase "occult science." Lacey's group viewed the advances of atomic science in a predominantly positive light, believing that this power could ultimately lead to the rejuvenation of human bodies and perhaps even human immortality. This optimism about the ability of modern science and technology to ameliorate the human condition recurs throughout the chapters in this section, as in the sections that follow.

In 1974, the editors of a primary-source reader on technology and society in Canadian history averred that "Canadians know more of prime ministers than they do of the creation of the technological structure that is the framework for the nation's economic and social life."[81] Just as a generation of social historians has broadened Canada's knowledge of past actors beyond the prime ministerial, so too historians of technology have begun to put flesh on the infrastructural and mechanical bones that support Canadian society. The chapters in the second section approach the specific question of technology and the modern in Canada in various ways. Some situate practice in place, examining the emergence of distinctive technoscientific cultures at the regional or national level in Canada. Others illustrate how technologies could reinforce or challenge what it meant to be modern in Canada, or to be Canadian in the modern era. Still others trace the circuits of technicians, technical knowledge, and technological artifacts between Canadian sites and those located elsewhere, extending the history of Canadian technology beyond the Canadian ecumene.

In the section's opening chapter, James Hull questions the supposed causal connection between urbanization, first-wave industrialization, and modernization in Canada. He argues that modernization actually emerged from Canada's "Second Industrial Revolution," or the move toward industrial research and applications grounded in science and technology and based upon an improved understanding of the natural world. His chapter discusses this concept with precision, considering the nature of production and organization in this revolution; the roles played by gender, class, and the state; the connections between Canadian and American industries; and the distinctive and outstanding features of the Canadian experience of this revolution. Hull seeks to trouble supposed turning points in Canadian modernization, such as the Laurier boom and the First World War, and gives Canadian actors full credit for technological success where credit is due. He concludes that the Second Industrial Revolution brought significant change to Canadian society by 1914 and that certain second-wave industrial sectors in Canada really were world class, in the sense of making important international contributions.

As Jan Hadlaw notes, the ability to use new technologies was not automatically acquired, but little information on technical education in past eras has survived.[82] Having uncovered such sources, Hadlaw is able to tell a relatively rare kind of technological story, about Bell's educational campaign in central Canadian cities to ease its subscribers' transition to dial telephony between the wars. The introduction of automatic dialling threatened well-established cultures of sociability surrounding the telephone and stoked all-too-modern fears about the dark side of mechanization, including social atomization and technological unemployment. Deftly defusing these modern anxieties by drawing upon pre-existing telephonic relationships and practices, Bell strove to convince its customers that dialling for oneself was "preferable and progressive." The company deployed a plethora of strategies, including window displays, print manuals, and person-to-person demonstrations, to teach customers how to operate the new instruments properly. Hadlaw argues that it was by learning how to use modern technologies such as dial telephones that Canadians acquired the ability, both practically and imaginatively speaking, to think of themselves as modern.

David Theodore offers a novel perspective on postwar science in Canada, one situated firmly in a specific place: the Montreal Neurological Institute at McGill University. Using the career of Christopher Thompson and his PDP-12 minicomputer as his narrative and analytical vector, Theodore develops the notion of "small science." In contradistinction to "big science," often considered the dominant or default mode of practice after the Second World War, small science was characterized by several key factors: a one-man research team, the idea of trained acquaintance, and bounded experiences of space (built and unbuilt) and time. A focus on small science, Theodore argues, may help researchers to glimpse and capture telling stories that often fly beneath the radar of historians of science attuned to bigger, multidisciplinary iterations of this activity. It also brings to the fore the importance of biographical or smaller-scale factors in the creation and maintenance of particular scientific workplaces. Finally, if "normal science in Canada is small science," as Theodore asserts, then his chapter may point the way to a better overall understanding of modern scientific practice in this country.

A man and his machine(s) also lie at the heart of Edward Jones-Imhotep's chapter, which traces the curious rise and fall of Gerald Bull and his outsized cannons. Bull's story exemplifies many of the tensions surrounding science, technology, and the modern in mid-to-late twentieth-century Canada. Like the optical illusion of Rubin's vase, depending on one's perspective, Bull's cannons could appear either as scientific instruments designed to forward atmospheric research and secure Canada's place in the global space race or as superguns designed to intimidate one's neighbours and thus recalibrate regional Cold War geographies of power. Place mattered: something that clearly facilitated research when positioned on the Quebec-Vermont border assumed a more threatening mien when embedded in Barbados or Iraq. But the cannons and the modern sentiments that encircled them also travelled well. As they moved from mid-twentieth-century Canada to late twentieth-century Iraq, they continued to embody a particular technological panacea for fears of insufficiently rapid national modernization. Yet, as Jones-Imhotep shows, Bull's technocratic, rational arguments about the peaceful nature of his superguns neither crossed space so easily nor wore so well over time, something mirrored in Bull's own transition from research scientist to arms dealer to assassination target.

Similar themes of violence and disorder permeate Eda Kranakis's chapter about Monsanto's patented Roundup Ready® strain of canola and the legal wrangles that ensued after its introduction to the Prairies near the end of the last century. Wielding biotechnological patents expertly, Monsanto challenged at long-standing modes of rural agrarian sociability and patterns of land use in its quest to saturate the Prairies with its new, weed-resistant seed regime. Monsanto approached the space of the court with equal expertise, redrawing the human-nonhuman boundary for its own benefit and relying on widespread scientific illiteracy among members of the legal profession and the Canadian public to press home its advantage. Canadian farmers had long operated according to the tenets of high modernist agriculture, in which science and technology represented the possibility of controlling and ordering agricultural landscapes. Kranakis illustrates how that ideological compact unravelled rapidly under Monsanto's onslaught; science was used as a means of control *against* farmers and as a tool to foment social *disorder*. Percy Schmeiser, the Saskatchewan farmer who bore the brunt of Monsanto's wrath, now stands as a global symbol of the power of late modern technoscience to harm, rather than help, farmers and agricultural landscapes.

Kranakis's chapter leads naturally into the volume's final section on environments, which reflects the nonhuman world's propensity to loom large in Canadian histories of all stripes, including those dealing with science and technology. These disciplines, practices, and artifacts have found an equally natural home in the pages of Canadian environmental historical scholarship, an especially fertile (and still growing) field within the past decade or so. The strengthening rapprochement between scholars of science and technology and those of the environment is evident outside Canada as well, perhaps most notably in the hybrid field of "envirotech," or envirotechnical history.[83]

Within the Canadian academy, Stephen Bocking has been among the scholars most committed to studying science and the environment in tandem. He continues that work in his contribution to this volume, which surveys what he terms the "landscapes of science" in modern Canada.[84] Bocking charts the history of modern scientific and technological interventions in Canadian landscapes, organizing his narrative around four key themes: the extension of state authority over territory, the transformation of landscapes for profit, the administration and regulation of environments and human activities therein, and the disruption of views about science and human interactions with the nonhuman world in the late modern era, most notably through Indigenous traditional ecological knowledge and citizen science. The symbiotic relationship between science and modernity helped to tame and rationalize unruly Canadian landscapes and their inhabitants. Yet, as Bocking illustrates humans and nonhumans outside this privileged coupling continually challenged and confounded the best-laid plans of scientists and modernizers alike.

Andrew Stuhl picks a different path through the landscapes of Arctic, Canadian, and global science. He follows the transnational production and circulation of one set of scientific texts, the fourteen-volume *Report of the Canadian Arctic Expedition 1913–1918*, throughout the interwar years. Although this expedition is often depicted, not without reason, as a story of scientific adventure in the service of the nation, its reports were produced with the assistance of and according to the standards of the international scientific community. Once complete, they helped to burnish the profile of Canadian science on the world stage. As the *Report* circulated throughout the Western world, it informed subsequent scientific endeavours in projects and disciplines sometimes quite distant from the field site and nation of its birth. Closer to home, as Stuhl illustrates, the *Report* produced new economic and environmental visions of the Arctic that enhanced, yet also subverted, established notions of Canada's northernness. In narrating this object biography and tracing its transnational geographies, Stuhl's chapter nicely complements other chapters' investigations of individual scientific actors and their thought-provoking travels through more-than-national networks.

In her chapter, Blair Stein, like Sera-Shriar, Adcock, and Stuhl, turns her gaze on Canada's cold climes; unlike her fellow authors, however, she is primarily concerned with questions of technology. Stein studies Trans Canada Air Lines/Air Canada's innovations in aviation between the 1940s and the 1970s, including the introduction of the Canadair DC-4M North Star and "sun destination" routes, and considers how they were conceived and marketed as solutions to Canadians' midcentury anxieties concerning winter mobility. Even as these new technologies and routes helped Canadians literally transcend their country's unique geographical and seasonal challenges, their characteristically modern disruption of space and time began to destabilize national identities rooted in common experiences of Canada's vast distances and cold winters. Climatic claims of triumphalism in the face of inevitable Canadian winters gave way to climatic expressions of ambivalence once Canadians could fly south on jetliners to the Caribbean and swap Canadian winters for "Canadian summers," or escape Canadian winters altogether. As with northern science in Stuhl's chapter, midcentury technologies of aviation both propped up and destabilized Canadians' long-held assertions of nordicity. The ability to spend the winter on sunny, warm beaches was cold comfort for some climatic patriots, leading to John Crosbie's call in 1977 to "ban the tan."

Daniel Macfarlane returns our gaze to Canada's heartland and, more specifically, to the St. Lawrence River, which became host to the eponymous Seaway and Power Project in the 1960s. This megaproject becomes the crucible in which Macfarlane develops the idea of "negotiated high modernism," a version of this ideological program tailored to liberal democratic states such as Canada. Negotiated high modernism bore many of the same hallmarks as high modernism, at least on the St. Lawrence. Engineers full of enthusiasm for rational, quantitative ways of understanding complex environments gathered local knowledge, modelled landscapes as they were and would soon become, and created whole new envirotechnical systems on the ground. People were relocated into modernized villages, but only after the local power authority had undertaken elaborate, multifaceted measures to manufacture their consent to do so. Knowledge travelled as well. The St. Lawrence Seaway and Power Project drew on personnel and expertise from other engineering megaprojects around the world, and it became a pedagogical site of high modernity to which non-Canadian hydraulic engineers flocked. Although negotiated high modernism favoured the carrot rather than the stick to bring about its desired outcomes, it still demonstrated the capacious power of governments and companies to permanently reshape Canadian lives and landscapes along the St. Lawrence River.

In 1980, Bruce Sinclair predicted that "we shall discover science and technology to be close to the centre of Canadian experience."[85] Since then, historians of these activities have done much to reveal the truth of this assertion. We now know that science offered settler-colonial Canadians the ability to comprehend and thus claim a sprawling territory through techniques of inventory and description.[86] Technology offered them the ability to "collapse" space and time, helping to knit a disparate and far-flung citizenry together. The discovery of insulin in the Banting laboratory brought hope to diabetes patients everywhere; the discovery of uranium in the Subarctic made Canada complicit in the real and potential destruction of the atomic age. Modern science and technology were perceived as a conduit for Canada's national maturation, holding out the prospect of pride and glory on an international stage. But they could also bring shame, as when doctors performed nutritional experiments on Indigenous children on reserves and in residential schools.[87] This volume affirms the central role that science and technology played in the creation of modern Canada and the value of circulating Canadian case studies to scholars of science, technology, and modernity beyond Canada's borders.

NOTES

The authors would like to thank Nicolas Kenny and the two anonymous reviewers for their thoughtful comments on earlier versions of this introduction.

1 For representative works, see David S. Landes, *The Unbound Prometheus: Technological Change and Industrial Development in Western Europe from 1750 to the Present* (London: Cambridge University Press, 1969); J.D. Chambers, *The Workshop of the World* (London: Oxford University Press, 1968); E.J. Hobsbawm, *Industry and Empire* (New York: Random House, 1968); E.P. Thompson, *The Making of the English Working Class* (New York: Pantheon, 1964); and A. Toynbee, *The Industrial Revolution* (Boston: Beacon Press, 1957).
2 Marshall Berman, *All that Is Solid Melts into Air: The Experience of Modernity* (New York: Viking Penguin, 1988), 13.
3 Herbert Butterfield, *The Origins of Modern Science: 1300–1800* (New York: Free Press, 1965), 8; Edwin Arthur Burtt, *The Metaphysical Foundations of Modern Science* (Garden City, NY: Doubleday, 1954), 15–24; Alexandre Koyré, *From the Closed World to the Infinite Universe* (Baltimore: Johns Hopkins University Press, 1979), 1–3; Katharine Park and Lorraine Daston, "Introduction: The Age of the New," in *The Cambridge History of Science*, vol. 3, *Early Modern Science*, ed. Katharine Park and Lorraine Daston (Cambridge, UK: Cambridge University Press, 2006), 15–16.
4 Steven Shapin, *The Scientific Revolution* (Chicago: University of Chicago Press, 1996).
5 Charles Sabel and Jonathan Zeitlin, "Historical Alternatives to Mass Production: Politics, Markets, and Technology in Nineteenth-Century Industrialization," *Past and Present* 108, 1 (1985): 133–76; Daryl Hafter, *European Women and Preindustrial Craft* (Bloomington: Indiana University Press, 1995); Francis Sejersted, "An Old Production Method Mobilizes for Self-Defense," in *Technological Revolutions in Europe: Historical Perspectives*, ed. Maxine Berg (Northampton, MA: Edward Elgar, 1998); Thomas Max Safley, *The Workplace before the Factory: Artisans and Proletarians, 1500–1800* (Ithaca, NY: Cornell University Press, 1993).
6 See, for example, John Tresch, *The Romantic Machine: Utopian Science and Technology after Napoleon* (Chicago: University of Chicago Press, 2012).
7 For the specific case of the history of science and technology, see, for example, David Edgerton, *The Shock of the Old: Technology and Global History since 1900* (Oxford:

Oxford University Press, 2011); Vanessa Ogle, *The Global Transformation of Time: 1870–1950* (Cambridge, MA: Harvard University Press, 2015); Avner Wishnitzer, *Reading Clocks, Alla Turca: Time and Society in the Late Ottoman Empire* (Chicago: University of Chicago Press, 2015); Gyan Prakash, *Another Reason: Science and the Imagination of Modern India* (Princeton, NJ: Princeton University Press, 1999); Joel Wolfe, *Autos and Progress: The Brazilian Search for Modernity* (Oxford: Oxford University Press, 2010); David Arnold, *Everyday Technology: Machines and the Making of India's Modernity* (Chicago: University of Chicago Press, 2013); Marian Aguiar, *Tracking Modernity: India's Railway and the Culture of Mobility* (Minneapolis: University of Minnesota Press, 2011); On Barak, *On Time: Technology and Temporality in Modern Egypt* (Berkeley: University of California Press, 2013); and Timothy Mitchell, *Rule of Experts: Egypt, Techno-Politics, Modernity* (Berkeley: University of California Press, 2002).

8   Steven Shapin and Simon Schaffer, *Leviathan and the Air-Pump: Hobbes, Boyle, and the Experimental Life* (Princeton, NJ: Princeton University Press, 1985).

9   Although the contributions to this volume partly interrogate these various forms, our working assumption is that modernization represents the institutional transformations that characterize the emergence of modernity, whereas modernism represents either implicit or explicit expressions of what it means to be modern. For a persuasive criticism of modernity as a historical construct, see, for example, Kathleen Davis, *Periodization and Sovereignty: How Ideas of Feudalism and Secularization Govern the Politics of Time* (Philadelphia: University of Pennsylvania Press, 2008); and Kathleen Davis and Nadia Altschul, eds., *Medievalisms in the Postcolonial World: The Idea of "the Middle Ages" outside Europe* (Baltimore: Johns Hopkins University Press, 2009).

10 Richard Foster Jones, *Ancients and Moderns: A Study of the Rise of the Scientific Movement in Seventeenth-Century England*, rev. ed. (New York: Dover, 1982); Joseph M. Levine, *Between the Ancients and the Moderns: Baroque Culture in Restoration England* (New Haven, CT: Yale University Press, 1999).

11 The literature on modernity is vast. On the controversy surrounding it, see David Lyon, *Postmodernity* (Minneapolis: University of Minnesota Press, 1999); P. Osborne, Michael Payne, and Jessica Rae Barbera, "Modernity," in *A Dictionary of Cultural and Critical Theory*, ed. Michael Payne (West Sussex, UK: Wiley-Blackwell, 2010), 456–59; and "AHR Roundtable: Historians and the Question of 'Modernity,'" *American Historical Review* 116, 3 (2011): 631–751.

12 C.A. Bayly, *The Birth of the Modern World, 1780–1914: Global Connections and Comparisons* (Oxford: Blackwell, 2004), 11.

13 Dipesh Chakrabarty, "The Muddle of Modernity," *American Historical Review* 116, 3 (2011): 663–75. On modernity as a set of contradictions, see Berman, *All that Is Solid Melts into Air*; and Torbjorn Wandel, "Too Late for Modernity," *Journal of Historical Sociology* 18, 3 (2005): 255–68. On modernity as loss, see Sumathi Ramaswamy, *The Lost Land of Lemuria: Fabulous Geographies, Catastrophic Histories* (Berkeley: University of California Press, 2004). See also Zvi Ben-Dor Benite, *The Ten Lost Tribes: A World History* (Oxford: Oxford University Press, 2009); Dick Teresi, *Lost Discoveries: The Ancient Roots of Modern Science – from the Babylonians*

to the Maya (New York: Simon and Schuster, 2002); Michael Hamilton Morgan, *Lost History: The Enduring Legacy of Muslim Scientists, Thinkers, and Artists* (Washington, DC: National Geographic, 2007); Elizabeth McHenry, *Forgotten Readers: Recovering the Lost History of African American Literary Societies* (Durham, NC: Duke University Press, 2002); and Jonardon Ganeri, *The Lost Age of Reason: Philosophy in Early Modern India, 1450–1700* (Oxford: Oxford University Press, 2011).

14 Chakrabarty, "The Muddle of Modernity," 663; Gurminder K. Bhambra, "Historical Sociology, Modernity, and Postcolonial Critique," *American Historical Review* 116, 3 (2011): 653–62; Dipesh Chakrabarty, *Provincializing Europe: Postcolonial Thought and Historical Difference* (Princeton, NJ: Princeton University Press, 2000), 8. As Lynn Thomas notes, African history came into being partly as a way of challenging the racist, teleological, and condescending presumptions of this understanding of modernity. Lynn Thomas, "Modernity's Failings, Political Claims, and Intermediate Concepts," *American Historical Review* 116, 3 (2011): 727–40. Achille Mbembe has argued that the burden of responding to this racism has limited writing and thinking about Africa. Achille Mbembe, "African Modes of Self-Writing," trans. Steven Rendall, *Public Culture* 14, 1 (2002): 239–73; Achille Mbembe, "On the Power of the False," trans. Judith Inggs, *Public Culture* 14, 3 (2002): 629–41.

15 Quoted in Alexander Woodside, *Lost Modernities: China, Vietnam, Korea, and the Hazards of World History* (Cambridge, MA: Harvard University Press, 2006), 18. For a discussion of the relationship between science, technology, and competing understandings of "nature," see the introduction to Edward Jones-Imhotep, *The Unreliable Nation: Hostile Nature and Technological Failure in the Cold War* (Cambridge, MA: MIT Press, 2017).

16 On the history of science, see Park and Daston, "Introduction," 2. On the emergence of technology, see Leo Marx, "Technology: the Emergence of a Hazardous Concept," *Technology and Culture* 51, 3 (1997): 561–77; and Eric Schatzberg, "'Technik' Comes to America: Changing Meanings of 'Technology' before 1930," *Technology and Culture* 47, 3 (2006): 486–512.

17 See Michael Adas, *Machines as the Measure of Men: Science, Technology, and Ideologies of Western Dominance* (Ithaca, NY: Cornell University Press, 1990). For an extended discussion of modern concerns about purity and hybridity, see Bruno Latour, *We Have Never Been Modern*, trans. Catherine Porter (Cambridge, MA: Harvard University Press, 1993).

18 See S.N. Eisenstadt, "Transformation of Social, Political, and Cultural Orders in Modernization," *American Sociological Review* 30, 5 (1965): 659–73; *Early Modernities*, special issue of *Daedalus* 127, 3 (1998), including Shmuel N. Eisenstadt and Wolfgang Schluchter, "Introduction: Paths to Early Modernities – a Comparative View," 1–18; *Multiple Modernities*, special issue of *Daedalus* 129, 1 (2000); Shmuel N. Eisenstadt, ed., *Multiple Modernities* (New Brunswick, NJ: Transaction Publishers, 2002); Dominic Sachsenmaier and Jens Riedel, with Shmuel N. Eisenstadt, eds., *Reflections on Multiple Modernities: European, Chinese, and Other Interpretations* (Leiden: Brill, 2002); Charles Taylor, *Modern Social Imaginaries* (Durham, NC: Duke University Press, 2004); and Ibrahim Kaya, *Social Theory and Later Modernities: The*

Turkish Experience (Liverpool: Liverpool University Press, 2004). For a summary of the multiple modernities approach, see Bhambra, "Historical Sociology, Modernity, and Postcolonial Critique."

19 Zvi Ben-Dor Benite, "Modernity: The Sphinx and the Historian," *American Historical Review* 116, 3 (2011): 638–52. For debate regarding the distinction between global and world history, see Bruce Mazlish, "Comparing Global History to World History," *Journal of Interdisciplinary History* 28, 3 (1998): 385–95; Maxine Berg, "From Globalization to Global History," *History Workshop Journal* 64, 1 (2007): 335–40; and William Gervase Clarence-Smith, Kenneth Pomeranz, and Peer Vries, "Editorial," *Journal of Global History* 1, 1 (2006): 1.

20 Carol Symes, "When We Talk about Modernity," *American Historical Review* 116, 3 (2011): 715–26.

21 Frederick Cooper, *Colonialism in Question: Theory, Knowledge, History* (Berkeley: University of California Press, 2005), 127; Chakrabarty, "The Muddle of Modernity," 665; James Vernon, *Distant Strangers: How Britain Became Modern* (Berkeley: University of California Press, 2014).

22 Carol Gluck, "The End of Elsewhere: Writing Modernity Now," *American Historical Review* 116, 3 (2011): 676–87. For a criticism of this usage, see Cooper, *Colonialism in Question*, 113–49.

23 Gluck, "The End of Elsewhere."

24 Sanjay Subrahmanyam, "Hearing Voices: Vignettes of Early Modernity in South Asia, 1400–1750," *Daedalus* 127, 3 (1998): 99–100.

25 Ibid.

26 Ibid.

27 On the wider self-conscious modernity of societies, see Bayly, *The Birth of the Modern World*.

28 Dorothy Ross, "American Modernities, Past and Present," *American Historical Review* 116, 3 (2011): 702–14. For an analysis of how American modernity differed from European modernity, see Jürgen Heideking, "The Pattern of American Modernity from the Revolution to the Civil War," in Eisenstadt, *Multiple Modernities*, 219–47.

29 Thomas, "Modernity's Failings, Political Claims, and Intermediate Concepts," 737.

30 Cynthia R. Comacchio, *The Dominion of Youth: Adolescence and the Making of a Modern Canada, 1920–1950* (Waterloo, ON: Wilfrid Laurier University Press, 2006), 12.

31 Although Quebeckers' experiences of modernity and modernization align in many respects with those of other Canadians and citizens of other industrializing nations, they have also been shaped, both in the past and present, by political and religious trends distinctive to that province. A full discussion of Quebec's historical and historiographical encounters with the modern lies outside the scope of this introduction. For pathways into this subject, see, for example, Jean-Philippe Warren, "Petite typologie philologique du 'moderne' au Québec (1850–1950). Moderne, modernisation, modernisme, modernité," *Recherches sociographiques* 46, 3 (2005): 495–525; Martin Petitclerc, "Notre maître le passé?: Le projet critique de l'histoire

sociale et l'émergence d'une nouvelle sensibilité historiographique," *Revue d'histoire de l'Amérique française* 63, 1 (2009): 83–113; and Peter Gossage and J.I. Little, *An Illustrated History of Quebec: Tradition and Modernity* (Don Mills, ON: Oxford University Press, 2012).

32 Keith Walden, *Becoming Modern in Toronto: The Industrial Exhibition and the Shaping of a Late Victorian Culture* (Toronto: University of Toronto Press, 1997); Christopher Dummitt, *The Manly Modern: Masculinity in Postwar Canada* (Vancouver: UBC Press, 2007); Jarrett Rudy, "Do You Have the Time? Modernity, Democracy, and the Beginnings of Daylight Savings Time in Montreal, 1907–1928," *Canadian Historical Review* 93, 4 (2012): 531–54; Nicolas Kenny, *The Feel of the City: Experiences of Urban Transformation* (Toronto: University of Toronto Press, 2014); Jane Nicholas, *The Modern Girl: Feminine Modernities, the Body, and Commodities in the 1920s* (Toronto: University of Toronto Press, 2015).

33 A decade ago, Cecilia Morgan noted Canadian historians' fascination with antimodernism and their concomitant lack of attention to modernity. Cecilia Morgan, *"A Happy Holiday": English Canadians and Transatlantic Tourism, 1870–1930* (Toronto: University of Toronto Press, 2008), 18–19.

34 Craig Heron, *Lunch-Bucket Lives: Remaking the Workers' City* (Toronto: Between the Lines, 2015), and Dummitt, *The Manly Modern*, provide good place-specific overviews of these broader changes. Placed side by side, these two narratives span most of the modern era.

35 In addition to other works cited in this section, see, for example, Katherine Arnup, *Education for Motherhood: Advice for Mothers in Twentieth-Century Canada* (Toronto: University of Toronto Press, 1994); Mary Louise Adams, *The Trouble with Normal: Postwar Youth and the Making of Heterosexuality* (Toronto: University of Toronto Press, 1997); Mona Gleason, *Normalizing the Ideal: Psychology, Schooling, and the Family in Postwar Canada* (Toronto: University of Toronto Press, 1999); Wendy Mitchinson, *Giving Birth in Canada, 1900–1950* (Toronto: University of Toronto Press, 2002); Tamara Myers, *Caught: Montreal's Modern Girls and the Law, 1869–1945* (Toronto: University of Toronto Press, 2006); and Kristine Alexander, *Guiding Modern Girls: Girlhood, Empire, and Internationalism in the 1920s and 1930s* (Vancouver: UBC Press, 2017).

36 See, for example, Andrew Parnaby, "'The Best Men that Ever Worked the Lumber': Aboriginal Longshoremen on Burrard Inlet, BC, 1863–1939," *Canadian Historical Review* 87, 1 (2006): 53–78; Mary-Ellen Kelm, "Flu Stories: Engaging with Disease, Death, and Modernity in British Columbia, 1918–19," in *Epidemic Encounters: Influenza, Society, and Culture in Canada, 1918–20*, ed. Magda Fahrni and Esyllt W. Jones (Vancouver: UBC Press, 2012), 167–92; Mary Jane Logan McCallum, *Indigenous Women, Work, and History, 1940–1980* (Winnipeg: University of Manitoba Press, 2014); Frank James Tester and Peter Kulchyski, *Tammarniit (Mistakes): Inuit Relocation in the Eastern Arctic, 1939–63* (Vancouver: UBC Press, 1994); and Caroline Desbiens, *Power from the North: Territory, Identity, and the Culture of Hydroelectricity in Quebec* (Vancouver: UBC Press, 2013).

37 Carolyn Strange, *Toronto's Girl Problem: The Perils and Pleasures of the City, 1880–1930* (Toronto: University of Toronto Press, 1995).

38 Comacchio, *The Dominion of Youth*. This rhetoric could apply to constituent parts of Canada, too. See, for example, Arn Keeling and Robert McDonald, "The Profligate Province: Roderick Haig-Brown and the Modernizing of British Columbia," *Journal of Canadian Studies* 36, 3 (2001): 7–23.
39 Cynthia R. Comacchio, *Nations Are Built of Babies: Saving Ontario's Mothers and Children, 1900–1940* (Montreal/Kingston: McGill-Queen's University Press, 1993), 14.
40 Donica Belisle, *Retail Nation: Department Stores and the Making of Modern Canada* (Vancouver: UBC Press, 2011); Sharon Wall, *The Nurture of Nature: Childhood, Antimodernism, and Ontario Summer Camps, 1920–55* (Vancouver: UBC Press, 2009); Walden, *Becoming Modern in Toronto*.
41 The classic work on antimodernism is T.J. Jackson Lears, *No Place of Grace: Antimodernism and the Transformation of American Culture, 1880–1920* (New York: Pantheon Books, 1981). For the Canadian context, see also Lynda Jessup, ed., *Antimodernism and Artistic Experience: Policing the Boundaries of Modernity* (Toronto: University of Toronto Press, 2001).
42 Ian McKay, *The Quest of the Folk: Antimodernism and Cultural Selection in Twentieth-Century Nova Scotia* (Montreal/Kingston: McGill-Queen's University Press, 1994); Tina Adcock, "Many Tiny Traces: Antimodernism and Northern Exploration between the Wars," in *Ice Blink: Navigating Northern Environmental History*, ed. Stephen Bocking and Brad Martin (Calgary: University of Calgary Press, 2017), 131–77.
43 Tina Loo, "Of Moose and Men: Hunting for Masculinities in British Columbia, 1880–1939," *Western Historical Quarterly* 32, 3 (2001): 296–319; Wall, *The Nurture of Nature*; Dummitt, *The Manly Modern*, Chapter 4; Ross Cameron, "Tom Thomson, Antimodernism, and the Ideal of Manhood," *Journal of the Canadian Historical Association* 10 (1999): 185–208; McKay, *The Quest of the Folk*; Candida Rifkind, "Too Close to Home: Middlebrow Anti-Modernism and the Sentimental Poetry of Edna Jaques," *Journal of Canadian Studies* 39, 1 (2005): 90–114; Michael Dawson, "'That Nice Red Coat Goes to My Head like Champagne': Gender, Antimodernism, and the Mountie Image, 1880–1960," *Journal of Canadian Studies* 32, 3 (1997): 119–39; Patricia Jasen, *Wild Things: Nature, Culture, and Tourism in Ontario, 1790–1914* (Toronto: University of Toronto Press, 1995); Mark Moss, *Manliness and Militarism: Educating Young Boys in Ontario for War* (Don Mills, ON: Oxford University Press, 2001), Chapter 5.
44 For a critical analysis of the Group of Seven's historical and contemporary legacies, see Lynda Jessup, "Prospectors, Bushwhackers, Painters: Antimodernism and the Group of Seven," *International Journal of Canadian Studies* 17 (1998): 193–214.
45 Tina Loo, *States of Nature: Conserving Canada's Wildlife in the Twentieth Century* (Vancouver: UBC Press, 2006); John Sandlos, *Hunters at the Margin: Native Peoples and Wildlife Conservation in the Northwest Territories* (Vancouver: UBC Press, 2007).
46 James C. Scott, *Seeing like a State: How Certain Schemes to Improve the Human Condition Have Failed* (New Haven, CT: Yale University Press, 1998).
47 Tina Loo and Meg Stanley, "An Environmental History of Progress: Damming the Peace and Columbia Rivers," *Canadian Historical Review* 92, 3 (2011): 407, 414.

48 Daniel Macfarlane, *Negotiating a River: Canada, the US, and the Creation of the St. Lawrence Seaway* (Vancouver: UBC Press, 2014). See also his chapter in this volume.
49 Tina Loo, "People in the Way: Modernity, Environment, and Society on the Arrow Lakes," *BC Studies* 142–43 (2004): 161–96.
50 Joy Parr, *Sensing Changes: Technologies, Environments, and the Everyday, 1953–2003* (Vancouver: UBC Press, 2010); Tina Loo, "Disturbing the Peace: Environmental Change and the Scales of Justice on a Northern River," *Environmental History* 12, 4 (2007): 895–919.
51 Comacchio, *Nations Are Built of Babies*; Cynthia Comacchio, "Mechanomorphosis: Science, Management, and 'Human Machinery' in Industrial Canada, 1900–45," *Labour/Le travail* 41 (1998): 35–67; Jones-Imhotep, *The Unreliable Nation*.
52 Tina Loo, "High Modernism, Conflict and the Nature of Change in Canada: A Look at *Seeing Like a State*," *Canadian Historical Review* 97, 1 (2016): 34–58; Parr, *Sensing Changes*, 189–98.
53 See William Cronon, "The Uses of Environmental History," *Environmental History Review* 17, 3 (1993): 1–22.
54 For a thorough discussion of the shift away from universalist histories of science and technology, see Lissa Roberts, "Situating Science in Global History: Local Exchanges and Networks of Circulation," *Itinerario* 33, 1 (2009): 9–30.
55 See Trevor H. Levere and Richard A. Jarrell, eds., *A Curious Field-Book: Science and Society in Canadian History* (Toronto: Oxford University Press, 1974); Wilfrid Eggleston, *National Research in Canada: The NRC, 1916–1966* (Toronto: Clarke, Irwin, 1978); and Vittorio de Vecchi, "The Dawning of a National Scientific Community in Canada, 1878–1896," *HSTC Bulletin: Journal of the History of Canadian Science, Technology, and Medicine* 8, 1 (1984): 32–58.
56 Representative works include M. Christine King, *E.W.R. Steacie and Science in Canada* (Toronto: University of Toronto Press, 1989); J.J. Brown, *Ideas in Exile* (Toronto: McClelland and Stewart, 1967); David Zimmerman, "The Organization of Science for War: The Management of Canadian Radar Development, 1939–45," *Scientia Canadensis: Canadian Journal of the History of Science, Technology, and Medicine* 10, 2 (1986): 93–108; Norman R. Ball and John N. Vardalas, *Ferranti-Packard: Pioneers in Canadian Electrical Manufacturing* (Montreal/Kingston: McGill-Queen's University Press, 1994); Arthur Kroker, *Technology and the Canadian Mind: Innis/McLuhan/Grant* (New York: St. Martin's Press, 1985); John N. Vardalas, *The Computer Revolution in Canada: Building National Technological Competence* (Cambridge, MA: MIT Press, 2001); Richard A. Jarrell, *The Cold Light of Dawn: A History of Canadian Astronomy* (Toronto: University of Toronto Press, 1988); and Richard A. Jarrell and James Hull, eds., *Science, Technology, and Medicine in Canada's Past: Selections from Scientia Canadensis* (Thornhill, ON: Scientia Press, 1991). For examples of the broader historiographical discussion of "national styles" to which these works contributed, see Mary Jo Nye, "National Styles? French and English Chemistry in the Nineteenth and Early Twentieth Centuries," *Osiris* 8 (1993): 30–49; and Jonathan Harwood, "National Styles in Science: Genetics in Germany and the United States between the World Wars," *Isis* 78, 3 (1987): 390–414.

57 See, for example, Richard A. Jarrell and Norman R. Ball, eds., *Science, Technology, and Canadian History* (Waterloo, ON: Wilfrid Laurier University Press, 1980).
58 Levere and Jarrell, *A Curious Field-Book*; Jarrell, *The Cold Light of Dawn*. See also Richard A. Jarrell and Roy MacLeod, eds., *Dominions Apart: Reflections on the Culture of Science and Technology in Canada and Australia, 1850–1945* (Ottawa: Canadian Science and Technology Historical Association, 1994).
59 See Adas, *Machines as the Measure of Men*.
60 Jarrell and MacLeod, *Dominions Apart*; Suzanne Zeller, *Inventing Canada: Early Victorian Science and the Idea of a Transcontinental Nation* (Montreal/Kingston: McGill-Queen's University Press, 1987); Suzanne Zeller, "Environment, Culture, and the Reception of Darwin in Canada, 1859–1909," in *Disseminating Darwinism: The Role of Place, Race, Religion, and Gender*, ed. Ronald L. Numbers and John Stenhouse (Cambridge, UK: Cambridge University Press, 1999), 91–122; Carl Berger, *Science, God, and Nature in Victorian Canada: The 1982 Joanne Goodman Lectures* (Toronto: University of Toronto Press, 1983); Robert Bothwell, *Nucleus: The History of Atomic Energy of Canada Limited* (Toronto: University of Toronto Press, 1988); Yves Gingras, *Physics and the Rise of Scientific Research in Canada* (Montreal/Kingston: McGill-Queen's University Press, 1991); Trevor H. Levere, "The History of Science of Canada," *British Journal of the History of Science* 21, 4 (1988): 419–25. On the imaginative role of technology, see R. Douglas Francis, *The Technological Imperative in Canada: An Intellectual History* (Vancouver: UBC Press, 2009); and A.A. den Otter, *The Philosophy of Railways: The Transcontinental Railway Idea in British North America* (Toronto: University of Toronto Press, 1997).
61 Ted Binnema, *Enlightened Zeal: The Hudson's Bay Company and Scientific Networks, 1670–1870* (Toronto: University of Toronto Press, 2014).
62 Kenneth Norrie and Doug Owram, *A History of the Canadian Economy*, 2nd ed. (Toronto: Harcourt Brace Canada, 1996), 245–50; Graham D. Taylor and Peter A. Baskerville, *A Concise History of Business in Canada* (Don Mills, ON: Oxford University Press, 1994), 264; Dorotea Gucciardo, "Wired! How Canada Became Electrified" (PhD diss., Western University, 2011). Ruth Schwartz Cowan has examined a similar intersection for the United States in her now-classic *More Work for Mother: The Ironies of Household Technology from the Open Hearth to the Microwave* (New York: Basic Books, 1983).
63 Pierre Bélanger, ed., *Extraction Empire: Sourcing the Scales, Systems, and States of Canada's Global Resource Empire* (Cambridge, MA: MIT Press, 2017). On Canada's "extractive empire" within its borders, namely the North, see Arn Keeling and John Sandlos, eds., *Mining and Communities in Northern Canada: History, Politics, and Memory* (Calgary: University of Calgary Press, 2015).
64 Key works include David Wade Chambers and Richard Gillespie, "Locality in the History of Science: Colonial Science, Technoscience, and Indigenous Knowledge," *Osiris* 15 (2000): 221–40; Kapil Raj, *Relocating Modern Science: Circulation and the Construction of Knowledge in South Asia and Europe, 1650–1900* (Basingstoke, UK: Palgrave Macmillan, 2007); James Delbourgo and Nicholas Dew, eds., *Science and Empire in the Atlantic World* (New York: Routledge, 2008); Simon Schaffer, Lissa Roberts, Kapil Raj, and James Delbourgo, eds., *The Brokered World: Go-Betweens*

and Global Intelligence, 1770–1820 (Sagamore Beach, MA: Science History Publications, 2009); and Sujit Sivasundaram, ed., "Focus: Global Histories of Science," *Isis* 101, 1 (2010): 95–158.

65 See, for example, Karen Dubinsky, Adele Perry, and Henry Yu, eds., *Within and without the Nation: Canadian History as Transnational History* (Toronto: University of Toronto Press, 2015); Benjamin H. Johnson and Andrew R. Graybill, eds., *Bridging National Borders in North America: Transnational and Comparative Histories* (Durham, NC: Duke University Press, 2010); Nancy Christie, ed., *Transatlantic Subjects: Ideas, Institutions, and Social Experience in Post-Revolutionary British North America* (Montreal/Kingston: McGill-Queen's University Press, 2008); Allan Greer, "National, Transnational, and Hypernational Historiographies: New France Meets Early American History," *Canadian Historical Review* 91, 4 (2010): 695–724; and Adele Perry, *Colonial Relations: The Douglas-Connelly Family and the Nineteenth-Century Imperial World* (Cambridge, UK: Cambridge University Press, 2015).

66 See Adi Ophir and Steven Shapin, "The Place of Knowledge: A Methodological Survey," *Science in Context* 4, 1 (1991): 3–21. On the "localist" (sometimes also termed the "spatial" or "geographical") turn, see Diarmid A. Finnegan, "The Spatial Turn: Geographical Approaches in the History of Science," *Journal of the History of Biology* 41, 2 (2008): 369–88.

67 We regret, for instance, that this volume does not explore science, technology, and modernity in Atlantic Canada. For recent scholarship about this region that engages these and cognate themes, including weather and climate, see, for example, Jennifer Hubbard, *A Science on the Scales: The Rise of Canadian Atlantic Fisheries Biology, 1898–1939* (Toronto: University of Toronto Press, 2006); Suzanne Zeller, "Reflections on Time and Place: The Nova Scotian Institute of Science in Its First 150 Years," *Proceedings of the Nova Scotian Institute of Science* 48, 1 (2015): 5–61; Liza Piper, "Backward Seasons and Remarkable Cold: The Weather over Long Reach, New Brunswick, 1812–21," *Acadiensis* 34, 1 (2004): 31–55; Liza Piper, "Colloquial Meteorology," in *Method and Meaning in Canadian Environmental History*, ed. Alan MacEachern and William J. Turkel (Toronto: Nelson Education, 2009), 102–23; and Teresa Devor, "The Explanatory Power of Climate History for the 19th-Century Maritimes and Newfoundland: A Prospectus," *Acadiensis* 43, 2 (2014): 57–78.

68 The first three points have been made most recently by contributors to the special forum on Canadian environmental history in the *Canadian Historical Review*'s December 2014 issue; see especially the essays by Stephen J. Pyne and Sverker Sörlin. On technology, see also, for example, the introduction to Bruce Sinclair, Norman R. Ball, and James O. Petersen, eds., *Let Us Be Honest and Modest: Technology and Society in Canadian History* (Toronto: Oxford University Press, 1974), 1–3. On the state, see, for example, Richard Jarrell, "Measuring Scientific Activity in Canada and Australia before 1915: Exploring Some Possibilities," in Jarrell and MacLeod, *Dominions Apart*, 27–52; and Philip Enros, ed., "Science in Government," *Scientia Canadensis: Canadian Journal of the History of Science, Technology, and Medicine* 35, 1–2 (2012): 1–149.

69 For a good overview of the rise of mobility in the history of science, see James A. Secord, "Knowledge in Transit," *Isis* 95, 4 (2004): 654–72.
70 See Paula Hastings and Jacob A.C. Remes, "Empire, Continent, and Transnationalism in Canada: Essays in Honor of John Herd Thompson," *American Review of Canadian Studies* 45, 1 (2015): 1–7.
71 Other scholarship on connections between Canada, the Caribbean, and Central and South America includes Karen Dubinsky, *Babies without Borders: Adoption and Migration across the Americas* (Toronto: University of Toronto Press, 2010); Sean Mills, *A Place in the Sun: Haiti, Haitians, and the Remaking of Quebec* (Montreal/Kingston: McGill-Queen's University Press, 2016); and a number of the essays in Dubinsky, Perry, and Yu, *Within and without the Nation*.
72 See, for example, Stephen Bocking, "A Disciplined Geography: Aviation, Science, and the Cold War in Northern Canada, 1945–1960," *Technology and Culture* 50, 2 (2009): 265–90; and Matthew Farish, "Frontier Engineering: From the Globe to the Body in the Cold War Arctic," *Canadian Geographer* 50, 2 (2006): 177–96.
73 For "life geographies," see David N. Livingstone, *Putting Science in Its Place: Geographies of Scientific Knowledge* (Chicago: University of Chicago Press, 2003), 182–83. Historians of the early modern world have recently demonstrated how a specific kind of travelling actor, the go-between, facilitated the translation and circulation of knowledge in that era. See Schaffer et al., *The Brokered World*.
74 Perry, *Colonial Relations*, 255.
75 Sverker Sörlin, "National and International Aspects of Cross-Boundary Science: Scientific Travel in the 18th Century," in *Denationalizing Science: The Contexts of International Scientific Practice*, ed. Elizabeth Crawford, Terry Shinn, and Sverker Sörlin (Dordrecht: Kluwer, 1993), 45, as quoted in Chambers and Gillespie, "Locality in the History of Science," 223.
76 Percy Bysshe Shelley, "A Defence of Poetry," in *Shelley's Poetry and Prose: Authoritative Texts, Criticism*, 2nd ed., ed. Donald H. Reiman and Neil Fraistat (New York: Norton, 2002), 538.
77 Recent examples of this work include Joy Parr, "Notes for a More Sensuous History of Twentieth-Century Canada: The Timely, the Tacit, and the Material Body," *Canadian Historical Review* 82, 4 (2001): 719–45; Parr, *Sensing Changes*; Kenny, *The Feel of the City*; and Jarrett Rudy, Nicolas Kenny, and Magda Fahrni, "'An Ocean of Noise': H.E. Reilley and the Making of a Legitimate Social Problem, 1911–45," *Journal of Canadian Studies* 51, 2 (2018): 261–88.
78 The history of medicine in Canada is a vibrant, distinct field of inquiry, with its own scholarly journal and association. A discussion of its central themes and concerns lies outside the scope of this chapter. For good introductions to this body of literature, see, for example, S.E.D. Shortt, *Medicine in Canadian Society: Historical Perspectives* (Montreal/Kingston: McGill-Queen's University Press, 1981); Wendy Mitchinson and Janice Dickin McGinnis, eds., *Essays in the History of Canadian Medicine* (Toronto: McClelland and Stewart, 1988); Jacques Bernier, *Disease, Medicine, and Society in Canada: A Historical Overview* (Ottawa: Canadian Historical Association, 2003); E.A. Heaman, Alison Li, and Shelley McKellar, eds.,

Essays in Honour of Michael Bliss: Figuring the Social (Toronto: University of Toronto Press, 2008); and Wendy Mitchinson, *Body Failure: Medical Views of Women, 1900–1950* (Toronto: University of Toronto Press, 2013).
79 Patrizia Gentile and Jane Nicholas, eds., *Contesting Bodies and Nation in Canadian History* (Toronto: University of Toronto Press, 2013).
80 See Steven Shapin, "Placing the View from Nowhere: Historical and Sociological Problems in the Location of Science," *Transactions of the Institute of British Geographers* NS 23, 1 (1998): 5–12.
81 Sinclair, Ball, and Petersen, *Let Us Be Honest and Modest*, 3.
82 An invaluable contribution here is Richard Jarrell's posthumously published monograph *Educating the Neglected Majority: The Struggle for Agricultural and Technical Education in Nineteenth-Century Ontario and Quebec* (Montreal/Kingston: McGill-Queen's University Press, 2016).
83 See, for example, *Scientia Canadensis: Canadian Journal of the History of Science, Technology, and Medicine* 40, 1 (2018), a special issue on environmental history and the history of technology co-edited by Daniel Macfarlane and William Knight; Dolly Jørgensen, Finn Arne Jørgensen, and Sara B. Pritchard, eds., *New Natures: Joining Environmental History with Science and Technology Studies* (Pittsburgh: University of Pittsburgh Press, 2013); and Dolly Jørgensen and Sverker Sörlin, eds., *Northscapes: History, Technology, and the Making of Northern Environments* (Vancouver: UBC Press, 2013). The introduction of Sara Pritchard's monograph *Confluence: The Nature of Technology and the Remaking of the Rhône* (Cambridge, MA: Harvard University Press, 2011) also provides a useful primer in envirotechnical theory and methods.
84 See also Bocking's related but distinct blog post on this topic: "Landscapes of Science," *The Otter~La loutre* (blog), Network in Canadian History and Environment (NiCHE), January 14, 2015, http://niche-canada.org/2015/01/14/landscapes-of-science/.
85 See Sinclair, "Preface," in Jarrell and Ball, *Science, Technology, and Canadian History*, xiii.
86 Zeller, *Inventing Canada*.
87 See Ian Mosby, "Administering Colonial Science: Nutrition Research and Human Biomedical Experimentation in Aboriginal Communities and Residential Schools, 1942–1952," *Histoire sociale/Social History* 46, 91 (2013): 145–72.

PART 1

BODIES

1

Civilizing the Natives
Richard King and His Ethnographic Writings on Indigenous Northerners

EFRAM SERA-SHRIAR

In 1836, Richard King (1810–76), the Arctic explorer and surgeon turned ethnologist, published an account of his experiences travelling across northern Canada as a member of the Great Fish River Expedition (later known as the George Back Expedition) between 1833 and 1835. During this trip, he engaged extensively with various groups of Indigenous northerners and began writing detailed ethnographic reports on their physical appearances, customs, habits, and belief systems. Dissatisfied with how Europeans had been examining and representing these peoples in published works, images, and exhibitions, King sought to transform British race science by founding a modern discipline built upon more rigorous and standardized criteria for producing reliable results. This effort coincided with the rise of scientific institutions and a growth in print culture.[1] King complained that, "in the absence of the necessary facts, late writers have had to trust to compilations, and defective authorities."[2] King's ethnographic writings on Indigenous northerners sparked the beginning of what would become a major preoccupation for the rest of his life – the scientific study of human diversity.

This chapter examines King's ethnographic reports from his voyage to the Canadian Arctic in the 1830s. This experience represented a formative moment in King's ethnological and ethnographic career and drove him to establish the Ethnological Society of London (ESL) in 1843.[3] Not only was it the first time that he encountered and engaged directly with extra-Europeans, but also it helped him to recognize some of the shortcomings that ethnographies of Indigenous peoples in British North America contained. Moreover, his time living among different Indigenous groups opened his eyes to the atrocities created by imperialism in British North America. He witnessed first-hand the effects that European settlement in Canada had on Indigenous populations, and as a result of seeing the dire conditions of these peoples he began championing the civilizing mission. King believed that the scientific study of extra-European peoples went hand in hand with humanitarianism. Science provided concrete evidence to support the notion that all humans, regardless of race, had the potential to be equal and therefore deserved fair treatment.[4] This preoccupation can be framed as a modernist turn in the development of the discipline, one indebted to the enlightened, rationalist thinking from the eighteenth century that was slowly displacing the racial hierarchies of the premodern era.[5]

King's ethnographic authority derived from his first-hand knowledge of Indigenous peoples in British North America. For many early ethnologists, travel narratives – such as the one produced by King – provided a crucial source of data for their studies. These narratives contributed to a growing archive that made possible the verification, expansion, and correction of ethnographic knowledge.[6] The physician and ethnologist Thomas Hodgkin (1798–1866), a cofounder of the ESL, believed that King's ethnographic work on Indigenous peoples in British North America was one of the best examples available and provided a sound framework for others at the ESL to emulate. Hodgkin wrote that "the Society has already excellent examples in the communications of Dr Richard King ... in which personal knowledge of the people was seconded by a careful reference to original authorities and observations ... which must be of the greatest value to those who may take up similar investigations."[7] King's travel narrative provided a detailed historical picture of Indigenous peoples in British North America. By combining his personal insights with the authoritative texts of other eyewitnesses, King could corroborate the accuracy of older reports, alter and improve ethnographic records that contained inconsistencies or ambiguities, and correct misinformation. At the core of his ethnographic writings was an ambition to promote the civilizing of northern Indigenous peoples through the acquisition of European knowledge and to advocate for their rights to land, fair pay, access to resources, and ethical treatment.

For the most part, King's contributions to the disciplinary history of British ethnology and ethnography have been overlooked. His most detailed biography, written by Hugh Wallace, focuses on his exploits as an Arctic explorer. However, during the first half of the nineteenth century, King played a central role in both improving and modernizing how ethnographic reporting was done by highlighting what he believed to be its main weaknesses and by helping to establish an ethnological community in Britain. He should therefore be positioned as one of the progenitors of the modern discipline.[8] Given that his interest in ethnology and ethnography originated during his voyage to the Canadian Arctic, it seems to be apropos to examine in detail his writing on Indigenous northerners in the *Narrative of a Journey to the Shores of the Arctic Ocean*.

There are three central points to underscore. First, King wanted to expand ethnographic knowledge of northern Indigenous peoples in his travel narrative by providing detailed reports on their physical appearance, material culture, and customs and habits. Second, he was critical of a lot of ethnographic reporting because he believed that it misrepresented Indigenous peoples. King therefore wanted to improve how information on extra-Europeans was recorded during encounters with them. Third, the civilizing mission was an integral component of his ethnographic research, and it is a main theme throughout his writings. The 1830s were pre-Confederation, and Canada was part of the British Empire. Although King was working as an agent for the British Navy, he was conducting his research in Canada. Therefore, this case study is best characterized as an example of British Canadian science, and it adds another dimension to our understanding of the making of modern science in Canada. King's ethnographic efforts provided researchers with previously unknown or neglected information on Indigenous peoples living in the emerging nation. As Suzanne Zeller argues, "science became the gauge by which Canadians assessed what their country ... could one day become."[9] Expeditions, such as the one that King embarked on during the early 1830s, were an essential part of the story that created the ethos of Canada's scientific identity in the years ahead.

King's Ethnographic Writing from the Great Fish River Expedition

Richard King grew up in London and came from a middling-sort background. His father, Richard King Sr., worked for the Ordnance Office in London. At the age of fourteen, King began training as a surgeon and apothecary, and in 1824 he commenced a seven-year apprenticeship with the Society of Apothecaries. He qualified as a licentiate in 1832 while working as a student probationer at Guy's Hospital in London.[10] Soon after completing his medical training, King joined the Royal Navy, and he was commissioned to be both the medical officer and the second-in-command under the veteran Arctic explorer George Back (1796–1878) for an expedition to the Canadian Arctic. Back's expedition was part of an ongoing effort by the British Navy to locate the Northwest Passage. Back himself was part of John Franklin's Coppermine River Expedition (1819–22) that nearly led to the deaths of the whole crew. Franklin led a slightly more successful mission to the Mackenzie River in the mid-1820s, and much of this improvement was the result of the support that he received from the Hudson's Bay Company (HBC).[11] The British Navy wanted to build upon this positive momentum and further explore the region.

Generally speaking, Back's voyage was primarily overland and comprised about sixteen men – though that number fluctuated depending on health issues and the needs of the expedition. Members of the team set sail in February 1833 with three main tasks. The first was to locate, and render assistance to, the missing British explorer John Ross (1777–1856) and his crew. The second was to collect natural history data and survey the landscape along the Great Fish River, which is known today as the Back River and is located in the Northwest Territories and Nunavut. The third was to find – if possible – a route to the Northwest Passage. The team embarked on their long trek from Montreal, travelling across Quebec, Ontario, and Manitoba en route to their winter lodgings at Fort Reliance, on the eastern arm of Great Slave Lake in the Northwest Territories. These lodgings were purpose-built for the expedition. By the time that Back's team reached Fort Reliance, though, Ross had returned to England. Thus, they turned their attention solely to natural history collecting and surveying.[12]

Although King was engaged in several duties, such as surveying the landscape, acting as medical officer, and collecting natural history specimens, ethnography seems to have been his preoccupation. Ethnographic observations are sprinkled throughout his travel account, but his most significant ethnographic analysis is Chapter 12 in the second volume of his *Narrative of a Journey*, which focuses almost exclusively on Indigenous peoples in North America, with an emphasis on the peoples living in the North.[13] Taken as a whole, the *Narrative of a Journey* is an impressive ethnographic source, containing detailed descriptions of several Indigenous groups, including Iroquois, Saulteaux, Cree, Chipewyan, Dene, and Inuit. King's descriptions outline the physical appearance of each group and their intelligence, and the account discusses their material cultures, customs, and habits. As a resource for early ethnologists, it was invaluable. King was not only highly attuned to the needs of researchers interested in the scientific study of races, but he was also trained in natural history and medicine – two fields regarded as essential to the emerging discipline's research program.[14]

Travel narratives, such as the one produced by King, were important evidentiary resources for armchair scholars in Britain, who were reliant on eyewitness accounts to substantiate the credibility of their theories.[15] King's ethnographic authority derived from his direct engagement with Indigenous populations in their local habitats, and he was seen as a trustworthy reporter.[16] Understanding how he composed his descriptions of Indigenous northerners is important, therefore, because it shows how modern constructions of Indigenous peoples were made and the sociopolitical influences that shaped them.[17] Within these "contact zones," as Mary Louise Pratt has called them, Europeans created racial characteristics by juxtaposing their own languages, customs, habits, and physical features with those of Indigenous populations.[18] King's *Narrative of a Journey* was no different, and to maximize the research value of this text, it is important to understand the politics of writing embedded in it. The racial identities that King constructed in his writings were entrenched in European preconceptions of northern Indigenous peoples and not representative of these peoples' self-perceptions.[19]

Although King's narrative is a highly sophisticated ethnographic source, in the first few chapters his racial descriptions are fairly basic, with general references to the physical appearances and clothing of Indigenous peoples. In many respects, these descriptions reflect the superficial nature of his encounters with different Indigenous groups during the early stages of the expedition. Most of his meetings en route to Fort Reliance were brief. For instance, one of his earliest encounters with Indigenous peoples in North America was with a group of Saulteaux men near Fort William in the modern-day Thunder Bay area.[20] King gave a brief description of their physicality as being "warlike, athletic and bold," and he included a passing reference to their jewellery and hair ornaments. There was no attempt to learn more about the significance of these decorations in Saulteaux culture. Nor did King try to build a rapport with them. The two groups simply exchanged some goods and then parted ways.[21] This sort of ethnographic description was common in early-nineteenth-century travel narratives, and the surgeon, medical lecturer, and ethnologist William Lawrence (1783–1867) lamented in his lecture on the *Natural History of Man* (1819) that scientific travellers had too often wasted their opportunities to expand ethnographic knowledge during interactions with Indigenous peoples by focusing on mere descriptions of their appearance and material culture, without prodding into the significance of tattoos, piercings, clothes, ornaments, or any other material object. What an ethnographer required was a more profound understanding of the culture under view.[22] King developed this approach as the expedition progressed.

As his knowledge of extra-Europeans grew through further interactions with various Indigenous peoples, his ethnographic observations improved and deepened. For instance, King soon learned the standard values of trade items among Indigenous populations, and he explained the customary trading system in his *Narrative of a Journey*:

> The beaver-skin is the standard of exchange in all transactions with the Indians, for which a coarse butcher's knife, or small file, is considered equivalent. A gun, worth about twenty shillings in England, is valued at fifteen beaver-skins; and a fathom of coarse cloth, or a small woollen blanket, at eight skins. Three marten, eight musk-rat, or a single wolverine skin, are reckoned one skin; a silver fox, or otter, two; and a black fox, or black bear, four skins.[23]

This sort of general overview of the currency value of furs and European objects among Indigenous communities was useful information not only to ethnologists and ethnographers interested in extra-European economic systems but also to merchants and traders, colonial agents, and military officials. As Michael Bravo has noted, "the success of ethnological research was dependent on gaining reliable access to information about the people of other nations ... In order to gain access to this kind of knowledge, it was vital to find reliable and productive sources throughout the world, and particularly in areas where there were British political and economic interests at stake."[24] Linking ethnographic knowledge to imperialist concerns was beneficial to both sides. Government officials learned about the subtleties of extra-European markets, and ethnographers gained further access to Indigenous communities by collecting information during government-sponsored explorations into little-known lands.[25]

King's engagement with Indigenous peoples, though, soon transformed substantially when he arrived at Fort Reliance in September 1833. Within this setting, he lived in close quarters with groups of Cree, Chipewyan, and Dene. This proximity afforded him an opportunity to learn about the details of their cultures. For example, King discussed at length how these Indigenous communities hunted caribou and used the entire carcass of the animal to fashion different items. He also examined how Indigenous northerners prepared the skins for clothing, made instruments and weapons out of the bones and antlers, and turned tendons and dorsal muscles into sewing thread.[26] In addition, King provided comprehensive descriptions of other Indigenous objects, such as sledges.[27] In short, these types of ethnographic observations painted a thorough portrait of Indigenous northerners' lives – albeit from a highly Europeanized perspective. They were records of the minute details of Indigenous peoples' social organization and material culture. For researchers in Britain interested in the Indigenous peoples of British North America, these were invaluable sets of data.[28]

As King became more confident in his own ethnographic reporting, he began to correct the observations of other travellers who had visited the region and commented on Indigenous populations. For example, King argued that a lack of orthographical knowledge among European explorers had created many inconsistencies when identifying geographical landmarks named by Indigenous northerners:

> The Great Fish and Fish Rivers have a host of names, because travellers will not give themselves the trouble to make use of their sense of hearing. The Great Fish River is compounded of three words, and should be written Thlĕwŷ-cho-dĕzză, Fish-Great-River: instead of which, the word for river is written dezeth, desh, and even tessy; while the word for fish is compounded into thlew-ee, threw-ey, thelew-eye, – meaning nothing at all, like twoy-to. The Fish River, again, is more abused by the terms, The-lew, The-lon, Thelew-ey-aze, Thlew-y-aze' whereas it should be written Thlĕwŷ-dĕzză.[29]

These discrepancies in the observations of explorers meant that it was not always clear whether the correct locations were being discussed in different narratives. In addition, it made it more difficult to cross-compare them for accuracy.[30] To counteract this problem, King argued that it was better to use English terms for naming geographic landmarks.[31]

There were also instances of heightened awareness in King's ethnographic observations. Before encountering Inuit for the first time, King had a preconceived view of them as savage brutes. One reason that he held an extreme view of these people was his prior communication with other Indigenous northerners. While wintering at Fort Reliance, King built a strong rapport with a local chief of the Yellowknives named Akaitcho (1786–1838), who had been supplying the British with meat.[32] Through his friendship with Akaitcho, King learned a great deal about Indigenous northerners. However, the information that he gathered on Inuit was prejudiced because

Akaitcho's people were competing with some Inuit communities for the scarce resources on neighbouring barren lands to the north. As a result of this feuding, Akaitcho depicted Inuit in a negative light.[33] Although King discussed a feud between Yellowknives and Inuit, it is possible that he was mistaken, and that Akaitcho was speaking of his rivalry with a local Dogrib chief named Edzo.[34] Confusing different northern Indigenous communities was common in nineteenth-century ethnographic reporting, and it often led to problematic misrepresentations of extra-Europeans that had long-term consequences. For instance, it could affect subsequent European interactions with Indigenous communities if a traveller read an account that portrayed a group as hostile or aggressive. Future dealings could be prejudiced by these false claims.[35]

The first meeting between Back's party and a group of Inuit was a tense affair. King had assumed, because of his discussions with Akaitcho, that the group of Inuit would be hostile to the British. However, King's self-proclaimed composure and sensitivity to Indigenous customs helped him to calm the situation and reassess the cultural meaning of the encounter. When the British approached the shore where the party of Inuit was encamped, armed Inuit quickly surrounded them. Because the British did not have an interpreter with them to speak to the group of Inuit, they struggled to determine the meaning of this action. King recounted that

> the Esquimaux, about nine in number, perceiving that it was our intention to land, approached the boat, brandishing their spears tipped with bone; and having formed themselves into a semicircle, they commenced an address in a loud tone of voice, during the whole of which time they continued alternately elevating [and] depressing both their arms. They motioned us to put off from the shore, and at the same time uttered some unintelligible words with a wildness of gesticulation that clearly show[ed] they were under the highest state of excitement.[36]

Back's party possessed a vocabulary of Inuit language, recorded by the British explorer William Edward Parry (1790–1855) during his various expeditions to the Arctic in the 1810s and 1820s.[37] However, because of their unfamiliarity with the spoken language, King and the others found it impossible to use the document when trying to understand the party of Inuit. Instead, King stated that they resorted to uttering basic words in an attempt to pacify the situation, and he wrote that, "at the sound of tĭmā (peace), kăblōōns (white people), they ceased yelling; and after repeating those words, they one and all laid down their spears, and commenced alternately patting their breasts and pointing to the heavens."[38] This is a wonderful example of how Back's team relied on the reports of earlier travellers for information on Indigenous peoples. Moreover, it highlights one of the shortcomings of the available literature on groups such as Inuit. There was no ethnographic manual for engaging with extra-European peoples until later in the century, and the information that the British had was not always helpful.[39] Yet, as King argued, it was at least effective in communicating the necessary information to prevent a fight.

The lack of a skilled interpreter also hindered the team's ability to collect information on the party of Inuit, and King bemoaned the loss of Augustus, an Inuk interpreter whom Back had met during his previous expedition to the Arctic under the command of Franklin (1786–1847) in the early 1820s. Augustus had intended to join Back's expedition but sadly perished en route to Fort Reliance. King stated that

> here we felt the want of poor Augustus, who could have explained to us, had his life been spared, many important facts relative to these interesting members of the human race. Numberless uncertainties as regards the line of coast might have been definitively set at rest, and our progress very much assisted, from the information we should have been able to glean from them.[40]

With a skilled interpreter present, King believed, the British could have gained more information – especially geographical knowledge of local terrain. Back's team had already acquired some information on the landscape along the coast of the Great Fish River from discussions with Akaitcho and his people; however, given how much more familiar Inuit were with the river, their knowledge would have greatly helped the British with their endeavour.[41] There was substantial evidence in Arctic travel narratives of situations in which Europeans learned about the natural history and geography of the Arctic through their interactions with Inuit. This sort of knowledge exchange was vital for European expeditions into unknown spaces.[42]

King's apparent reflectiveness as an ethnographer sensitized him to some of the limitations of his observations. During his first meeting with Inuit, he recognized that, because it was a short interaction, he was unable to provide a comprehensive portrait of the community: "These arrangements, occupying about half an hour, were no sooner completed than we embarked: our observations, therefore, as regards those people were necessarily very limited."[43] King also reassessed the meaning of the initial engagement of Back's party with the male members of the Inuit group after another elderly Inuk man joined them later. As he recounted,

> that they were labouring under the greatest alarm when the boat first grounded, there cannot be a shadow of doubt; but as the same alternate elevation and depression of their arms was made with a cheerful countenance, during the interview, to an elderly man at some distance in the rear of the party, who thereupon immediately laid down his arms, and with a fearless and quick step joined the younger warriors, I am inclined to believe those motions [were] emblematic of peace.[44]

King continued to describe the physical appearance of the Inuit party, their material culture, and their social practices. He concluded by stating that, contrary to Akaitcho's opinion, they were a friendly and helpful people.[45]

Because many early ethnologists never left the shores of Europe, they highly prized visual representations of Indigenous peoples. These images were used to cross-compare the physical features of different groups. By identifying similar and dissimilar characteristics among the races of the world, ethnological researchers could build arguments in support of either monogenesis or polygenesis, or reject them, depending on their theoretical orientation.[46] Ethnographers who collected this type of data took great care in constructing the alleged accuracy of their drawings.[47] To sketch his subjects, King used a camera lucida, an optical device that superimposed a projection of the subject onto a drawing surface.

The use of a camera lucida added further authenticity to King's illustrations and allowed him to make an objective claim because he could argue that they were verisimilar representations of Indigenous peoples based upon projections created through an instrument.[48] What is interesting about his attempts to draw the Indigenous peoples with whom he interacted during the expedition is that, in many cases, the subjects did not like how they were being depicted in the illustrations – even though they were done through the camera lucida. In several cases, King's subjects actively negotiated how they were being depicted, and this shows an interesting case of Indigenous agency in the travel narrative.

As an example, King attempted to draw a young Indigenous woman who had a damaged eye, which she hid with her hair. However, when he used the camera lucida to project her image onto a drawing service, the eye appeared in view. King recounted that, seeing this, "she was so exceedingly mortified as for a long time to refuse sitting for her portrait, and then persisted in covering that imperfection."[49] Similarly, when drawing Akaitcho, King remarked that

> Akaitcho, who had an excrescence about the size of a pea upon his forehead, seemed amused in the highest degree as long as he thought its appearance on paper was intended as a caricature; but finding it remained so, he placed his finger over the representation, observing, with a smiling countenance, in that way it was năzōō (good); but withdrawing his finger, he said, in a contemptuous manner, năzōōlăh (bad).[50]

Both the young woman and Akaitcho wanted to be represented in an idealized manner, whereas King wanted to produce authentic representations of them for ethnographic purposes. In the end, neither illustration featured in the travel narrative. This suggests either that the images were never completed because of the protests or that King might have decided against their inclusion because of his respect for the young woman and Akaitcho.

In sum, King's ethnographic descriptions were included throughout the two volumes of *Narrative of a Journey* and provided detailed portraits of several Indigenous groups in British North America, including Iroquois, Saulteaux, Cree, Chipewyan, Dene, and Inuit. King recorded information on their physical appearances, material cultures, and customs and habits. He was aware of the limitations of his observations – openly stating them in some places – and made suggestions on how other travellers to British North America could strengthen and expand their reports. But there was another dimension to his ethnographic writings on Indigenous northerners, one critical of how European settlement and trade in the region had adversely affected Indigenous life. In response to the negative impacts that Europeans had on Indigenous peoples in British North America, King promoted a civilizing mission.

King and the Civilizing Mission

Richard King's promotion of the civilizing mission in British North America transformed after he witnessed first-hand the dire conditions of Indigenous northerners. Although steeped in Eurocentric biases that prioritized European culture over those of extra-Europeans, the mission was still a highly sympathetic undertaking. According to King, the situation at Fort Reliance during the winter of 1833–34 was severe. Many parties of northern Indigenous peoples arrived at the encampment in search of food. As the months grew colder, the situation worsened. King stated that it was a particularly harsh winter, and the landscape became increasingly barren. With few resources available, the British could do little to help the local population. A large percentage of the Cree, Chipewyan, and Dene residing at Fort Reliance died of starvation. This extreme circumstance heavily influenced his ethnographic reporting, sensitizing him to the plight of the Indigenous inhabitants. For instance, King remarked in January 1834 that "between forty and fifty human beings lay dead around us; and so scattered that it was impossible to walk in any direction within twenty miles of the house without stumbling against a frozen body."[51] He was overwhelmed by the death toll and wanted to introduce a system in British North America to improve the livelihoods of Indigenous peoples. King linked the decline of Indigenous peoples in North America to the arrival and settlement in the region of Europeans who upset the fragile balance of the local environment. He devoted nearly an entire chapter in the second volume of *Narrative of a Journey* to a discussion of how to improve the grim situation. It was an impassioned analysis – one that promoted his civilizing mission.

A key aim of King's ethnographic portrait of northern Indigenous peoples was to frame them in a positive light, demonstrating both their kindness and their intellectual prowess. This portrait was significant because, if Indigenous populations in British North America were shown to be kindhearted, then they had the potential to become moral citizens through European enculturation. This sort of reappraisal of extra-European identities was indebted to the enlightened, rationalist thinking of the eighteenth century and marked a shift from premodern and derogatory conceptions of humans heavily influenced by the racism of medieval Christian thought.[52] In contrast, for many nineteenth-century Britons, morality was seen as a symbol of modern civilized life. Similarly, if Indigenous peoples demonstrated advanced intelligence through their actions or craftsmanship, then this was evidence of rationalism, which meant that they could adapt to European customs, habits, and values. Again this suggested to King that North American Indigenous peoples could become modern, civilized subjects.

To highlight their moralism, King discussed several historical episodes in which Indigenous northerners assisted the British during severe moments of need. For instance, using Franklin's disastrous Coppermine River Expedition (on which over 50 percent of the crew died of starvation), King stated that the local Dene population bestowed a "delicate and humane attention towards Captain Sir John Franklin and his party." Franklin and the other survivors escaped death because the Dene had provided the British with food and shelter, despite being short themselves on supplies and space.[53] As King argued, their actions "would have done honour to the most civilised people" and illustrated that Indigenous northerners had the potential to be equal to Europeans through their kindness and generosity.[54] Reflecting on how he and the other members of Back's team had treated the Indigenous groups seeking food and shelter at Fort Reliance during the harsh winter, he stated, "I wish we could record a reciprocal kindness on our part."[55]

When it came to showing instances of advanced intelligence among Indigenous northerners, King included a wide range of examples to illustrate their comprehensive knowledge of the natural world and skill in craftsmanship. His aim was to show that northern Indigenous peoples could become as rational and civilized as modern Europeans. King argued, for instance, that Indigenous peoples' knowledge of physiology was extensive, and "with the effect of wounds I found them particularly familiar: they were aware that an injury done to some of the organs of the animal frame caused either an instantaneous or lingering death, while a more severe wound to others would merely produce temporary inconvenience."[56] This depth of physiological understanding, according to King, was comparable to that of Europeans. Moreover, it suggested that northern Indigenous peoples could feasibly learn natural history and medicine.

When it came to constructing material objects such as canoes, smoking pipes, or snowshoes, King argued that "they have shown themselves by no means inferior mechanics" to Europeans, and he believed that this was especially impressive given that Indigenous peoples had only a small selection of tools with which to work, including hatchets, knives, files, and awls.[57] King was equally impressed with how easily Indigenous northerners mastered new languages: "They also possess a great facility for acquiring different languages; for they very commonly speak three, and I have met with some that could converse in four, viz. the English, French, Cree, and Chipewyan tongues."[58] The ability to learn multiple languages, according to King, was another example of advanced intelligence among Indigenous northerners, and it provided further evidence to support the notion that they could be civilized and acculturated into European society.

Despite this potential to become civilized, King argued that European settlement in British North America both hindered the social growth of Indigenous peoples and often led to the decline (or disappearance) of different Indigenous communities:

It is a matter of melancholy reflection that the civilisation of the North American Indians, a numerous race, gifted with the finest qualities that human nature is capable of displaying, should have been obstructed, rather than promoted, by their communication with Europeans: but so it is. They have, by force of example, been taught every vice that could tend to their degradation; while they have not been instructed in those arts which would have added to their comforts and conveniences.[59]

Instead of trying to improve the lives of northern Indigenous peoples by providing them with access to new types of knowledge, medicine, or technology, European settlers had been fostering their denigration through the introduction of vices such as alcohol for their own gains. The HBC, for example, used spirits as a way of controlling and exploiting Indigenous peoples. By encouraging the more skilled hunters to indulge in drinking, its aim was to indenture them into the service of the HBC by supplying them with liquor and burdening them with debt.[60] However, if a hunter became overly dependent on alcohol, he was seen as expendable, and the HBC would cut him off – leaving him in abject poverty.[61]

The dependence on firearms also created problems among Indigenous communities, and King remarked that

> the introduction of fire-arms may be assigned as one cause: for as long as they could obtain a supply of ammunition, they neglected the use of the bow and arrow, the spear, and various modes of trapping and snaring their game; which, from constant disuse, they have now wholly forgotten. That of granting on credit, both in the spring and autumn, a larger outfit of clothing and ammunition than the Indians are able to repay by their winter and summer hunting excursions, places them so completely in the power of the trader by the debt thus incurred, that this must be considered another cause of their decline.[62]

Firearms proved to be highly effective tools for hunting and quickly became the preferred weapon of choice for chasing prey. However, as Indigenous huntsmen became increasingly dependent on the technology, they began losing their proficiency with traditional hunting tools because of disuse. Indigenous hunters were becoming wholly reliant on European weaponry, and their debts quickly grew with the HBC. This system, according to King, created further major problems in North American Indigenous communities because, as hunters advanced in age and became less capable of acquiring furs, the HBC stopped giving them credit, and without the ability to pay for ammunition many of these elderly huntsmen died in destitution.[63]

There were other imbalances in the relationship between the HBC and Indigenous peoples who were collecting furs for the company. Most notably, these hunters were not given fair payments for their labours compared with the profits that the HBC was generating from the sales of these items across Europe. As King stated, "three marten-skins are obtained for a coarse knife, the utmost value of which, including the expense of conveying it to those distant regions, cannot be estimated at more than sixpence; and three of these skins were sold last January in London for five guineas."[64] He continued by outlining that, "with the more expensive furs, such as the black fox, or sea-otter, the profit is more than tripled; and but a few years ago a single skin of the former species sold for fifty guineas, while the native obtained in exchange the value of two shillings."[65] This was a huge monetary gain for the HBC, and King asserted that the "honourable company which by royal charter is permitted to reap such golden harvests might appropriate a small fund to rescue from starvation the decrepit and diseased, who in their youthful days have contributed to its wealth."[66] King believed that the HBC had a moral obligation to render help to Indigenous populations suffering tremendously under this exploitative economic system.

The solution to the plight of Indigenous peoples in British North America, according to King, was to civilize them and afford them the same opportunities as Europeans. However, the problem, King argued, was that Indigenous communities no longer trusted Europeans because of how they had exploited and subjugated them since their arrival in British North America. King wrote that it would be an

> arduous task of making them forget the impressions we have already given to them. If it were possible to eradicate from the mind of the North American Indian all knowledge or traditionary [sic] remembrance of the interference of the whites, which has been exerted with fearful demoralisation for two centuries, and place him in the same state as when first discovered, it would be far easier to effect his civilisation.[67]

If it were possible to re-establish some trust between Indigenous communities and European settlers, King believed that it would then be possible to civilize them. During the 1830s, Europeans promoted two standard ways of civilizing Indigenous peoples. The first method was through conversion, but King argued that such a method would only fuel more distrust among

Indigenous peoples because it would seem like another case of Europeans asserting themselves over Indigenous populations. The second method, which King supported, was through acculturation. He believed that, if Indigenous communities were introduced first to European customs, values, knowledge, and so forth, then there was a better chance that they would civilize. For such a method to be successful, though, he argued that it was essential to improve their current situation.[68]

King's support of this civilizing method was indebted to the writings of Thomas Hodgkin. The two were close friends and corresponded frequently with each other from the 1830s on.[69] Hodgkin was a critic of Christian missionaries trying to civilize Indigenous peoples throughout the world through conversion. He believed that the acceptance of Christianity should be a rational choice, but to make a rational choice the person had to be acculturated by European values, customs, habits, and so forth. The acquisition of European culture was therefore a necessary first step in this model.[70] King forwarded a similar plan: "I think they should be both civilised and converted; but, from my own experience, I certainly am of [the] opinion that it will be found a far easier and more successful task to commence by the former; and if proper attention be paid to the rising generation, conversion would promptly and happily spread amongst old and young."[71] The success of this civilizing mission, according to King, depended on whether the younger generation of Indigenous peoples in North America subscribed to the system. If they did, proponents of this approach argued that it had a higher chance of working on the ensuing generations.

Conclusion

In this chapter, I have examined the ethnographic writings of Richard King from his expedition to the Great Fish River between 1833 and 1835. The travel narrative that he produced during this voyage was an important resource for many different types of readers, including ethnologists, philanthropists, military officials, colonial agents, and fur traders.[72] His ethnographic research represented a transformative moment in the history of the discipline that saw a shift toward a modern view of race that was indebted to enlightened, rational thinking, incorporated new forms of technology, and was grounded in the latest scientific and medical knowledge available. Not only did it contain detailed, sophisticated, and sensitive ethnographic pictures of different Indigenous peoples, but it also highlighted the effects that European imperialism in British North America had on Indigenous peoples living there. King identified the root of the problem (Europeans exploiting them) and offered a solution to the problem (civilizing them).

Nevertheless, despite its broadmindedness and charitable tone, King's narrative was a highly subjective account embedded in European biases of the day that still viewed Indigenous societies as inferior – even if they had the potential to be equal to modern European societies. Studying how King constructed his narrative and racial categories is an important historical exercise because it affords an opportunity to view scientific understandings of race in the early nineteenth century. It also presents opportunities to see the legacies of these historical discourses today.[73] As an ethnographic resource, King's narrative transformed British understandings of Indigenous peoples living in what is today northern Canada, and it represented an early attempt to persuade colonizers to reflect on the implications of their imperial projects. Canadian science – especially in the 1830s, when King was active – was not working in isolation but was part of a much larger global network. Linking this story to a larger British imperial perspective adds another dimension to the modern history of science in Canada. King was an important founding member of British race science. Therefore, his ethnographic writing on Indigenous northerners forms an important part of British anthropology's past. In a way, these Indigenous peoples provided an important sample group for further research, and King's narrative is a classic example of British Canadian science during its prime.

NOTES

1 Janet Browne, "Biogeography and Empire," in *Cultures of Natural History*, ed. N. Jardine, J.A. Secord, and E.C. Spary (Cambridge, UK: Cambridge University Press, 1996), 305–21.
2 Richard King, "Address to the Ethnological Society of London Delivered at the Anniversary, 25th May 1844," *Journal of the Ethnological Society of London* 2 (1850): 18.
3 For more on the Ethnological Society of London, see George Stocking, *Victorian Anthropology* (New York: Free Press, 1987), 244–57; Geoffrey Cantor, *Quakers, Jews, and Science: Religious Responses to Modernity and the Sciences in Britain, 1650–1900* (Oxford: Oxford University Press, 2005), 133–38; Robert Kenny, "From the Curse of Ham to the Curse of Nature: The Influence of Natural Selection on the Debate on Human Unity before the Publication of the *Descent of Man*," *British Journal for the History of Science* 40, 3 (2007): 363–88; Sadiah Qureshi, *Peoples on Parade: Exhibitions, Empire, and Anthropology in Nineteenth-Century Britain* (Chicago: University of Chicago Press, 2011), 186–87; and Efram Sera-Shriar, *The Making of British Anthropology, 1813–1871* (London: Pickering and Chatto, 2013), 53–79.

4 Efram Sera-Shriar, "Arctic Observers: Richard King, Monogenism, and the Historicisation of Inuit through Travel Narratives," *Studies in History and Philosophy of Biological and Biomedical Sciences* 51 (June 2015): 25. King discusses his civilizing mission in detail in the twelfth chapter of the second volume of his narrative. Richard King, *Narrative of a Journey to the Shores of the Arctic Ocean in 1833, 1834, and 1835*, 2 vols. (London: Richard Bentley, 1836), 2: 30–64.
5 For more on shifting notions of race in the modern age, see David N. Livingstone, *Adam's Ancestors: Race, Religion, and the Politics of Human Origins* (Baltimore: Johns Hopkins University Press, 2008), 11–25; and Paul Gillen and Devleena Ghosh, *Colonialism and Modernity* (Sydney: University of New South Wales, 2007), 156–58. For more on King's reconfiguration of Inuit racial constructions, see Sera-Shriar, "Arctic Observers," 23–31.
6 Michael T. Bravo, "Ethnological Encounters," in Jardine, Secord, and Spary, *Cultures of Natural History*, 344; and Sera-Shriar, "Arctic Observers," 23–31. For more on the importance of travel literature in the making of natural sciences generally, see Peter Hulme and Tim Youngs, "Introduction," in *The Cambridge Companion to Travel Writing*, ed. Peter Hulme and Tim Youngs (Cambridge, UK: Cambridge University Press, 2002), 1–16; Janet Browne, "A Science of Empire: British Biogeography before Darwin," *Revue d'histoire des sciences* 45, 4 (1992): 453–75; Lisbet Koerner, "Purposes of Linnaean Travel: A Preliminary Research Report," in *Visions of Empire: Voyages, Botany, and Representation in Nature*, ed. David Phillip Miller and Peter Hanns Reill (Cambridge, UK: Cambridge University Press, 1996), 117–52; and Daniel Carey, "Compiling Nature's History: Travellers and Travel Narratives in the Early Royal Society," *Annals of Science* 54, 3 (1997): 269–92.
7 Thomas Hodgkin, "The Progress of Ethnology," *Journal of the Ethnological Society of London* 1 (1848): 43.
8 Hugh N. Wallace, *The Navy, the Company, and Richard King: British Exploration in the Canadian Arctic, 1829–1860* (Montreal/Kingston: McGill-Queen's University Press, 1980). The most extensive attention that King's ethnology has received in the secondary literature is in Sera-Shriar, *The Making of British Anthropology*, 53–63; and Sera-Shriar, "Arctic Observers," 23–31. King is briefly mentioned in Stocking, *Victorian Anthropology*, 244, 255; and Browne, "A Science of Empire," 465.
9 Suzanne Zeller, *Inventing Canada: Early Victorian Science and the Idea of a Transcontinental Nation* (Montreal/Kingston: McGill-Queen's University Press, 2009), 6.
10 Wallace, *The Navy, the Company, and Richard King*, 20–21.
11 Nanna Kaalund, "From Science in the Arctic to Arctic Science: A Transnational Study of Arctic Travel Narratives, 1818–1883" (PhD diss., York University, 2017), 89–104. For more on Franklin's voyages to the Arctic, see Anthony Brandt, *The Man Who Ate His Boots: Sir John Franklin and the Tragic History of the Northwest Passage* (New York: Knopf, 2010).
12 King, *Narrative of a Journey*, 1: v. See also M.J. Ross, *Polar Pioneers: John Ross and James Clark Ross* (Montreal/Kingston: McGill-Queen's University Press, 1994), 192–99.
13 King, *Narrative of a Journey*, 2: 30–64.

14 Sera-Shriar, *The Making of British Anthropology*, 21–52.
15 For more on armchair anthropology in the nineteenth century, see Efram Sera-Shriar, "What Is Armchair Anthropology? Observational Practices in Nineteenth-Century British Human Sciences," *History of the Human Sciences* 27, 2 (2014): 26–40.
16 For more on the relationship between travel narratives and early ethnology, see Stocking, *Victorian Anthropology*, 79–86; Janet Browne, "Natural History Collecting and the Biogeographical Tradition," *História, ciêcias, saúde – Manguinhos* 8 (2001): 960–61; and Efram Sera-Shriar, "Tales from Patagonia: Phillip Parker King and Early Ethnographic Observation in British Ethnology, 1826–1830," *Studies in Travel Writing* 19, 3 (2015): 204–23.
17 There is an extensive body of literature on European constructions of Indigenous peoples in Canada. For examples, see Elizabeth Vibert, *Traders' Tales: Narratives of Cultural Encounters in the Columbian Plateau, 1807–1846* (Norman: University of Oklahoma Press, 1997); Carolyn Podruchny, *Making the Voyageur World: Travellers and Traders in the North American Fur Trade* (Lincoln: University of Nebraska Press, 2006); and Carolyn Podruchny and Laura Peers, eds., *Gathering Places: Aboriginal and Fur Trade Histories* (Vancouver: UBC Press, 2010).
18 Mary Louise Pratt, *Imperial Eyes: Travel Writing and Transculturation* (London: Routledge, 1992), 5–8.
19 For more on constructions of North American Indigenous peoples in European texts, see Michael Witgen, "The Rituals of Possession: Native Identity and the Invention of Empire in Seventeenth-Century Western North America," *Ethnohistory* 54, 4 (2007): 639–68. See also Livingstone, *Adam's Ancestors*, 16–25.
20 Saulteaux are part of the Ojibwa nation of Anishinaabe-speaking peoples.
21 King, *Narrative of a Journey*, 1: 32–33.
22 William Lawrence, *Lectures on Physiology, Zoology, and the Natural History of Man, Delivered at the Royal College of Surgeons* (Salem, NY: Foote and Brown, 1828), 109.
23 King, *Narrative of a Journey*, 1: 64–65.
24 Bravo, "Ethnological Encounters," 342.
25 For more on the link between anthropology and British imperialism in the nineteenth century, see Sera-Shriar, *The Making of British Anthropology*, Chapter 2.
26 King, *Narrative of a Journey*, 1: 152–54.
27 Ibid., 1: 182–83.
28 For more on the application of travel reports in early ethnological writings, see ibid., 43–50; and Bravo, "Ethnological Encounters," 341–49.
29 King, *Narrative of a Journey*, 1: 260.
30 Sera-Shriar, "Tales from Patagonia," 209.
31 King, *Narrative of a Journey*, 1: 260.
32 Much has been written on Akaitcho, and he is most famous for his involvement in Franklin's Coppermine River Expedition between 1819 and 1822. See, for example, Keith Crowe, *A History of the Original Peoples of Northern Canada* (Montreal/Kingston: McGill-Queen's University Press, 1991), 79–80; June Helm, *The People of Denendeh: Ethnohistory of the Indians of Canada's Northwest Territories* (Iowa City: University of Iowa Press, 2000), 231–34; and Catherine Lanone, "Arctic Romance

under a Cloud: Franklin's Second Expedition by Land," in *Arctic Exploration in the Nineteenth Century*, ed. Frédéric Regard (London: Pickering and Chatto, 2013), 105–6.
33 King, *Narrative of a Journey*, 1: 264. For more on the war between Akaitcho's people and nearby Inuit groups, see Richard Clarke Davis, *Lobsticks and Stone Cairns: Human Landmarks in the Arctic* (Calgary: University of Calgary Press, 1996), 144.
34 June Helm and Beryl Gillespie, "Dogrib Oral Tradition as History: War and Peace in the 1820s," *Journal of Anthropological Research* 37, 1 (1981): 8–27.
35 Sera-Shriar, "Arctic Observers," 27.
36 King, *Narrative of a Journey*, 2: 5.
37 For more on Parry, see Trevor H. Levere, *Science and the Canadian Arctic: A Century of Exploration, 1818–1918* (Cambridge, UK: Cambridge University Press, 1993), 63–83.
38 King, *Narrative of a Journey*, 2: 5–6.
39 The earliest British ethnographic guidebooks were produced in the late 1840s and early 1850s. They were based upon French instructions for travellers from the beginning of the nineteenth century. See James Cowles Prichard, "Ethnology," in *A Manual of Scientific Enquiry: Prepared for the Use of Officers in Her Majesty's Navy and Travellers in General*, ed. John Herschel (London: John Murray, 1849), 423–40; and Thomas Hodgkin and Richard Cull, "A Manual of Ethnological Inquiry," *Journal of the Ethnological Society of London* 3 (1854): 193–208.
40 King, *Narrative of a Journey*, 2: 11.
41 Ibid., 1: 131–32.
42 There are multiple examples of Europeans gathering geographic information from Inuit within the narratives of voyagers in the Arctic. See Levere, *Science and the Canadian Arctic*, 248–49; and Michael T. Bravo, "Ethnographic Navigation and the Geographical Gift," in *Geography and Enlightenment*, ed. David N. Livingstone and Charles W.J. Withers (Chicago: University of Chicago Press, 1999), 199–223.
43 King, *Narrative of a Journey*, 2: 6–7.
44 Ibid., 2: 7.
45 Ibid., 2: 7–11.
46 Sera-Shriar, *The Making of British Anthropology*, 88–89. There is much literature on the history of visual anthropology in the nineteenth century. For examples, see John M. MacKenzie, "Art and the Empire," in *The Cambridge Illustrated History of the British Empire*, ed. P.J. Marshall (Cambridge, UK: Cambridge University Press, 1996), 296–317; Elizabeth Edwards, *Raw Histories: Photographs, Anthropology, and Museums* (Oxford: Berg, 2001); Qureshi, *Peoples on Parade*; and Marcus Banks and Jay Ruby, *Made to Be Seen: Perspectives on the History of Visual Anthropology* (Chicago: University of Chicago Press, 2011).
47 For more on instruments as technologies of precision, see Fraser Macdonald and Charles Withers, eds., *Geography, Technology, and Instruments of Exploration* (Farnham, UK: Ashgate Publishers, 2015).
48 Similar objective claims were used later in the century by researchers using photography. See David Green, "Veins of Resemblance: Photography and Eugenics," *Oxford Art Journal* 7, 2 (1984): 4; Kelley Wilder, *Photography and Science* (London: Reaktion

Books, 2009), 19–20, 32–35; and Efram Sera-Shriar, "Anthropometric Portraiture and Victorian Anthropology: Situating Francis Galton's Photographic Work in the Late 1870s," *History of Anthropology* 53, 2 (2015): 158.
49 King, *Narrative of a Journey*, 2: 112.
50 Ibid., 2: 112–13.
51 Ibid., 1: 171.
52 Gillen and Ghosh, *Colonialism and Modernity*, 156–58; Livingstone, *Adam's Ancestors*, 11–25; Sera-Shriar, "Tales from Patagonia," 216–17.
53 Brandt, *The Man Who Ate His Boots*.
54 King, *Narrative of a Journey*, 2: 40–41.
55 Ibid., 2: 41.
56 Ibid., 2: 59–60.
57 Ibid., 2: 60.
58 Ibid., 2: 61.
59 Ibid., 2: 49.
60 For more on King's criticism of the Hudson's Bay Company, see Ted Binnema, *Enlightened Zeal: The Hudson's Bay Company and Scientific Networks, 1670–1870* (Toronto: University of Toronto Press, 2014), 144–46. See also Elle Andra-Warner, *Hudson's Bay Company Adventures: Tales of Canada's Fur Traders* (Victoria: Heritage House, 2011).
61 King, *Narrative of a Journey*, 2: 50–51.
62 Ibid., 2: 52–53.
63 Ibid., 2: 53.
64 Ibid.
65 Ibid., 2: 53–54.
66 Ibid., 2: 54.
67 Ibid., 2: 56.
68 Ibid., 2: 57.
69 Ibid., 2: 33–34, 63.
70 For more on Hodgkin's civilizing mission, see Zoë Laidlaw, "Heathens, Slaves, and Aborigines: Thomas Hodgkin's Critique of Missions and Anti-Slavery," *History Workshop Journal* 64, 1 (2007): 133–61.
71 King, *Narrative of a Journey*, 2: 58.
72 For more on the relationship between travel, print, and readership, see Innes N. Keighren, Charles W.J. Withers, and Bill Bell, *Travels into Print: Exploration, Writing, and Publishing with John Murray, 1773–1859* (Chicago: University of Chicago Press, 2015).
73 For more on the relationship between science, race, and politics, see Douglas Lorimer, *Science, Race Relations, and Resistance: Britain 1870–1914* (Manchester: Manchester University Press, 2013).

2

Scientist Tourist Sportsman Spy
Boundary-Work and the
Putnam Eastern Arctic Expeditions

TINA ADCOCK

> *Exploring is neither hazard nor sacrifice, but great fun.*
> *And killing big game is about as risky as shooting cows.*
> *These two theses, to which (with individual modification)*
> *I subscribe, have got me into considerable trouble.*
>
> – George Palmer Putnam[1]

On August 19, 1925, the governor general of Canada gave royal assent to an amendment to Section 8 of the Northwest Territories Act. Known as the Scientists and Explorers Ordinance, it stipulated that all such fieldworkers had to obtain permission from the federal government before carrying out investigations in the Northwest Territories. To date, historians have rightly emphasized the political significance of the ordinance. By asserting state control over foreign scientists' and explorers' access to and movements through the High Arctic, it was intended to enhance Canada's claims to sovereignty over that still-contested region.[2] In solving this political problem, however, federal bureaucrats unwittingly created an epistemic one. They now had to specify the qualities of scientists and explorers in order to ascertain which applicants merited permits and which did not. Although civil servants interpreted these occupational categories liberally at first, clause 4a of the ordinance explicitly forbade fieldworkers from pursuing commercial or political activities.[3] Sportsmen or spies masquerading as scientists or explorers were not wanted in the Northwest Territories. Yet, as bureaucrats discovered in the late twenties, the boundaries between legitimate and illegitimate actors and practices were not only surprisingly difficult to locate but also disquietingly easy to transgress.

This epistemic problem was not unique to the early twentieth-century Canadian North. Geoffrey Bowker and Susan Leigh Star have identified classification as a key practice of modernity and an important vector by which bureaucrats shaped late nineteenth- and early twentieth-century civil societies. They reveal classification to be an inherently political act that empowers certain individuals and groups and effaces or marginalizes others.[4] In like vein, Thomas Gieryn has demonstrated the centrality of boundary-work to science, meaning the sociopolitical processes by which certain actors, activities, and places are deemed credible producers of scientific knowledge, and others are excluded from the terrain of science and devalued accordingly. Boundary-work is localized, contingent, and often pragmatic. How epistemic credibility is contested, and the borders of knowledge drawn and defended, determines what science looks like and how it functions in any given time and place.[5]

Of all the venues in which science occurs, the field has been deemed the one in greatest need of boundary-work; it is also the place most resistant to such work in practice. It poses special material and intellectual challenges to those wishing to erect firm boundaries between scientific and nonscientific personnel and activities. Unlike the private space of the laboratory, the field's borders are public and therefore permeable. A plethora of human and nonhuman actors may enter, circulate through, and use these spaces and their resources toward diverse ends. Yet those ends are often achieved by means that look remarkably similar to the casual observer. "In the field," Henrika Kuklick and Robert Kohler note, "scientists' practices shade, sometimes almost imperceptibly, into those of other activities that are also carried on there, such as travel, sport, or resource harvesting."[6]

Behavioural similitude could, and did, lead to occupational misidentification in the field. Northern bureaucrats' fears that they would mistake scientists for sportsmen, tourists, or poachers, or vice versa, were perfectly legitimate. Indeed, it was especially difficult to distinguish between scientific collectors and sport hunters in early twentieth-century North America. The skills and knowledge required to succeed at either task, including fieldcraft, competent marksmanship, and intimate familiarity with animal habits, overlapped substantially.[7] Making intraoccupational distinctions – between university-trained scientists and self-taught collectors, for example – could prove just as challenging. The field's characteristic blurring of the professional-amateur boundary may well have been heightened in the Arctic, where the professionalization of the field sciences seems to have occurred later in the twentieth century, if at all.[8] Northern administrators in the late twenties took an instrumental approach to epistemic boundary-work, drawing the boundaries of northern scientific and exploratory praxis generously and then narrowing their remit as subsequent events warranted.

Applications made under the Scientists and Explorers Ordinance display the evolution of bureaucratic thought regarding how best to define, and thus control, science and exploration in the modern Canadian North.[9] Bowker and Star observe that classifications "become more visible ... when they break down or become objects of contention."[10] Indeed, expeditions that seemed scientifically legitimate, but that went on to violate the conditions of the ordinance, most clearly demonstrate the entwined political and epistemic boundary-work required for the state's effective regulation of scientific exploration in the Northwest Territories after 1925. This chapter draws on two such expeditions, those led by George Palmer Putnam to the eastern Canadian Arctic in the summers of 1926 and 1927, to illustrate how the boundaries of scientific exploration in northern Canada were negotiated, constructed, breached, and rebuilt in practice between the wars. By attending closely to the beliefs, desires, and relationships that animated this episode, this chapter also begins to delineate the emotional and affective contours of field science and exploration in this time and place. Ronald E. Doel, Urban Wråkberg, and Suzanne Zeller have recently affirmed "the importance of micro-social and individual experience in the larger institutional and ideological dynamics of modern field science."[11] Questions of affect and emotion are, moreover, assuming increasing prominence in the history of science more broadly.[12] Analyzing people's strong but disparate reactions to the birds killed on Putnam's second expedition helps illuminate the evanescent fault line between science and sport in the early twentieth-century Canadian North and helps clarify the practical ramifications of not minding this gap.

The ordinance enabled the government to control the admission of foreign fieldworkers to the sovereign space of the Northwest Territories. Its practical power, however, relied on the mending and maintenance of good neighbourly boundaries between Canada and other nations with Arctic interests, particularly the United States. The MacMillan-Byrd expedition of 1925 had badly damaged diplomatic fences along the forty-ninth parallel. In its wake, as the Putnam expeditions illustrate, American explorers, diplomats, and representatives of scientific institutions and Canadian bureaucrats cultivated friendly bilateral ties to ensure that American fieldworkers retained access to Arctic field sites and that they would respect Canadian laws while there. Yet when the Putnam Baffin Island Expedition breached those legal boundaries in 1927, the border between the two countries hardened once more. Canadians then used formal diplomatic means to demand that future scientific and exploratory guests behave appropriately in northern spaces that the American government tacitly, if not explicitly, recognized as being within Canada's sovereign ambit. Ironically, the resolution of the Putnam affair exposed the limits of Canadian power over its Arctic territories.

The Canadian government had viewed American activities in the North American Arctic with rising concern since the late nineteenth century. Following the Alaska Boundary Dispute in 1903, when British diplomats sacrificed disputed lands in what became the Alaskan panhandle on the altar of Anglo-American amity, Canadian politicians became all the more determined to protect the country's other Arctic borders.[13] In the early 1920s, as American and European explorers began to conduct fieldwork in Davis Strait and the eastern Arctic Archipelago, civil servants in the Department of the Interior's Northwest Territories and Yukon Branch took measures to make the Canadian state more visible in this region. From 1922 on, the Eastern Arctic Patrols carried government representatives annually to settlements in Labrador, Hudson Bay, and the eastern Arctic Archipelago. The Royal Canadian Mounted Police (RCMP) founded posts at strategic places on Devon, Baffin, and Ellesmere Islands and stationed officers there throughout the year.

Officials in Ottawa were well aware, however, that it suited American explorers to act as though the Arctic were an international no-man's land. In May 1925, Robert A. Logan, a former officer in the Royal and Canadian Air Forces, reported hearing "a great deal of loose talk among explorers generally as to sovereignty in the Arctic regions" at the Explorers Club in New York City. "Men like [Donald B.] MacMillan, for their own reasons, did not desire effective control by any state," he said.[14] Indeed, an expedition led by MacMillan and sponsored in part by the United States Navy brought the question of Canadian sovereignty in the Arctic to a head later that year.[15] MacMillan and his second-in-command, Richard Byrd, proposed to establish a supply base on Axel Heiberg Island to support their program of aerial exploration in the High North. Canadian officials worried that the American government might dispute Canada's claim to this island, based upon earlier Norwegian exploratory activities there. They were also concerned, not without justification, that the expedition might discover new islands in the archipelago to which the United States would lay claim in anticipation of future commercial activity there.

Canadian bureaucrats first learned of MacMillan and Byrd's plans through reports published in American newspapers, rather than through formal diplomatic channels. They quickly revived an idea that had first occurred to them in 1921: legislation that would require foreign scientists and explorers to apply for permission to carry out fieldwork in the Northwest Territories.[16] The necessary amendment was introduced in the House of Commons on June 1, and Canadian officials notified the State Department by way of the British Embassy in Washington that MacMillan and Byrd were expected to apply for permits. Initially, American bureaucrats demurred, not wishing to commit to any course of action that might constitute official recognition of Canadian sovereignty over Axel Heiberg or other High Arctic islands. By mid-August, however, State Department officials had softened their stance. They wired MacMillan instructions to informally request a permit from the Canadian government – albeit one that only covered activities south of Ellesmere Island. Neither MacMillan nor Byrd ever did so, even though Byrd told George P. Mackenzie, the commander of the 1925 Eastern Arctic Patrol, that the necessary permits had been obtained prior to their departure. Mackenzie, having no immediate means of verifying this claim, backed down. But Canada was the ultimate victor. This was the last time that American explorers backed by the American government would contest Canadian sovereignty in the High Arctic so brazenly. For Ottawa's northern bureaucrats, however, the incident cast a long shadow on subsequent American expeditions to the Canadian Arctic, especially those that followed closely on its heels.[17]

The MacMillan-Byrd expedition constituted a watershed moment in Canada's political boundary-work in the Arctic and in its external relations more broadly. One of the key outcomes of this affair, the Scientists and Explorers Ordinance, became a prime vector of both political and epistemic boundary-work in the Canadian Arctic after 1925. Motivated by an urgent need to assert Canadian sovereignty in the North, the ordinance materially strengthened the border that separated non-British fieldworkers from this field site writ large. It granted northern bureaucrats the ability to screen aliens wishing to conduct scientific work in the Northwest Territories, to deny access to fieldworkers with suspect credentials or motivations, to eject those found to be conducting unauthorized fieldwork there, and to exact a fine of up to $1,000 upon conviction of such an offence.[18]

The ordinance simultaneously buttressed the state's epistemic authority in the North. After the Canadian Arctic Expedition of 1913–18, which Andrew Stuhl discusses later in this volume, the federal government neither undertook nor sponsored any more large-scale expeditions to the Arctic or Subarctic. The Geological Survey of Canada placed an effective moratorium on reconnaissance expeditions to remote and economically "unproductive" areas of the country.[19] Even when money could be found to send a scientist on the Eastern Arctic Patrol, civil servants initially had trouble finding someone both technically and temperamentally equipped for the rigours of Arctic fieldwork.[20] Meanwhile, by the mid-twenties, non-British expeditions to the Northwest Territories had become an annual occurrence. Officials feared that these interlopers were exerting significant pressure on the North's natural and cultural resources, despite the existence of protective legislation such as the Northwest Game Act. While game and fish disappeared down the gullets of explorers and their dogs, zoological and botanical specimens and archaeological and ethnological objects were being spirited into the vaults of American and European natural history museums.[21] "We are very glad to have scientific work carried on in our Territory," O.S. Finnie, the director of the Northwest Territories and Yukon Branch, noted in 1924, "but it seems regrettable that our Canadian museums cannot reap some benefit from all this research work."[22]

Given the strong relationship between science and sovereignty in northern Canada, as in other intemperate spaces of fieldwork, the federal government needed to assert control over foreign scientists' production of northern knowledge and consumption of northern resources. The ordinance placed these scientists and explorers in a relationship of compulsory reciprocity with the Canadian government. To gain access to biological specimens and cultural artifacts, fieldworkers had to agree to submit a report of their scientific findings, a list of specimens taken, and a record of their travels in the Northwest Territories. By arrogating to itself the knowledge produced by foreign fieldworkers, the government enhanced and consolidated its intellectual as well as physical control over its Arctic territories. The ordinance also enabled civil servants to track the circulation and final resting places of northern objects beyond Canadian borders and compelled fieldworkers to mind the protective boundaries that sister pieces of legislation, including the Northwest Game Act and the Migratory Birds Convention Act, drew around certain nonhuman populations.

The ordinance did not, however, offer precise definitions of the occupations that it purported to control. When RCMP Commissioner Cortlandt Starnes asked Finnie to "kindly define the meaning of the words 'scientific' and 'exploration,' so that there may be no misunderstanding," Finnie merely referred him to the ordinance's wording, particularly the exclusion of political and commercial activities specified in clause 4a.[23] The boundaries of "science" and "exploration" remained both capacious and diffuse in the ordinance's early years. This was almost certainly a strategic move intended to maximize its potential reach as a political and epistemic device of control over visitors to the Northwest Territories. At the same time, civil servants recognized the need to ensure a basic level of scientific or exploratory competence in applicants for permits. The bureaucratic history of the 1926 American Museum Greenland Expedition provides a representative snapshot of the process by which applicants' bona fides were verified and permits issued in the years directly following the ordinance's passage.[24]

This enterprise was the brainchild of George Palmer Putnam, a prominent New York publisher by profession and an explorer by avocation. An avid outdoorsman from his youth, Putnam appeared most interested in a northern pleasure cruise aboard the *Effie M. Morrissey*, a former whaling schooner refitted for travel in ice-infested waters. "I don't like to think of [the Arctic] as an icy wilderness where men die," he declared. "It is better to compare it with a magnificent natural park quite within daily touch of those at home."[25] But the expedition had scientific components as well as touristic ones. It aimed to collect zoological material on Ellesmere, Devon, and Baffin Islands, as well as lands farther east, for the American Museum of Natural History (AMNH).[26] G.W.H. Sherwood, the museum's acting director, wrote to J.B. Harkin, the parks commissioner in the Department of the Interior, to request both scientist's and explorer's licences and permits to collect species protected under the Migratory Birds Convention Act. Of the application, Harkin's colleague commented that "there would appear to be no question as to the bona fides of the Museum."[27] The support of a recognized scientific institution and the inclusion of suitably trained fieldworkers were sufficient proof of its credibility in the eyes of Interior officials.

Coming so early in the life of the ordinance, the American Museum Greenland Expedition helped to refine the process by which scientific and exploratory permits were dispensed, in ways that reflected civil servants' predilection for inclusiveness. Officials decided that masters of vessels carrying expeditions entering waters adjacent to the Northwest Territories required licences, as did scientific and exploratory personnel entering the Northwest Territories by land.[28] Moreover, though very few members of the expedition had formal scientific training, the deputy minister of the interior thought it best to issue individual scientist's and explorer's permits to all of its personnel.[29] By such means were the expedition's cinematographer, artist, radio operator, and professional cowboy/would-be muskox wrangler transformed into "scientists" and/or "explorers" and placed in the same category as its professional taxidermist, ichthyologist, and zoologist. The Advisory Board for Wild Life Protection awarded permits to collect specimens of mammals, game birds, nests, and eggs separately, and more selectively. Even in this case, however, the freelance big-game hunter Daniel W. Streeter and Putnam's thirteen-year-old son David Binney Putnam were among the recipients.[30]

The political boundary-work between American expedition personnel and Canadian civil servants reflected a similarly pragmatic deployment of generosity, likely intended to mend borders frayed only months before by the actions of MacMillan and Byrd. Just prior to the expedition's departure, Putnam *père* expressed his appreciation for the courtesy and cooperation demonstrated by Canadian officials. "It will be my pleasure and responsibility to see that ... the expedition is conducted in a way to merit the full approval of yourself and the Canadian authorities," he told Finnie. He promised to freely share any data collected, including motion picture footage, and invited Finnie to be his guest at the annual dinner of the Explorers Club in November as a personal token of gratitude.[31] While the expedition was in northern waters, Commodore Fitzhugh Green, its agent, struck a similarly warm note in his own letter to Finnie. Noting the "happy ... contact" between expedition members, RCMP, and Hudson's Bay Company personnel at Pond Inlet, he pronounced this evidence of "a growing concord between the spirits of those from both our countries who are interested in Northern matters."[32]

These seemingly banal pleasantries can be read as microdiplomatic attempts to reset Arctic relations between Canada and the United States in the immediate wake of the MacMillan-Byrd affair. As members of the Explorers Club, Putnam and Green knew very well that Canadian officials now viewed American exploration in the Arctic with fresh and justifiable suspicion. To ensure their continued access to this region, they explicitly affirmed that *this* American expedition, unlike others, would respect Canadian laws and sovereignty while in Canada's Arctic territories. Such affirmations were made publicly as well as privately. One of Putnam's dispatches to the *New York Times* described their reconnaissance of Jones Sound on behalf of Byrd, who hoped to establish a base there as part of a future aerial expedition. In the same paper of record in which MacMillan had declared his intent to place newly discovered Arctic islands under the Stars and Stripes, Putnam now stressed that Byrd "would be working on Canadian territory. Canada would have to give him permission to use it. The saying, 'No law of God or man runs north of fifty-three' holds no longer."[33] Although Byrd had dined under Putnam's roof, Putnam did not hesitate to lay aside personal ties, at least temporarily, to strengthen bilateral ones.[34]

Finnie responded to Putnam with genuine but perhaps not uncalculated warmth. He praised the Boys' Own–style narrative in which David Binney Putnam chronicled his summer experiences aboard the *Morrissey*, published under his father's imprint as *David Goes to Greenland* (1926). Upon receiving a complimentary copy, Finnie wrote that "David has produced a very interesting and readable story ... The volume will stand on the shelves of our library on Northern Exploration with the works of all the other 'Great' scientists and explorers."[35] In a subsequent letter, Finnie revealed that his own son, Richard, had also made several trips to the Northwest Territories. "I want to meet that boy of yours who has been north in the Arctic," Putnam responded enthusiastically. "I didn't know we had this brand of parenthood in common."[36] These affective exchanges doubled as political boundary-work, reflecting the ability of personal relationships to build as well as detract from cross-border amity. Bureaucrats such as Finnie were keenly aware of the state's limited practical ability to monitor foreign scientists' and explorers' activities and to punish any breaches of the ordinance they might commit. They cultivated friendly individual relationships with foreign fieldworkers, hoping that the latter would comply with Canadian regulations out of personal trust and respect for the civil servants with whom they were now directly acquainted.

These carefully nurtured ties came undone in the wake of Putnam's second Arctic expedition, undertaken the next summer on the *Morrissey* in the vicinity of Foxe Basin and southwestern Baffin Island. Everything began auspiciously. Putnam asked Finnie to renew the licences granted to their party in 1926, reminding him that, because of troubles with the *Morrissey*, they had had little opportunity to collect zoological specimens in Canadian territory on that occasion. He assured Finnie of the scientific propriety of his expedition, explicitly stating that there would be no indiscriminate killing of mammals or birds. "My primary interest is to hunt with the camera," he affirmed.[37] Given the expedition's previous record of compliance with Canadian laws and regulations, its continued association with the American Museum of Natural History, and its new relationships with other reputable scientific institutions such as the American Geographical Society (AGS), Finnie had no cause to doubt Putnam's intentions. He replied that the Canadian government would happily issue the expedition with permits to conduct scientific fieldwork and to collect specimens of mammals and nonmigratory birds in the Northwest Territories. He also forwarded a copy of Putnam's letter to Harkin so that permits to collect migratory birds might be granted as well. Finnie's letter concluded on a warm personal note: "I am glad to see that your son, David, is again on the personnel of the expedition and will look forward with great interest to another work from his pen."[38]

The Putnam Baffin Island Expedition, as its leader later wrote, "evolved almost entirely into a geographical undertaking, with practically no zoological collection."[39] Under the leadership of its geographer and second-in-command, the University of Michigan's Laurence M. Gould, the expedition literally redrew the southwestern coastline of Baffin Island, converting over 5,000 square miles of presumed land into water on southern maps. "It was a unique experience, to say the least," marvelled Putnam, "to find that we actually were cruising, comfortably afloat, along a course that showed clearly on the chart as being fifty miles and more from salt water!"[40] Although Canadian authorities were surprised to receive this information, they extended to Putnam the traditional exploratory prerogative of appending his preferred names to various newly legible topographical features. Putnam Island and Bowman Bay – the latter named after Isaiah Bowman, the AGS's president at the time, in acknowledgment of his support – appear on maps of the Arctic Archipelago even today.[41]

In addition to the geographical exploration undertaken on behalf of the AGS, members of the expedition also performed oceanographical fieldwork for the Smithsonian Institution, the United States Bureau of Fisheries, and the Buffalo Society of Natural Sciences; archaeological and ethnological fieldwork for the Museum of the American Indian, Heye Foundation; zoological fieldwork for the American Museum of Natural History; and medical experiments on Inuit at Amadjuak and Cape Dorset.[42] The only zoological specimen brought south, however, was the skin of a blue goose shot in Foxe Basin. In his narrative of that summer's expedition, *David Goes to Baffin*

Land (1927), David noted that "the hunting was very disappointing. We'd expected great things from the Hudson Bay country, but it didn't work out well at all."⁴³

As he had done the previous year, Putnam sent a copy of David's newest book to Finnie. It was presumably lodged in the library of the Northwest Territories and Yukon Branch, to which civil servants in the Department of the Interior had access. It may well have been this copy that found its way into the hands of Harrison F. Lewis, the chief migratory bird officer of Ontario and Quebec, the following spring. Lewis was an ardent amateur ornithologist and an indefatigable tracker of poachers in the field – and, in this case, on the page.⁴⁴ David's narrative revealed, to Lewis's horror, that members of the Putnam Baffin Island Expedition had killed numerous protected migratory birds, including ducks, geese, and shore birds, as well as other animals, during their sojourn in Canadian territory. "The killing of these creatures is often mentioned incidentally in the narrative," Lewis lamented, "giving the impression that such killings were common and many of them went unrecorded."⁴⁵ Having checked the return on permit number 1488, which had given the expedition licence to secure scientific specimens of migratory birds, Lewis confirmed that Putnam had declared only the aforementioned blue goose. All of the other victims, it appeared, had been intended for the expedition's consumption while in the field and thus had been killed unlawfully in closed season.

Their fate was all the more tragic for its seeming purposelessness. "This expedition was in the field only one short season, was not invited or sponsored by Canada, and appears to have been undertaken chiefly as a money-making and advertising venture," Lewis noted. "There was no cause for its not having been provisioned adequately before sailing, yet the record shows the killing of ducks soon after Baffin Island was reached." The expedition's "selfish and ruthless activities" had both immediate and potentially wide-ranging political consequences. As American guests on Canadian soil and in Canadian waters, their actions violated the Migratory Birds Convention Act, a bilateral treaty between the two nations, in a manner that threatened to inflame Canadian-American relations in the Arctic once more. Moreover, in so blithely and publicly flouting both Canadian and international laws, these killings had the potential to revive the myth of the "open Arctic" that Canada was doing its best to quell.

These dead birds can be usefully read as boundary objects, or things that have the capacity to unite different communities of practice. Boundary objects have certain fixed properties that render them intelligible and meaningful to a range of intersecting actors and groups but retain enough malleability to respond effectively to individual and situated desires.[46] The values attached to these birds' bodies by American and Canadian members of different but overlapping epistemic communities shed additional light on the tenuous boundary between field science and field sport in the interwar Canadian North. Both Putnam and Canadian civil servants subscribed to an elite conservationist ethos that mandated the wise and rational use of natural resources so as to ensure their longevity. Where they parted company was on the appropriate nature and scale of consumption.

Putnam carefully distinguished himself from an earlier generation of Arctic sport hunters and explorers who had wantonly decimated wildlife populations. Referring to himself and Bob Bartlett, the ship's captain for his two expeditions, he wrote that "neither of us had much enthusiasm for killing, except for the larder or scientific use."[47] But for Putnam, the ability to consume wildlife freely, though not excessively, through hunting was an essential component of the strenuous leisure in "wild" surrounds that this expedition proffered, in typical antimodern fashion, as a physical and intellectual panacea to his usual hectic and enervating urban lifestyle. It was important to Putnam both as a father and publisher that David have the opportunity to refine his tracking, stalking, shooting, and field-dressing skills through encounters with various species, often facilitated by Inuit guides.

Learning to hunt successfully was a key rite of passage in elite Anglo-American cultures of sport and masculinity. David's hunting episodes provided valuable didactic material for adult and juvenile readers of his 1927 narrative. On one occasion, having fired five bullets at a hare that "seemed to enjoy the show," David was then told by two adults (one Inuk, one Euro-American) that he could have only one more shot. This critique of David's wasteful and careless shooting hits the proverbial target; his last shot kills the hare.[48] Meanwhile, watching Inuit methods of hunting Arctic wildlife, such as harpooning bull walrus from skin kayaks, taught David and his readers that gentlemanly sporting conduct transcended cultural boundaries. Hunting practices that afforded animals a reasonable chance of escape were not exclusively the preserve of elite Anglo-American sportsmen. Such lessons served as necessary moral complements to more efficient, but less thrilling modes of hunting for scientific collection. Putnam averred that bagging walrus with high-powered rifles for the halls of natural history museums was about as challenging and dangerous as shooting cows in a pasture.[49] Yet even these deaths provided occasions for Putnam to model gentlemanly hunting practices to his son and to civil servants in Ottawa. The meat from scientific specimens was always given to Inuit and their dogs to ensure that it was not wasted.[50]

Hunting and fishing anecdotes were also a generic convention of expeditionary narratives that enhanced their appeal in the eyes of the reading public. Such moments provided a suitable *mise en scène* for subtle product placement: the superior properties of Parker shotguns, Remington rifles, Thomas fishing rods, and Armours pemmican are woven deftly into David's sketches of sporting excursions (see Figure 2.1).[51] For David's book to sell well among adolescent (and adult) armchair travellers, and to attract corporate sponsorship for future expeditions, its protagonist had to be seen hunting and fishing – the more frequently, the better.

Members of the Putnam Eastern Arctic Expeditions appear to have considered the literal consumption of Arctic wildlife a privileged perquisite of polar exploration – a *droit d'explorateur*, if you will. Even on well-supplied expeditions such as Putnam's, many fieldworkers savoured the opportunity to eat country food, which leavened and enlivened the nourishing but uninteresting imported fare on which they normally subsisted.[52] After returning from Greenland in 1926, Putnam hosted a dinner to celebrate the publication of Arctic expeditionary narratives by Knud Rasmussen, Bob Bartlett, and, of course, David Binney Putnam. Nearly every dish featured some animal – polar bear, ptarmigan, narwhal, walrus, white whale, or capelin – from the field in which they had lately laboured. Naturally, it was capped off by Eskimo pie.[53]

On Baffin Island in 1927, David enjoyed feasts of fried caribou steak, young goose meat, and snipes. With gustatory relish, he wrote of this last species that

> they certainly taste good. You can either pick them and then singe off the little feathers that are left, or skin them entirely. The frying pan with some bacon and bacon fat does the rest, and the little birds are so tender you can eat them all, crunching up the bones and the rest of them in about four decent mouthfuls.[54]

David's evident delight damned him further in the eyes of Lewis, who reproduced this passage verbatim in his letter with, one fancies, a kind of disgusted fascination. Like their counterparts in the United States, Canadian conservationists decried and sought to legislatively circumscribe "pothunting," or the rural and working-class practice of killing animals and birds

Figure 2.1 One of the Putnam Baffin Island Expedition's commercial tie-ins: a Daisy air rifle advertisement featuring the "14-year-old explorer" David Binney Putnam and promoting his narrative *David Goes to Baffin Land* (1927). The *Morrissey* was equipped with twelve of these rifles, according to John A. Pope. | *American Boy*, March 1928.

for food to supplement family larders.[55] By contrast, conspicuously – even gaudily – consuming the flesh of Arctic critters enhanced Putnam's status as an elite sportsman and tourist, even as it put the lie to his personal scientific credibility in a way that civil servants could scarcely ignore.

For Lewis and his colleagues in the Department of the Interior, these slain birds were personal objects of grief as well as political time bombs. All were natural historians by avocation, if not profession, and passionate advocates of wildlife conservation.[56] Both J.B. Harkin and Maxwell Graham had been prime movers in the bilateral negotiations that led to the passage of the Convention for the Protection of Migratory Birds in 1916, and the drafting of its Canadian corollary, the Migratory Birds Convention Act, which came into force a year later.[57] The Advisory Board of Wild Life Protection, to which Harrison Lewis owed his position, was created specifically to ensure Canadian compliance with this treaty and legislation, but it quickly became a centralizing and driving force behind federal legislation and policy on wildlife conservation.[58] Natural resources had traditionally been a provincial rather than a federal responsibility in Canada; wildlife in the Northwest Territories, however, fell under the purview of the Department of the Interior. Managing northern fauna therefore received a disproportionate amount of the board's time and energy from the late teens into the twenties.[59] But laws pertaining to wildlife were particularly challenging to enforce in the North, given its vast physical dimensions and scant number of game officers, whether appointed wardens or *ex officio* ones (as in the case of the police).

Yet Canadian bureaucrats realized the necessity of trying to secure at least a handful of convictions to demonstrate publicly the efficacy of such legislation. This was even more the case when it came to a "solemn international treaty," to quote Lewis.[60] Canada could not be seen to be falling down on its neighbourly obligations; nor could it allow its neighbours to take advantage of its hospitality, as Putnam and his expedition had done. Fortunately for Canadian officials, the North American diplomatic landscape had shifted since 1925 in ways that enhanced their ability to take up this matter with the American government. At the time of the MacMillan-Byrd expedition, Canada had lacked independent diplomatic relations with the United States. Ottawa's ability to converse with State Department officials had been hampered by the need to channel missives through the British Embassy in Washington, which did not always accord high priority to Canadian communiqués. The next year, following the Imperial Conference in London, Prime Minister William Lyon Mackenzie King established direct diplomatic relations with the United States. Informal diplomatic ties having failed in the case of the Putnam expedition, Canadian officials now had better formal means to compel these Americans and their government to explain and atone for their actions.

Undersecretary of State for External Affairs O.D. Skelton used Lewis's letter as the basis for Canada's official communiqué to the American government, a document routed through the Canadian Legation in Washington in August 1928.[61] William R. Castle Jr., Skelton's counterpart in the State Department, soon assured Canadian officials that it would be "a matter of serious concern" if an American expedition had "abused the privileges accorded it by your Government." He forwarded the communiqué to the American Museum of Natural History, the expedition's chief sponsor, so that its administrators could inquire into the matter.[62] Meanwhile, Putnam, apparently still unaware of the growing diplomatic brouhaha, sent a final friendly note to Finnie in mid-September to ensure that all of the administrative t's had been crossed and i's dotted in connection with the previous summer's expedition. "I very genuinely want to live up to every obligation, to evidence my great appreciation for your own kindly courtesy and the cooperation extended to us by all concerned," he wrote.[63] Given that his note was dated three weeks after Castle's, Canadian officials found it difficult to believe that Putnam had not yet been contacted by museum or state personnel.[64] Even if Putnam's note constituted a last-ditch attempt to resurrect the friendly ties that had once existed between him and Finnie, the latter's hands were now tied by the formal diplomatic machinery set in motion. Skelton advised Finnie to make no reply until the matter had been officially investigated and resolved.

By the end of September, AMNH officials had spoken with Putnam, and Castle was able to report informally to Vincent Massey, Canada's envoy to the United States, and formally to the Canadian Legation in Washington.[65] A greatly embarrassed Putnam had admitted to the infractions, which he characterized as arising from his misunderstanding of Canadian game laws and regulations. He had mistakenly believed that only game killed and taken out of the country as scientific specimens needed to be reported to the government, so had not reported game killed for food. Claiming, incorrectly, that Inuit living on Baffin Island killed animals freely for consumption in violation of game regulations, Putnam said that the expedition might have unconsciously, if improperly (to his mind), followed their lead while hunting in that region. Although his actions had inadvertently revealed him to be more sportsman than explorer, a spy he was not. He denied any secret or covert intent, pointing out that the expedition's official report, published in the AGS's *Geographical Review*, mentioned their killing game for food, as had David's narrative.[66] Putnam also foregrounded his long-standing respect for wildlife legislation, noting that in his twenty-year hunting career he had "grasped every opportunity of upholding game laws."[67] Museum officials attested that Putnam was a "gentleman of the highest character," one who would never intentionally give officials any cause for complaint.[68]

Putnam's boundary-work had now come full circle. In 1926, he had pre-emptively sought to mend the breaches in Canadian-American relations concerning the High Arctic, the better to smooth his first expedition's path north. Two years later, he turned once more to this task – this time retroactively, and in response to damage of his own making. First, Putnam sought to contain the epistemic damage to the AMNH and to future Arctic expeditions that it might sponsor. He noted once more that the expedition, having developed into an essentially geographical one, had moved outside the museum's remit and expertise while in the field. He also emphasized that no member of the museum's staff had accompanied the expedition and that "the fault and responsibility for violation of Canadian game laws was entirely his."[69] Next, Putnam sought to repair the political damage that he had caused. Not only did he provide a full explanation to museum and State Department officials, but he also requested an audience with the Advisory Board of Wild Life Protection to atone in person, a gesture that had US Secretary of State Frank B. Kellogg's explicit support.[70] "In light of this," Massey wrote to Skelton, "I think it unwise to make any further representations."[71] The board accepted Putnam's explanation and apology at a special meeting held on the afternoon of December 10, 1928. Earlier that day, at a meeting arranged by the American minister to Canada, Putnam had also tendered a personal apology to Prime Minister Mackenzie King. A communiqué detailing the outcome of this visit went forth to the State Department, and there the matter rested.[72]

The American government may have been involved in some political boundary-work of its own, both inside and outside its borders, in connection with this affair. Someone in Washington with inside knowledge of the affair and access to official correspondence may have leaked the story to the Washington bureau of the *Ottawa Morning Journal*.[73] Reports of "how the writings of a 14-year-old boy have brought about an international diplomatic controversy" duly appeared in newspapers on both sides of the border.[74] Although the publicity came too late to alter the outcome of the investigation, the widespread coverage of the expedition's misdeeds, particularly its contravention of the Migratory Birds Convention Act, made it especially difficult for officials to overlook its actions.[75] But why would the American government wish to cast its own Arctic explorers' misdeeds in high relief? By publicizing Putnam's transgression and its consequences, officials in Washington may have wanted to openly affirm American as well as Canadian legal presence in the eastern Arctic Archipelago, further debunking the myth of the open Arctic. Regardless of the leak's exact origin, the State Department's handling of the Putnam matter signalled that American explorers would henceforth be held accountable for their actions in the Arctic and that their government would not ignore breaches of treaties to which it was party there. Although this was a recognition of bilateral rather than Canadian sovereignty over Arctic space, the American government's willingness not only to pursue such claims according to Canadian wishes, but also, perhaps, to expose such cases to public scrutiny, demonstrated a new measure of respect for Canadian claims to the High Arctic, even if the question of Canadian sovereignty there had not yet been settled to Americans' satisfaction.

This boundary-work on the part of American individuals and institutions appears to have been successful, at least in the short to medium term. Canadian officials resumed friendly micro/diplomatic relationships with their American counterparts. Finnie thanked Putnam for inviting him once more to the Explorers Club's annual dinner and reiterated his interest in viewing the expedition's motion pictures of Baffin Island, which Putnam had offered to screen for Canadian civil servants.[76] Meanwhile, few, if any, American expeditions to the Arctic triggered similar diplomatic exchanges between the Canadian and American governments in the years following Putnam's journeys. It was not until the mid-1930s that foreign applications for scientist's and explorer's permits became vetted by the Department of External Affairs as a matter of course.[77] In the end, the effects of the affair may have been felt most trenchantly by Putnam. Although he continued to publish exploration narratives and to act as patron to other explorers, he never organized another large-scale expedition, to the Arctic or elsewhere.

The parable of the Putnam Eastern Arctic Expeditions reveals much about the Canadian government's attempted, and sometimes failed, management of foreign fieldworkers directly after the passage of the Scientists and Explorers Ordinance. Civil servants drew generous epistemic boundaries around the activities of science and exploration in the Northwest Territories. After a cursory examination of credentials, they often found it convenient to designate expeditions scientific, the better to maintain the illusion of official control over their actions.[78] The Putnam expeditions, however, appeared scientifically sound. Of all the early applications for scientist's and explorer's permits that crossed officials' desks, those submitted by Putnam would have seemed among the least likely to cause any trouble. It was all the more surprising, then, when its members transgressed the boundary between science and sport in the summer of 1927. Putnam's Arctic cruises nicely illustrate the epistemic dangers associated with fieldwork identified at the beginning of this chapter: the difficulty of monitoring the boundaries of these open-ended spaces, the ease with which scientific practices slid into contiguous ones of sport hunting or resource exploitation, and the uncertainty surrounding occupational identities in such varied natural and social landscapes of labour.[79]

Canadian officials feared that sportsmen and spies would intentionally masquerade as scientists and explorers. The Putnam expeditions taught them that that slippage could occur unintentionally, too. To borrow from the language of Christian theologians, expeditions could sin out of ignorance as well as malice. Putnam's fall from grace was all the more disappointing to Canadian civil servants, given that both parties had taken such pains to rebuild neighbourly fences after the MacMillan-Byrd affair. American explorers cultivated good personal relationships with Canadian officials so as to be allowed to cross into the latter's territorial waters. Officials in Ottawa reciprocated, hoping that friendlier ties would convince American explorers to act appropriately while there. But when these hopes were dashed, Canadians could seek formal acknowledgment of wrongdoing through diplomatic channels. In Putnam's case, this action was triggered by the violation of a bilateral treaty and eased through recent structural developments in Canadian-American diplomacy. This was the first time that an American expedition to the Canadian Arctic merited such a reprimand, but it would not be the last.

In this story, dead birds serve not only as boundary objects, bringing different political and epistemic communities together while simultaneously exposing divergent beliefs about the appropriate use and consumption of natural resources. For Canadian officials, they also came to mark the political boundaries of redress. Dead birds could not be resurrected, despite the significant personal and professional sorrow occasioned by their demise. But contrite bird killers had to be absolved in the interests of maintaining good cross-border relations in Arctic matters. The diplomatic imperative to forgive the sins of scientific expeditions to the Northwest Territories, however grave, points to the pragmatic limits of Canadian sovereign power over foreigners in the interwar High Arctic.

NOTES

My thanks to Preston Lann and Leah Wiener for their assistance with archival research, and to Nancy Earle, Jasmine Nicholsfigueiredo, Jennifer Scott, and the two anonymous reviewers for their thoughtful comments on earlier drafts of this chapter.

1 George Palmer Putnam, *Wide Margins: A Publisher's Autobiography* (New York: Harcourt, Brace, 1942), 216.
2 See, for example, D.H. Dinwoodie, "Arctic Controversy: The 1925 Byrd-MacMillan Expedition Example," *Canadian Historical Review* 53, 1 (1972): 58–59; Morris

Zaslow, "Administering the Arctic Islands 1880–1940: Policemen, Missionaries, Fur Traders," in *A Century of Canada's Arctic Islands, 1880–1980*, ed. Morris Zaslow (Ottawa: Royal Society of Canada, 1981), 66–69; and Janice Cavell and Jeff Noakes, *Acts of Occupation: Canada and Arctic Sovereignty, 1918–25* (Vancouver: UBC Press, 2010), 222, 228–29.

3 "An Ordinance Respecting Scientists and Explorers," *Canada Gazette* 60, 3 (July 17, 1926): 209.
4 Geoffrey C. Bowker and Susan Leigh Star, *Sorting Things Out: Classification and Its Consequences* (Cambridge, MA: MIT Press, 1999), 13, 285, 312.
5 Thomas F. Gieryn, *Cultural Boundaries of Science: Credibility on the Line* (Chicago: University of Chicago Press, 1999).
6 Henrika Kuklick and Robert E. Kohler, introduction to "Science in the Field," *Osiris* 11, 2nd ser. (1996): 3.
7 Robert E. Kohler, *All Creatures: Naturalists, Collectors, and Biodiversity, 1850–1950* (Princeton, NJ: Princeton University Press, 2006), 69–70.
8 Felix Driver has suggested, provocatively, that polar exploration has never truly professionalized. Felix Driver, "Modern Explorers," in *New Spaces of Exploration: Geographies of Discovery in the Twentieth Century*, ed. Simon Naylor and James R. Ryan (London: I.B. Tauris, 2010), 246.
9 I trace this evolution in Tina Adcock, "'The Maximum of Mishap: Adventurous Tourists and the State in the Northwest Territories, 1926–1948," *Histoire sociale/Social History* 49, 99 (2016): 431–52.
10 Bowker and Star, *Sorting Things Out*, 2–3.
11 Ronald E. Doel, Urban Wråkberg, and Suzanne Zeller, "Science, Environment, and the New Arctic," *Journal of Historical Geography* 44, 1 (2014): 11.
12 Paul White, ed., "The Emotional Economy of Science," *Isis* 100, 4 (2009): 792–851; and Otniel E. Dror, Bettina Hitzer, Anja Laukötter, and Pilar León-Sanz, eds., *History of Science and the Emotions*, *Osiris* 31, 2nd ser. (2016).
13 Nancy Fogelson, *Arctic Exploration and International Relations 1900–1932: A Period of Expanding National Interests* (Fairbanks: University of Alaska Press, 1992), 54–56.
14 Memorandum by J.A. Wilson enclosed in G.J. Desbarats to W.W. Cory, May 12, 1925, Library and Archives Canada (hereafter LAC), RG 85, vol. 759, file 4831.
15 On the MacMillan-Byrd expedition, see Dinwoodie, "Arctic Controversy"; Fogelson, *Arctic Exploration*, 90–96; Cavell and Noakes, *Acts of Occupation*, Chapter 8; Shelagh D. Grant, *Polar Imperative: A History of Arctic Sovereignty in North America* (Vancouver: Douglas and McIntyre, 2010), 228–36; and Gordon W. Smith, *A Historical and Legal Study of Sovereignty in the Canadian North: Terrestrial Sovereignty, 1870–1939*, ed. P. Whitney Lackenbauer (Calgary: University of Calgary Press, 2014), Chapter 14.
16 Cavell and Noakes, *Acts of Occupation*, 222.
17 Ibid., 237–38.
18 "An Ordinance Respecting Scientists and Explorers."
19 Morris Zaslow, *Reading the Rocks: The Story of the Geological Survey of Canada, 1842–1972* (Toronto: Macmillan, 1975), 338, 341.
20 J.D. Craig to O.S. Finnie, October 30, 1924, LAC, RG 85, vol. 666, file 3918.

21 Explanatory note to An Act to Amend the Northwest Territories Act, May 1925, LAC, RG 85, vol. 85, file 202-2-1-1.
22 O.S. Finnie to W.W. Cory, April 14, 1924, ibid.
23 Cortlandt Starnes to O.S. Finnie, October 13, 1928, and Finnie to Starnes, October 22, 1928, ibid.
24 I provide more detail on the verification of scientific and exploratory bona fides and analyze another early expedition, led by John D. Fuller in 1927, that helped to shape this process in Adcock, "The Maximum of Mishap."
25 "Putnam in Arctic and Wife Here Talk," *New York Times*, August 25, 1926.
26 George Palmer Putnam, "Greenland Expedition Is about to Push Off," *New York Times*, June 13, 1926.
27 G.W.H. Sherwood to J.B. Harkin, March 12, 1926, and Maxwell Graham to O.S. Finnie, March 25, 1926, LAC, RG 85, vol. 766, file 5099-1.
28 Maxwell Graham to O.S. Finnie, May 21, 1926, and Finnie to Graham, May 21, 1926, ibid.
29 O.S. Finnie to W.W. Cory, May 28, 1926, ibid.
30 Hoyes Lloyd to O.S. Finnie, May 7, 1926, ibid.
31 George Palmer Putnam to O.S. Finnie, June 8, 1926, ibid.
32 Fitzhugh Green to O.S. Finnie, September 4, 1926, ibid.
33 "Wants American Flag Raised over Land Near North Pole," *New York Times*, March 31, 1925; George Palmer Putnam, "Decry Jones Sound as a Base for Byrd," *New York Times*, September 4, 1926.
34 Sally Putnam Chapman with Stephanie Mansfield, *Whistled like a Bird: The Untold Story of Dorothy Putnam, George Putnam, and Amelia Earhart* (New York: Warner Books, 1997), 40–41.
35 O.S. Finnie to George Palmer Putnam, December 11, 1926, LAC, RG 85, vol. 766, file 5099-1.
36 O.S. Finnie to George Palmer Putnam, November 21, 1927, and Putnam to Finnie, November 28, 1927, ibid.
37 George Palmer Putnam to O.S. Finnie, March 10, 1927, ibid. The expedition did, however, carry north a considerable number of guns (more than fifty) and a certain amount of ammunition, some of which would have been earmarked for trading with Inuit or paying Inuit hunters in kind for their services. John A. Pope, to whom Putnam assigned care of the ship's arsenal, was visibly impressed: "The stock on board makes V L + D's [Von Lengerke and Detmold, a sporting goods retailer] ammunition room look sick. Never handled so much 'powder + ball.'" John A. Pope, Diary of the *Morrissey* Expedition, June 13 and 14, 1927, National Archives at College Park, College Park, MD (hereafter NACP), Collection XJAP: John A. Pope Papers, box 1.
38 O.S. Finnie to George Palmer Putnam, March 15, 1927, LAC, RG 85, vol. 766, file 5099-1.
39 George Palmer Putnam to O.S. Finnie, April 27, 1928, ibid.
40 George Palmer Putnam, "5,000 Miles Clipped from Baffin Island," *New York Times*, September 13, 1927.
41 For details of that support, see the correspondence between Putnam and Bowman in "Putnam, George P., Baffin Land Expedition: 1928, Correspondence, 1927–1928,"

University of Wisconsin–Milwaukee Libraries, American Geographical Society Library, AGSNY Archival Collection 1, subseries 8B, box 264, folder 27.
42 George Palmer Putnam, "The Putnam Baffin Island Expedition," *Geographical Review* 18, 1 (1928): 3–4. For other scientific publications arising from the expedition, see Laurence M. Gould, Aug. F. Foerste, and Russell C. Hussey, "Contributions to the Geology of Foxe Land, Baffin Island," *Contributions from the Museum of Paleontology, University of Michigan* 3, 3 (1928): 19–76; Bruno Oetteking, "A Contribution to the Physical Anthropology of Baffin Island, Based on Somatometrical Data and Skeletal Material Collected by the Putnam Baffin Island Expedition of 1927," *American Journal of Physical Anthropology* 15, 3 (1931): 421–68; and Peter Heinbecker, "Studies in Hypersensitiveness XXXIV. The Susceptibility of Eskimos to an Extract from Toxicodendron Radicans (L.)," *Journal of Immunology* 15, 4 (1928): 365–67.
43 David Binney Putnam, *David Goes to Baffin Land* (New York: G.P. Putnam's Sons, 1927), 90.
44 J. Alexander Burnett, *A Passion for Wildlife: The History of the Canadian Wildlife Service* (Vancouver: UBC Press, 2003), 14–16.
45 All quotations in this and the subsequent paragraph derive from Harrison Lewis to J.B. Harkin, April 5, 1928, LAC, RG 85, vol. 766, file 5099-1. The "impression" given by David's narrative is confirmed by the other two contemporary records of the expedition that I have located thus far: the personal diaries of the expedition's assistant surveyors, John A. Pope (cited above) and Monroe Grey Barnard. For a digital transcription of the latter, see George G. Barnard II, "Dad's Diary," http://www.archive.ernestina.org/history/Mgblog1927.html.
46 Susan Leigh Star and James R. Grisemer, "Institutional Ecology, 'Translations,' and Boundary Objects: Amateurs and Professionals in Berkeley's Museum of Vertebrate Zoology, 1907–39," *Social Studies of Science* 19, 3 (1989): 387–420; Bowker and Star, *Sorting Things Out*, 296–98. For a recent discussion of birds as boundary objects, see Nancy J. Jacobs, *Birders of Africa: History of a Network* (New Haven, CT: Yale University Press, 2016), especially 110–12.
47 George Palmer Putnam, *Mariner of the North: The Life of Captain Bob Bartlett* (New York: Duell, Sloan and Pearce, 1947), 161.
48 D.B. Putnam, *David Goes to Baffin Land*, 155. Pope noted that David was "a good hunter and a good shot." Diary of the *Morrissey* Expedition, August 21, 1927.
49 Putnam, *Wide Margins*, 261.
50 George Palmer Putnam to O.S. Finnie, October 11, 1926, LAC, RG 85, vol. 766, file 5099-1.
51 D.B. Putnam, *David Goes to Baffin Land*, 42, 56, 60, 104.
52 Of a subsequent Arctic expedition, Bartlett wrote that "we were after walrus, for we wanted fresh meat even though we had plenty of tinned supplies. Stuff out of cans becomes tiresome, and you don't have to be living on it long to find it out either." Bob Bartlett, *Sails over Ice* (New York: Charles Scribner's Sons, 1934), 136–37.
53 Putnam, *Wide Margins*, 229–30.
54 D.B. Putnam, *David Goes to Baffin Land*, 141.

55 Karl Jacoby, *Crimes against Nature: Squatters, Poachers, Thieves, and the Hidden History of American Conservation* (Berkeley: University of California Press, 2001); Tina Loo, *States of Nature: Conserving Canada's Wildlife in the Twentieth Century* (Vancouver: UBC Press, 2006).
56 Burnett, *A Passion for Wildlife*, 13–16; Janet Foster, *Working for Wildlife: The Beginning of Preservation in Canada*, 2nd ed. (Toronto: University of Toronto Press, 1998), 161.
57 Mark Cioc, *The Game of Conservation: International Treaties to Protect the World's Migratory Animals* (Athens: Ohio University Press, 2009), 72; Foster, *Working for Wildlife*, Chapter 6.
58 Foster, *Working for Wildlife*, 147.
59 John Sandlos, *Hunters at the Margin: Native People and Wildlife Conservation in the Northwest Territories* (Vancouver: UBC Press, 2007), 167.
60 Harrison Lewis to J.B. Harkin, April 5, 1928, LAC, RG 85, vol. 766, file 5099-1.
61 O.D. Skelton to Chargé d'Affaires, Canadian Legation in Washington, August 8, 1928, ibid.; Laurent Beaudry to Frank B. Kellogg, August 13, 1928, NACP, General Records of the Department of State, Record Group 59, file series 031.11 (American Museum of Natural History/Baffin Land), box 292 (031.11–031.11C21/87).
62 W.R. Castle Jr. to Laurent Beaudry, August 23, 1928, LAC, RG 85, vol. 766, file 5099-1; W.R. Castle Jr. to Henry Fairfield Osborn, August 23, 1928, NACP, RG 59, file series 031.11, box 292.
63 George Palmer Putnam to O.S. Finnie, September 13, 1928, LAC, RG 85, vol. 766, file 5099-1.
64 See O.S. Finnie to J.B. Harkin, September 15, 1928; Harkin to Finnie, September 24, 1928; and O.D. Skelton to O.S. Finnie, October 2, 1928, ibid.
65 Vincent Massey to O.D. Skelton, October 26, 1928, and W.R. Castle Jr. to Laurent Beaudry, October 27, 1928, ibid.; W.R. Castle Jr. to John D. Hickerson, October 26, 1928, NACP, RG 59, file series 031.11, box 292.
66 George Palmer Putnam to G.W.H. Sherwood, October 10, 1928, NACP, RG 59, file series 031.11, box 292.
67 Extract from Minutes of Advisory Board on Wild Life Protection meeting, December 10, 1928, LAC, RG 85, vol. 766, file 5099-2.
68 G.W.H. Sherwood to Secretary of State, October 11, 1928, NACP, RG 59, file series 031.11, box 292.
69 George Palmer Putnam to G.W.H. Sherwood, October 10, 1928, ibid.
70 Ibid.; also see Frank B. Kellogg to William Phillips, November 3, 1928, NACP, RG 59, file series 031.11, box 292.
71 Vincent Massey to O.D. Skelton, October 26, 1928, LAC, RG 85, vol. 766, file 5099-1.
72 William Phillips to William R. Castle Jr., November 6, 1928, and December 12, 1928, NACP, RG 59, file series 031.11, box 292; Vincent Massey to Frank B. Kellogg, February 7, 1929, ibid.
73 D.L. McKeand to J. Lorne Turner, January 20, 1934, LAC, RG 85, vol. 766, file 5099-2.

74 Leland S. Conness, "Boy's Writings Cause Trouble for Diplomat," *Ottawa Morning Journal,* November 10, 1928, LAC, RG 85, vol. 766, file 5099-1; "Writings by Boy, 14, Stir Diplomatic Row," *Washington Post,* November 11, 1928.
75 Extract from Minutes of Advisory Board on Wild Life Protection meeting, December 10, 1928.
76 O.S. Finnie to George Palmer Putnam, December 22, 1928, ibid.
77 J.M. Wardle to Barnum Brown, June 10, 1936, LAC, RG 85, vol. 783, file 5958.
78 Adcock, "The Maximum of Mishap."
79 Kuklick and Kohler nicely delineate these dangers in their introduction to *Science in the Field,* 1–14.

Nature's Tonic
Electric Medicine in Urban Canada, 1880–1920

DOROTEA GUCCIARDO

The electrical era will broaden our vision and widen our hearts and out of all the chaos and unrest and disorder, then will come peace and rest and order. Cant and creed, doctrinal hypocrisy and religious and social narrowness and inequality will all give away to a bigger, broader, godlier conception and the electrical era will be the greatest era, and the happiest era the world has ever seen.[1]

Thus wrote journalist Guy Cathcart Pelton in 1919, expressing a vision of a positively transformed Canada thanks to the power of electricity. Inspired by recent electrical developments in medicine, communication, and transportation, Pelton wrote of his "growing impression ... that electricity is the source of life." He predicted that the sick would be healed, that machines would be powered, that the burdens of physical labour would be lifted, and that communication would become instantaneous because of the electrical energy of the air. Although his utopian visions of a Canadian society on the verge of a new, electrical era at times bordered on the absurd (he believed that the universe was "bound together by invisible electribands ... that would transform tomorrow into an age of hope and happiness and love"), his sentiment that electricity held the key to a prosperous future was a popular one among Canadians in the early twentieth century.[2]

Canada was indeed in the midst of an electrical era. Nineteenth-century advances in the science of electromagnetism in Europe and the United States had a direct impact on industrial development in Canada and left many feeling optimistic that electricity, especially hydroelectric power, would liberate central Canadian industry from the limitations posed by imported coal. In the first decade of the twentieth century, Canada's "installed hydro-electric capacity grew from 173,000 horsepower to almost one million horsepower. By 1920, it was up to two and a half million."[3] At the same time, electricity in the form of communication (telegraph and eventually telephone), transportation (electric streetcars), and illumination became staples in urban centres across the country. Aside from the very wealthy, who could afford to install electric lighting on their properties early on, between the 1880s and the 1920s the public's encounter with electricity tended to be in shared spaces: lighting along public streets, in shop windows, at fairs and exhibitions. The exaggerated distribution of electric light developed into the fashionable way to display civic pride, and city officials across the country incorporated "Great White Ways" into urban schemes during the interwar years.[4] Electricity fundamentally altered the experiences of these shared spaces; although the bright lights of city streets and the clanging sounds of electric streetcars might have been awe-inspiring to some, they remained perplexing to others. Medical practitioners in particular produced an impressive body of literature questioning what role this modern power source could play in human health.

The idea that electricity could act as a curative was not new, but interest in its medicinal qualities was heightened during this electrical age. Many contemporary observers, like Pelton, were optimistic that electricity would revolutionize Canadian life; perhaps nowhere were the transformative effects of electric technologies felt more acutely than in Canadian cities. It was in them that electricity became synonymous with modernity, providing, as historian Marshall Berman writes, a "dynamic new landscape in which modern experience takes place."[5] Part of that modern experience took place behind closed doors; at a time when most people's experiences with electricity were limited to public settings, some Canadian residents were exposing their bodies to a more intimate relationship with electric power via electrotherapy, a treatment based upon the concept that electricity was the body's natural life force and, if necessary, could be replenished – or, more accurately, recharged – by an external energy source. With electricity giving power to so many aspects of Canadian life, some believed that this "mysterious force" could give power to their own bodies as well. Those who sought electrical cures for countless symptoms and afflictions (both real and imagined) reflect an obscure layer in the public experience of electricity.

This chapter examines the practice of electrotherapy within the wider context of electrification in English-speaking, urban Canada between 1880 and 1920.[6] Electrotherapy, or "medical electricity," was offered by both licensed physicians and electrotherapeutists, the latter typically composed of enterprising individuals who centred the practice on the notion that electric current could revitalize a run-down body. Just as Tina Adcock's chapter illustrates that the lines between legitimate and illegitimate scientific practices in Canada's North were sometimes blurred, so too this chapter explores how the boundaries between a licensed professional and a self-styled physician practising electric medicine were often ill defined. This was partly because of a lack of regulatory oversight, which allowed anyone to extol the virtues of electric therapies through newspaper advertisements and pamphlets or by setting up their own practices and treating patients.

One self-proclaimed "medical electrician" from Toronto, identified only as J. Adams in his writings, presented electricity as the fundamental motive power in the universe, responsible for all natural phenomena, including the inner workings of the human body. It was a common motif: the nervous system and the brain in particular were seen as electric and vulnerable to "overtaxation" by external pressures – such as those that helped to define the modern Canadian city. Adams argued that, "as Electricity is the greatest electro-motive power in Health, it must be Nature's own most appropriate remedy for restoring the human frame."[7] An examination of the theories and practices of electrotherapy exposes an interesting paradox of modernity in Canada at this time: electricity, responsible for powering the modern world, was also seen as the culprit in depleting society of its energy. By using nature's own tonic as a therapeutic agent, Canadians redefined what it meant to be modern in this new electric age.

Understanding the Body as Battery

By 1900, the notion that electricity could act as a curative was accepted by mainstream medical practitioners and laypersons alike. Increasing numbers of experiments conducted in the nineteenth century seemed to confirm that the human body had electric qualities that could be manipulated with exposure to electric current and electric light. The backdrop to these studies was the Second Industrial Revolution, made possible thanks to advances in the science of electromagnetism and the subsequent electric technologies, such as the motor and the generator, that it produced. It was no coincidence that the rise of electrotherapy corresponded to this period of rapid industrialization. The degree to which industrialization helped or hindered standards of living is difficult to assess, but to contemporary observers "industrialism jeopardized health."[8] In his 1880 publication *American Nervousness*, George Beard wrote that modernity, in the guise of new technological devices such as the telegraph, railway, and electric light, was creating a crisis in human health. Beard maintained that the consequences of these new technologies – instant communication, speed, glaring brightness – posed more physical and emotional demands than any person could reasonably withstand.[9]

Readers today would be forgiven for dismissing Beard's writing as a historical oddity, irrelevant compared with the concurrent rise of germ theory, which fundamentally transformed understandings of health and disease. But Beard's observations would easily have been understood by a generation that believed in the body as a permeable entity. Environmental historians such as Linda Nash and Gregg Mitman have documented how, throughout the nineteenth century, many Americans understood health and illness in relation to their local environments. "Long before the advent of modern ecology," Nash writes, "they understood themselves as organisms that were connected to their environment in [a] multitude of ways."[10] And, for many years after the discovery of bacteria, people continued to worry about the vulnerability of their bodies to local conditions; the nineteenth-century city, with all of its technological marvels, presented heretofore unknown dangers to personal health. Although Beard described electrical overstimulation as a uniquely American experience, industrialized life also seemed to leave Canadians in emotional distress, with more and more complaining of nervous disorders such as "neurasthenia," a term coined by Beard.[11] Many urban Canadians could not seem to acclimatize their bodies to the modern city.[12] By 1888, Daniel Clark, the medical supervisor of an asylum in Toronto, remarked that his "class of patients is growing larger day by day in this nerve-exhausting age."[13]

With electricity giving power to so many aspects of Canadian life, some believed that this "mysterious force" could give power to their bodies as well. Canadian doctors, most of whom had been trained outside the country, were likely well versed in electrotherapeutic techniques. Ideas about medicine flowed freely across national borders, and – with the majority of medical texts being written by European and American practitioners – it should not be surprising that Canadian doctors were influenced by works produced outside Canada. Nineteenth-century Canadian physicians were not averse to the international influences in their field; as historian Wendy Mitchinson notes, they "allowed Canadian physicians to feel they were part of a world-wide scientific community and not a parochial profession."[14] The transnational flow of ideas was also evident in the pages of the country's medical journals, including the *Canada Lancet*, that republished foreign articles about methods and techniques for their Canadian audiences. Practitioners were likely aware of Beard's writings, along with those of his colleague Alphonso D. Rockwell, with whom Beard published *A Practical Treatise on the Medicinal and Surgical Uses of Electricity*.[15] Their book, which went through multiple editions, was received favourably by the medical profession in Europe and North America. Many practitioners embraced the argument that electricity acted as a powerful sedative, could treat most nervous disorders, and provided pain relief.[16]

To help explain electricity's role in the human body, doctors often turned to the "body as battery" metaphor, where diseases resulted from "disturbances breaking up the electrical polarities of the [human] system."[17] The most common complaint about a nineteenth-century battery was its polarization; thus, if the body was like a machine, then it stood to reason that the human system could become polarized and drain the body's battery.[18] Others described the body as operating by electrical circuits: the brain received and distributed electricity via the nerves, which acted like wires. MacKay Jordan, a Vancouver doctor, theorized that the eye is the "main artery through which light and life enter the body."[19] The medical community's use of technological euphemisms to explain the body trickled into mainstream society. Historian Carolyn Thomas de la Peña maintains that "folk beliefs, unsettled medical knowledge, and an ill-defined technology, combined to create a space for electrical enthusiasm where phrases such as 'recharging my batteries' and 'short circuiting' crept into everyday speech."[20]

It was a sign of the times that Canadians could find electric belts and vibrators on sale in the daily press, along with ads for nostrums claiming to possess all of the healing qualities of electricity.[21] Patients could also receive treatment from electrotherapeutic institutions or from physicians such as Abner Mulholland Rosebrugh, an eye and ear specialist who combined his regular services with electric therapy. A vocal proponent of electrotherapy, Rosebrugh wrote extensively about the need to teach the principles of electricity in medical schools. In the 1880s, he wrote a series of articles on electrotherapy for the *Canada Lancet*, the country's foremost medical journal, in which he argued that "the administration of electricity is quite within the power of every physician."[22] He also published at least one treatise, *A Handbook of Medical Electricity*, in 1885, and he designed a portable battery for electrotherapeutic uses.[23]

Despite interest from practitioners such as Rosebrugh and Dr. Robert L. MacDonnell, who claimed to have introduced electrotherapeutic technologies to Montreal, developments in electric medicine were primarily left to electrotherapeutists, such as Professor S. Vernoy, founder of the Vernoy Electro-Medical Institute in Toronto in 1876.[24] That Vernoy, like many of his electrotherapeutist counterparts, did not have a medical degree would likely have been considered irrelevant to those who sought his care. Because orthodox doctors were unable to treat most ailments successfully, many people turned to alternative medicines in their quest for cures. Electrotherapy thrived in this environment, in which the line between doctor and quack was blurred. Vernoy advertised "radical cures" for men suffering from "nervous diseases, sexual and spinal weakness," as well as chronic ailments such as acne, gout, and rheumatism. Sufferers could receive treatment at his Jarvis Street location or "be treated at their homes if desired."[25] Alternatively, patients could treat themselves by purchasing his "Improved Family Switch Battery," a wood-encased battery from which patients would apply current to their bodies using zinc plates.[26] All that sufferers had to do was flick a switch and they could become their own medical electricians. Vernoy was also a prolific writer, publishing countless pamphlets to teach interested parties about the healing powers of electricity. He also published the *Electric Age*, a quarterly that "extolled the virtues of electrotherapy."[27]

Female sufferers could receive treatment in the adjacent building, where Jenny Trout, Canada's first licensed female doctor, along with Amelia Tefft and Emily Stowe, offered electrical treatment to women suffering from countless diseases.[28] Their Toronto Electro-Therapeutic Institution offered patients general electrification through galvanic baths or more localized treatment for specific afflictions, such as nervous debility or rheumatism.[29] However, for Trout and Tefft, "no organs of the human system are more liable to derangement than those situated in the pelvic basin."[30] In 1877, the two women published a handbook titled *The Curative Powers of Electricity Demonstrated*, which explained how electrotherapy was the most appropriate way to restore "vital energy" to a run-down body. Patients could seek treatment at the institution – an impressive structure, with space for out-of-town patients, a dining room, and a parlour – or call for a doctor to treat them at home.[31]

Tefft also paired up with Vernoy, and they offered full courses on electrophysiology, electrodiagnosis, and electrotherapeutics to anyone seeking instruction in these areas in order to treat themselves. On graduation, students would receive a diploma as proof of their competence to treat disease with electricity.³² The lack of regulation apparent in these types of institutions was fiercely opposed by some licensed physicians, who pointed to them as prime examples of quack medicine administered by charlatans. Doctors such as Rosebrugh tried to disqualify self-styled specialists by offering electric treatment in conjunction with their regular services, but despite their best intentions they actually helped to legitimize electrotherapeutic institutions and electrotherapy.³³ The presence of a medical battery in a doctor's work space would have made it more acceptable to seek similar treatment elsewhere. Prospective patients would have had little reason to suspect that medical electricians were anything but legitimate, especially when electrotherapeutists themselves probably firmly believed in the "curative powers of electricity." But how, exactly, could electricity cure the ailing person?

Despite the proliferation of advertisements and pamphlets touting the benefits of electrotherapy, none defined precisely what it was. Electrotherapeutists often filled in this lack of information by using vague yet promising language to assure the public that electricity was "nature's greatest healing agent" and that electric current would "restore vigor to the system."³⁴ Medical textbooks provided the clearest explanation: the principle behind electrotherapy was to administer treatment by either exposing the patient's body to the heat of electric light or applying electric current to a patient's limbs.³⁵ Electric current could be used to treat almost any malady by either relieving pain or stimulating health through muscle contraction via static or frictional electricity, galvanism, or faradic electricity.³⁶ A patient undergoing galvanic treatment at a doctor's office, for example, might have been confronted with a mahogany-encased battery capable of administering more than fifty volts. These devices typically included a rheostat, which allowed the medical electrician to increase or decrease the voltage, as well as electrode cords attached to the patient. These devices were also available for home use. Historian John Senior notes that today "these forms of treatment would not be considered valid for the purposes [for which] they were originally developed."³⁷ But at the time, administering electrotherapy in these forms made perfect sense, especially if one believed that electricity occurred naturally in the body. Lapthorn Smith, a Montreal-based gynecologist, noted that his patients continued to experience relief from their symptoms long after treatment. He reasoned that "the tissues and fluids of the body act as an induction apparatus, or, rather, as a storage battery, which continues to emit an electric current for some time afterward."³⁸

An established form of electrotherapy in the 1870s was generalized treatment via the electric bath. It was used to treat, among other ailments, debility, anemia, rheumatism, gout, and fever, and it was considered the ideal treatment to remove metallic toxins, such as mercury or lead, from the body.[39] The electric bath was the technological alternative to the long-held custom of bathing in mineral spas, except that some electrotherapeutists claimed that the electric version was superior – indeed safer – because the patient was not breathing in hot air, thus reducing the chances of cardiac or respiratory problems.[40] There were two common forms of electric baths: the incandescent light bath and the electric water bath. In the former, the patient would remove his clothes and sit on an insulated stool. The medical electrician would encase him within a specially designed cabinet that would expose only his head. The walls of the cabinet would be lined with incandescent (or sometimes arc) lamps, and the patient would receive treatment from the heat emitted by the bulbs.[41] Robert Bartholow observed that the "patient is more or less highly charged with electricity, which is silently received without pain ... The face [is] flushed, the action of the heart is quickened, and the pulse is more rapid. A general sense of tingling in the skin is experienced, and an abundant perspiration breaks out over the body."[42] The body was considered a permeable entity; the method was designed to sweat out any toxins and to allow the light to penetrate the skin and purify the blood. Practitioners could order or build their own cabinets; special collapsible baths were also available to administer treatment in a patient's home.[43]

The incandescent light bath retained its usefulness for years. In a 1932 article for *Saturday Night*, Isabel Morgan described the "magnetic effects of ... [the] modern cabinet bath." Instead of sitting on an insulated stool, she was told to lie down on a cot that slid into the cabinet, with her head exposed and her face shielded from the heat and light with a cold towel. She wrote that the lights gave an intense heat and that

> the local exercisers (round flat pads of felt attached to a slow alternating current) bring a slightly prickly massaging movement[, and] as the heat begins to take effect the body becomes bathed in its own moisture – a sign that the pores are becoming active and that the body is throwing off the various impurities that lodge in the most immaculate skin.[44]

The principle behind the light bath retained its late-nineteenth-century meanings; however, the venue had changed. Instead of going to an electrotherapeutic institution focused on healing, Morgan received treatment at a spa, suggesting that this form of electrotherapy had evolved into a luxury treatment rather than a medical cure.

For the electric bath, a bathtub, typically constructed of porcelain or wood and usually five and a half feet long, was filled with enough water for the patient to be covered up to the shoulders, and the water was to be kept at a temperature of ninety degrees Fahrenheit.[45] The medical electrician would place one electrode on the back of the patient's head and another under the patient's feet. Occasionally, the medical electrician would place a smaller plate on the patient's hips or knees if the person had expressed any aches or pains specific to those areas.[46] Copper was the preferred metal for the plates since it would not corrode, but sometimes plates made from zinc were used instead. The principle of this bath was that the water would act as a conductor, and the patient would receive only a portion of the current that the medical electrician would supply from an electrostatic generator or galvanic battery.[47] According to one user manual, once the current was turned on, the patient would feel a prickling sensation at the ankles or knees.[48] There was also a chance that the patient would experience a slight metallic taste as the current grew stronger. Depending on the affliction, the treatment would usually last between ten and fifteen minutes, with twelve baths being given over the course of one month.[49] The generalized treatment of the electric water bath was mostly prescribed to soothe nervous disorders, though a May 1873 article in the *Canada Lancet* questioned, "if the curing of constipation were the only thing we could do with it, would it not be deserving of high praise?"[50]

By the 1880s, Canadians could seek more localized treatment. With the aid of special attachments, electric current could be passed over any part of the body where the patient experienced pain. Women were frequent recipients of this method of treatment within the relatively new medical specialty of gynecology; with the increased popularity in electricity as a form of treatment, gynecological electrotherapeutics became an important research topic.[51] Mitchinson maintains that gynecology was "based on the assumption that [a] woman was predisposed to disease and that the culprit was her reproductive system."[52] Most doctors during this period regarded a woman's menstrual cycle as a barometer of her health; in 1892, American physician Henry Chavasse noted that, with first menstruation, a woman's "mental capacity both enlarges and improves."[53] Despite that promise, the stress of modern life, according to contemporary social commentators, could lead to irregularities such as hysteria.

An article published in the *Canada Lancet* in April 1880 described hysteria as "a disease to which every woman is liable."[54] The symptoms of hysteria were difficult to pin down; it could manifest itself as sleeplessness, nervousness, irritability, pain in the lower abdomen, or irregular or absent menstrual cycles. Doctors took great effort to distinguish it from George Beard's concept of neurasthenia. Canadian physician J. Matthews tried to explain the difference in a July 1889 article: "A hysterical woman often shows great power and capacity of both mind and body. A neurasthenic has lost elasticity and power ... The nerves are weak."[55] Two main differentiators between neurasthenia and hysteria were modernity and gender; whereas contemporary pressures caused a body to be rundown, hysteria was linked to the centuries-old belief of female diseases originating in the womb.

However, some gynecologists did posit that hysteria could be triggered by external pressures. An 1868 article in the *Canada Lancet* observed that women who used sewing machines for prolonged periods could experience disorders of the reproductive system, such as dysmenorrhea (intense pain during menstruation) and leucorrhea (excess vaginal discharge), leading to hysteria. The article noted that "the motion of the limbs in working the machines occasions sexual excitement."[56] The labour of working the treadle was simply too much for the delicate frames of young women. For Lapthorn Smith, school was to blame. Girls' brains were "using up all the blood that their enfeebled appetite and digestion can supply[, which] means that the generative organs are being starved at the very time that they most require a plentiful supply of good blood for their development."[57] Smith's observation reflected a widespread societal belief that women were the weaker sex. It would not have been unusual at the time for Smith to extend his "involvement to encompass not only the physical causes but the social and moral ones as well."[58] This would have allowed him to reassert his own importance and offer remedies, many electrical, to reverse the ill effects of reproductive disorders and thereby cure women of their hysteria.[59]

George Apostoli helped to popularize the use of electricity in gynecology.[60] The Paris-based doctor encouraged the use of galvanic and faradic currents in the treatment of gynecological disorders, and his methods were adopted throughout Europe and North America. Apostoli popularized the use of specially designed electrodes that would introduce current into the uterus to treat female reproductive disorders, such as amenorrhea (the absence of regular periods) or dysmenorrhea.[61] In a typical session, the patient would be placed in a dorsal position, with her head and shoulders elevated and her feet held in supports. The gynecologist would insert an intrauterine electrode and establish the patient's tolerance of the current by gradually increasing it with the aid of a galvanometer.[62] The patient was likely told to keep perfectly still in order to avoid shock. If she complained of pain, then the physician would reduce the current for a moment before increasing it again in successive stages up to a maximum of 250 milliamperes.[63] The gynecologist would insert his bare hands into the vagina in order to adjust the electrode; some, such as Smith, stressed cleanliness – "the hands have to be well washed and the fingers scrubbed with sublimate solution" – but it is doubtful that intrauterine procedures were sanitary.[64]

Vibration and massage were also acceptable forms of treatment for hysteria at the turn of the century. Some gynecologists noted that, though it was a strenuous treatment, manual vaginal massage could at least "give the patient a measure of relief."[65] Either manually or with a specially designed tool, doctors would massage a woman's genitals until the woman reached climax. The process was often laborious, and few physicians favoured it. As a result, historian Rachel Maines notes, the vibrator evolved into an "electromechanical medical instrument ... in response to physicians' demands for more efficient physical therapies, particularly for hysteria."[66] Electrotherapeutists argued that, if the womb could be "brought into a healthy condition," then hysteria could be cured, and electricity could provide the motive power to ensure instant relief.[67] The electric vibrator, introduced in the 1880s, required little skill to operate and was a quick alternative to manual stimulation. Facetiously referring to it as a "capital-labour substitution option," Maines indicates that the vibrator "reduced the time it took physicians to produce results from up to an hour to about ten minutes."[68] Personal vibrators were also on sale in the daily press, though none of them explicitly stated their use in vaginal stimulation.[69] Interested consumers could also purchase a vibrator from Eaton's, which carried its own brand, designed for "muscular and nervous diseases," after 1916.[70]

There seemed to be a consensus among medical electricians and gynecologists adopting electrotherapeutics that intrauterine electrodes and vibrators were acceptable only for married women. Virgins required more noninvasive treatments. Smith argued that there was no form of reproductive disorder that warranted vaginal examinations in unmarried or virginal patients.[71] Instead, these women would receive treatment *over* their reproductive organs: "In virgins, the faradic current, one pole at the lumbar region and the other over the uterus, does remarkably well."[72] Another specially adapted treatment was the electric bidet. Similar to the incandescent light bath, the bidet took the form of a box stool, with a back and arms. In the middle of the seat was a hole, below which six lamps would provide treatment. Mirrors positioned on the floor would reflect light and heat. This form of local bath was considered good for the treatment of menstrual disorders and rectal issues in both sexes. Medical electricians claimed that it would equalize the body's circulation and relieve pain.[73]

These were solutions to age-old problems given a modern cast. Men also experienced a gender-specific version of electrotherapy, except their illnesses were often linked to Beard's concept of neurasthenia. The average age of a sufferer tended to range from twenty-one to thirty-five, when, according to electrotherapeutist Jeanne Cady Solis, "the strain and stress of life are the greatest."[74] These stresses were especially acute in men who suffered from one particular symptom that helped to establish neurasthenia as a predominantly male illness: impotence.[75]

Impotence was a taboo subject in Victorian Canada, where social propriety was often delineated by strict gender codes. Erectile difficulty was equated with moral weakness – but a physical imbalance caused by psychological pressures associated with neurasthenia helped to legitimize a man's loss of sexual power.[76] Beard and Rockwell, for example, encouraged the use of electricity as a remedial agent in sexual problems, especially those caused by "over-taxation."[77] Historian Angus McLaren notes that some doctors attempted to attribute impotence to physical and organic causes such as cancer, malformation, or venereal disease.[78] Canadian electrotherapeutists tended to couch the words in advertisements for electric procedures and devices in the neurasthenic language promoted by Beard and Rockwell, and in doing so they helped to shift attention away from men whose impotence was traditionally blamed on sexual excess and moral delinquency in the form of masturbation.

The enterprising medical electrician offered a variety of electric solutions to men's problems, and one of the earliest forms of treatment was the electric belt. De la Peña notes that in the United States electric belts were the most popular form of treatment for impotence well into the 1920s.[79] Statistics on belts sold in Canada are not available; however, a proliferation of advertisements for this product occurred from the 1890s through the early decades of the twentieth century, suggesting a similar, though delayed, trend in Canada. Belts tended to vary from simple constructions made from cloth and galvanized discs to high-end versions made from silk, with brass-capped batteries hidden inside leather liners.[80] Electric belts were relatively harmless, though some early versions had exposed zinc electrodes, which could cause burning or blistering of the skin.[81]

Instructions for electric belts varied, but generally patients were to remove the batteries, immerse them in acid or vinegar, wipe off any excess moisture, and replace them in the belt. This was meant to "charge" the batteries.

According to the directions for one belt, the patient was to "buckle the belt snugly on the body around the naked hips, with the back plates resting on the base of the spine ... and pass the testicles through the loop in front which contains small silver plates."[82] These belts could be worn during the day or at night to provide the "vitalizing power of electricity" while the patient slept.[83] Later versions of the belt included a cord and plug, and the patient would literally plug himself into the wall to receive treatment.[84] Advertisements for electric belts were targeted at men and framed in neurasthenic terms. "Worn-out" and "weak" men were encouraged to seek treatment from specialists or to purchase electric belts for home use. Manufacturers promised that their belts could reverse youthful indiscretions by providing the "vitalizing power of electricity direct to all weak parts, developing the full, natural vigor of manhood."[85] The images associated with these advertisements were often suggestive of the relationship between penile health and masculinity – men "cured" by electric belts were featured with broad shoulders, Herculean strength, and defined muscles. In contrast, "the debilitated man" was frail, often depicted in a seated or slouched position, with slumped shoulders and a worried facial expression.[86] The electric belt, as a restorer of the body's natural life force, was frequently shown with tiny lightning bolts shooting out from the batteries and occasionally from the wearer himself. These images not only reflected Victorian notions of masculinity but also helped to define the social meaning of electricity by promoting electrical medical technologies as the modern cure for the modern man.

Whither the Modern Cure

Canadians understood electricity in the context of the world around them. Electric medicine matured at a time when the telegraph was shrinking the distances between provinces and between countries; when electromagnetism was proving its worth in industrial machinery; when streetcars were transforming the urban landscape; when Thomas Edison was turning night into day. Electricity powered these innovations, but how it worked remained a mystery. Medical electricians presented themselves as voices of authority by providing a definition of electricity for a society largely ignorant of its properties. Electric technologies were still in their infancy, and the social meanings of electricity remained ill defined. Journalists touted it as a "mysterious agent" and a "powerful force." At least one reporter labelled it a "fluid," not unlike the blood that flows through human veins.[87] Some medical electricians capitalized on these descriptions and maintained that electricity was "food": just as it powered the modern world, so too it powered the human body.[88] The electric means to replenish the body at the turn of the century were readily available, but the Canadians who received treatment were not simply pawns in the process – they had a choice. If it was cheaper to visit an electrotherapeutist than a regular physician, the decision was an easy one. And, if they were dissatisfied with the service, it is unlikely that they would have returned.

Because most people at the turn of the century did not fully understand electricity, electrotherapeutists – sincere or otherwise – had an audience ready to be sold on the idea that it could cure them. One Ottawa woman reportedly believed that riding on a streetcar would cure her rheumatism.[89] Medical electricians sought to distance themselves from charges of quackery by asserting the scientific merits of their practice: "The greater the scientific literacy of the doctor and patient, the more acceptable would therapeutic electricity become."[90] Many practitioners published material on electrotherapy in the late nineteenth century, intended for general physicians, medical students, and the public. Most stressed that electrotherapeutists become electricians first and physicians second. "To undertake the practice of electro-therapeutics without a thorough knowledge of the fundamental principles and laws of the science of electricity," American physician Wellington Adams wrote in 1891, "is as ridiculous and impracticable as would be an effort to carry on chemical analysis without having first become conversant with the principles of chemistry."[91]

Despite their best efforts, medical electricians could not stop the decline of electrotherapy as they practised it. With the professionalization of medicine, the pluralistic medical society of the nineteenth century shifted to more concentrated health care in the twentieth century.[92] The influx of useless devices into the market, coupled with an increasing risk of harm to patients by unlicensed, untrained, and self-styled physicians, led to harsh criticism of electrotherapy. Doctors became increasingly intolerant of unsatisfactory treatment, and many shifted their attention to promising developments in bacteriology.[93] If electric machinery and modern life were not the causes of illness among Canadians, then the body did not need to be "recharged" with electricity in order to be healed. Although facets of electrotherapy continued to evolve (the well-publicized developments in shock therapy in the late 1930s for treating mental illness comprise a prime example), so did perceptions of electricity. Electric technologies still had potential, especially in the realm of domestic and rural electrification; however, by the time electricity became common in urban areas during the interwar years, it had lost some of its intrigue, making it difficult for the average Canadian to consider it a magical cure for the modern age.

NOTES

1 Guy Cathcart Pelton, "The Electrical Era," *Western Women's Weekly*, July 12, 1919, 2.
2 Guy Cathcart Pelton, "Electrical Wonders Yet to Come," *Western Women's Weekly*, July 26, 1919, 11.
3 John Negru, *The Electric Century: An Illustrated History of Electricity in Canada* (Toronto: Canadian Electrical Association, 1990), 28.
4 The *Canadian Electrical News* published many stories about civic adventures in ornamental street lighting. See, for example, "Ornamental Street Lighting in Hamilton," December 1910, 40; "Illumination in New Hamburg Town," April 1911, 51; "Waterloo's Ornamental Street Lighting," April 1912, 23; "Electrical Canada from Coast to Coast," June 1912, 96; "Decorative Lighting in Winnipeg," December 1912, 73; and "Street Illumination in Prince Albert, Saskatchewan," June 1913, 122. See also "Street Lighting Improvements," *Live Wire*, March–April 1925, 5; and "Modern Street Lighting in Windsor," *Hydro Bulletin* 19, 1 (1932): 11–13.
5 Marshall Berman, *All that Is Solid Melts into Air: The Experience of Modernity* (New York: Simon and Schuster, 1982), 18.
6 Thomas P. Hughes has demonstrated that the growth of electrical systems in the Western world was determined by a multitude of factors, ranging from geographical necessity, to political culture, to available technology. Thomas Hughes, *Networks of Power: Electrification in Western Society, 1880–1930* (Baltimore: Johns Hopkins University Press, 1983). Other scholars have adopted his method in studies of national and international electrical systems; see Vincent Lagendijk, *Electrifying Europe: The Power of Europe in the Construction of Electricity Networks* (Amsterdam: Aksant, 2008); and William J. Hausman, Peter Hertner, and Mira Wilkins, *Global Electrification: Multinational Enterprise and International Finance in the History of Light and Power, 1878–2007* (Cambridge, UK: Cambridge University Press, 2008). Although Hughes acknowledges the existence of consumers, he does not provide space for them. David E. Nye, in *Electrifying America: Social Meanings of a New Technology, 1880–1940* (Cambridge, MA: MIT Press, 1990), analyzes electrification as a social process. For in-depth studies of the social meanings of light, including the relationships between light and medicine, see Wolfgang Schivelbusch, *Disenchanted Night: The Industrialization of Light in the Nineteenth Century* (Berkeley: University of California Press, 1988); and Chris Otter, *The Victorian Eye: A Political History of Light and Vision in Britain, 1800–1910* (Chicago: University of Chicago Press, 2008).
7 J. Adams, *Electricity, Its Mode of Action upon the Human Frame, and the Diseases in Which It Has Proved Beneficial* (Toronto: Dudley and Burns, c. 1870), 5.

8 Roy Porter, *The Greatest Benefit to Mankind: A Medical History of Humanity from Antiquity to the Present* (London: HarperCollins, 1997). For a more in-depth analysis of standards of living during this time, consult Terry Copp, *The Anatomy of Poverty: The Condition of the Working Class in Montreal, 1897–1929* (Toronto: McClelland and Stewart, 1974).
9 Tim Armstrong, *Modernism: A Cultural History* (Cambridge, UK: Polity, 2005), 92; John E. Senior, "Rationalizing Electrotherapy in Neurology, 1860–1920" (PhD diss., University of Oxford, 1994), 5.
10 Linda Nash, *Inescapable Ecologies: A History of Environment, Disease, and Knowledge* (Berkeley: University of California Press, 2006), 210.
11 Wendy Mitchinson, "Hysteria and Insanity in Women: A Nineteenth Century Canadian Perspective," *Journal of Canadian Studies* 21, 3 (1986): 87.
12 Gregg Mitman, "In Search of Health: Landscape and Disease in American Environmental History," *Environmental History* 10, 2 (2005): 12.
13 Daniel Clark, "Neurasthenia," paper read to the Ontario Medical Association meeting, June 1888.
14 Mitchinson, "Hysteria and Insanity in Women," 88.
15 The *Canada Lancet* frequently reviewed and referenced foreign books and made numerous mentions of Beard and Rockwell's treatise. For more information, see "When Are Involuntary Seminal Emissions Pathological?," *Canada Lancet* 12, 4 (1879): 112; "Fallacies Regarding Electricity," *Canada Lancet* 13, 3 (1880): 66–69; "Electrotherapeutics," *Canada Lancet* 13, 6 (1881): 161–67; and "Electricity in the Treatment of Specific Diseases," *Canada Lancet* 14, 5 (1882): 129–32.
16 George M. Beard and Alphonso D. Rockwell, *A Practical Treatise on the Medical and Surgical Uses of Electricity*, 7th ed. (New York: William Wood, 1866), 216–25.
17 O.K. Chamberlain, *Electricity: Wonderful and Mysterious Agent* (New York: John F. Trow, 1862), 2.
18 Michael Brian Schiffer, *Power Struggles: Scientific Authority and the Creation of Practical Electricity before Edison* (Cambridge, MA: MIT Press, 2008), 75; Senior, "Rationalizing Electrotherapy in Neurology," 6.
19 "Electricity: The Source of Life," *Western Women's Weekly* 1, 14 (1918): 5.
20 Carolyn Thomas de la Peña, *The Body Electric: How Strange Machines Built the Modern American* (New York: New York University Press, 2003), 108.
21 Museum of Health Care at Kingston, file Patent Medicines, Advertisements: "Electricity Is Life" (electric pills) and "$100 Proclamation" (electric oil), n.d. See also "Briggs' Genuine Electric Oil," *Ottawa Citizen*, November 7, 1883, 1.
22 A.M. Rosebrugh, "Electrotherapeutics," *Canada Lancet* 13, 8 (1881): 232.
23 J.T.H. Connor, "'I Am in Love with Your Battery': Personal Electrotherapeutic Devices in 19th and Early 20th Century Canada," paper presented to the Ontario Museum Association, London, ON, October 26, 1986.
24 J.T.H. Connor and Felicity Pope, "A Shocking Business: The Technology and Practice of Electrotherapeutics in Canada, 1840s to 1940s," *Material History Review* 49 (Spring 1999): 61; Charles Pelham Mulvany et al., *History of Toronto and County of York, Ontario* (Toronto: C. Blackett Robinson, 1885), 7.

25 "Professor Vernoy's Electro-Therapeutic Institution," *Globe*, January 31, 1880, 4.
26 Western University Medical Artifact Collection, Inventory 2004.005.01.01, "Prof. Vernoy's Improved Family Switch Battery," accessed June 2010, http://rabbit.vm.its.uwo.ca/MedicalHistory/Default.aspx?type=showItemFrameset&itemID=2138.
27 Connor, "'I Am in Love with Your Battery,'" n.p.
28 Ibid.
29 Advertisement, *Globe*, December 15, 1877, 4.
30 Toronto Electro-Therapeutic Institution, *The Curative Powers of Electricity Demonstrated* (Toronto: Monetary Times, 1877), 17.
31 Ibid.
32 "Electricity, Nature's Chief Restorer," *Globe*, August 18, 1877, 4.
33 It was common for Canadians to seek treatment from doctors practising in a variety of fields of medicine: orthodox, eclectic, and homeopathic, for example. Even though it was illegal to practise medicine without a licence, governing bodies such as the College of Physicians and Surgeons were largely ineffective at enforcing rules and regulations until well into the twentieth century. For more information, see Colin D. Howell, "Elite Doctors and the Development of Scientific Medicine: The Halifax Medical Establishment and Nineteenth Century Medical Professionalism," in *Health, Disease, and Medicine: Essays in Canadian History*, ed. Charles G. Roland (Toronto: Hannah Institute for the History of Medicine, 1984), 106.
34 "The Vernoy Electro-Medical Battery," *Toronto Star*, June 23, 1905, 46.
35 S.H. Monell, *Rudiments of Modern Medical Electricity* (New York: Edward R. Pelton, 1900), 17.
36 J.O.N. Rutter, *Human Electricity: The Means of Its Development, Illustrated by Experiments* (London: John W. Parker, 1854), 36.
37 Senior, "Rationalizing Electrotherapy in Neurology," 1.
38 Lapthorn Smith, "Disorders of Menstruation," in *An International System of Electrotherapeutics: For Students, General Practitioners and Specialists*, Horatio R. Bigelow (Philadelphia: F.A. Davis, 1894), G-163.
39 George M. Schweig, *The Electric Bath: Its Medical Uses, Effects, and Appliance* (New York: G.P. Putnam's Sons, 1877), 27.
40 Thomas Dowse, *Lectures on Massage and Electricity in the Treatment of Disease* (Bristol: John Wright, n.d.), 440.
41 See, for example, Justin Hayes, *Therapeutic Use of Faradic and Galvanic Currents in the ElectroThermal Bath* (Chicago: Jansen, McLurg, 1877), 7–14; Homer Clark Bennett, *The Electro-therapeutics Guide, or A Thousand Questions Asked and Answered* (Lima, OH: National College of Electrotherapeutics, 1907), 94; and R.V. Pierce, *The People's Common Sense Medical Adviser in Plain English: Medicine Simplified* (Bridgeburg, ON: World's Dispensary Medical Association, 1924), 884–85.
42 Robert Bartholow, *Medical Electricity: A Practical Treatise on the Application of Electricity to Medicine and Surgery* (Philadelphia: Henry C. Lea's Son, 1881), 212.
43 Bennett, *Electrotherapeutics Guide*, 92.
44 Isabel Morgan, "Baths of Light," *Saturday Night*, May 14, 1932, 16.
45 George M. Schweig, *The Electric Bath: Its Medical Uses, Effects and Appliance* (New York: G.P. Putnam's Sons, 1877), 8, 14–15; Edward Trevert, *Electro-Therapeutic*

Handbook (New York: Manhattan Electrical Supply Company, 1900), 81; Bennett, *Electrotherapeutics Guide*, 96.

46 Trevert, *Electro-Therapeutic Handbook*, 82.
47 Bennett, *Electrotherapeutic Guide*, 96; George J. Engelmann, "The Faradic or Induced Current," in Bigelow, *An International System of Electrotherapeutics*, A-157.
48 Trevert, *Electro-Therapeutic Handbook*, 84.
49 Schweig, *The Electric Bath*, 21; Trevert, *Electro-Therapeutic Handbook*, 84.
50 "Medical Electricity," *Canada Lancet* 5, 9 (1873): 521.
51 Andrew F. Currier, *Under What Circumstances Can Electricity Be of Positive Service to the Gynecologist?* (New York: Danbury Medical, 1891); George J. Engelmann, "Fundamental Principles of Gynecological Electrotherapy," *Journal of Electrotherapeutics* (May 1891): 1–46; Willis E. Ford, "The Methods of Administering Galvanism in Gynecology," *Transactions of the Medical Society of the State of New York* (New York: Medical Society of the State of New York, 1892); Augustin H. Goelet, *The Electrotherapeutics of Gynecology* (Detroit: George S. Davis, 1892); Chauncy D. Palmer, *The Gynecological Uses of Electricity*" (booklet), c. 1894; Betton Massey, *Conservative Gynecology and Electrotherapeutics*, 3rd ed. (Philadelphia: F.A. Davis, 1900); Francis H. Bermingham, "Electrotherapeutics in Some of the Diseases of the Genito-Urinary Tract," *American Journal of Surgery* 23 (1909): 2–3; Herman E. Hayd, *Electricity in Gynecological Practice* (Buffalo: self-pub., n.d.).
52 Wendy Mitchinson, "Causes of Disease in Women: The Case of Late Nineteenth Century English Canada," in *Health, Disease, and Medicine: Essays in Canadian History*, ed. Charles G. Roland (Toronto: Hannah Institute for the History of Medicine, 1984), 381.
53 Henry Chavasse, *What Every Woman Should Know: Containing Facts of Vital Importance to Every Wife, Mother, and Maiden* (Chicago: Royal Publishing House, 1892), 33.
54 "The Treatment of Hysterics," *Canada Lancet* 12, 7 (1880): 244.
55 J. Matthews, "Clinical Lecture on Hysteria, Neurasthenia, and Anorexia Nervosa," *Canada Lancet* 21, 10 (1889): 335–37.
56 "Effect of Sewing Machines on Menstruation," *Canada Lancet* 1, 3 (1868): 56.
57 Smith, "Disorders of Menstruation," G-157.
58 Mitchinson, "Causes of Disease in Women," 392.
59 Goelet, *The Electrotherapeutics of Gynecology*, 1–16; Hayd, *Electricity in Gynecological Practice*, 1–5; Currier, *Under What Circumstances?*, 6–15.
60 Palmer, *The Gynecological Uses of Electricity*, 4.
61 J.H. Kellogg, "A Discussion of the Electrotherapeutic Methods of Apostoli and Others," in Bigelow, *An International System of Electrotherapeutics*, G53–G89.
62 Hayd, *Electricity in Gynecological Practice*, 3; Kellogg, "A Discussion of the Electrotherapeutic Methods," G65–G66; Engelmann, "The Faradic or Induced Current," A-175, A-180.
63 Massey, *Conservative Gynecology*, 58; see also Engelmann, "Fundamental Principles of Gynecological Electrotherapy," 6.
64 A. Lapthorn Smith, "Electricity in Gynecology," *Canada Lancet* 20, 4 (1887): 99.
65 Horatio Bigelow, *Gynaecological Electro-Therapeutics* (Philadelphia: J.B. Lippincott, 1889), 170.

66 Rachel P. Maines, *The Technology of Orgasm: "Hysteria," the Vibrator, and Women's Sexual Satisfaction* (Baltimore: Johns Hopkins University Press, 1999), 3.
67 Chavasse, *What Every Woman Should Know,* 48.
68 Maines, *The Technology of Orgasm,* 4.
69 Rachel P. Maines, "Situated Technology: Camouflage," in *Gender and Technology: A Reader,* ed. Nina E. Lerman, Ruth Oldenziel, and Arwen P. Mohun (Baltimore: Johns Hopkins University Press, 2003), 99.
70 "Eaton's Vibrator," *Eaton's Catalogue,* fall–winter 1916–17, 412.
71 Smith, "Disorders of Menstruation," G-151.
72 Bigelow, *Gynaecological Electro-Therapeutics,* 158.
73 Bennett, *Electrotherapeutics Guide,* 194–95.
74 Jeanne Cady Solis, "The Psychotherapeutics of Neurasthenia," *Canada Lancet* 39, 1 (1905): 93.
75 M.J. Grier, *The Treatment of Some Forms of Sexual Debility by Electricity* (Philadelphia: Medical Press, 1891), 5–15; A.R. Rainear, *Electricity in the Treatment of Male Sexual Disorders* (Philadelphia: n.p., 1896).
76 Angus McLaren, *Impotence: A Cultural History* (Chicago: University of Chicago Press, 2007), 116.
77 A.D. Rockwell, *Electrotherapeutics of the Male Genital Organs* (New York: William Wood, 1874), 8; Beard and Rockwell, *A Practical Treatise,* 619.
78 McLaren, *Impotence,* 131.
79 De la Peña, *The Body Electric,* 149.
80 Andrew Chrystal, *Catalogue of Professor Chrystal's Electric Belts and Appliances: Consisting of Electric Belts and Bands, Electric Belts with Suspensory Appliances* (pamphlet) (Michigan: n.p., c. 1899), 6–7.
81 "A Victim of the Bare Metal Electrode Belt," *Globe,* July 12, 1900, 5.
82 Museum of Health Care at Kingston, File: Pamphlets, Museum Catalogue No. 003.050.036f, "Directions and General Remarks," n.d.
83 "I Cure Varicocele!," *Globe,* September 18, 1900, 9.
84 Carolyn Thomas de la Peña, "Plugging In to Modernity: Wilshire's I-ON-A-CO and the Psychic Fix," in *The Technological Fix: How People Use Technology to Create and Solve Problems,* ed. Lisa Rosner (New York: Routledge, 2004), 44–53.
85 "No Cure, No Pay," *Globe,* May 29, 1900, 9.
86 "The Debilitated Man," *Globe,* January 2, 1900, 7.
87 "Electricity on Tap," *Globe,* June 21, 1913, 6.
88 C.H. Bolles and M.J. Galloway, *Electricity: Its Wonders as a Curative Agent* (Philadelphia: Philadelphia Electropathic Institution, n.d.), 2.
89 As reported in *Canadian Electrical News* 1, 5 (1891): ix.
90 Robert Newman, "The Want of College Instruction in Electrotherapeutics," *Electrical Journal* (October 1, 1896): 1–8; William J. Herdman, *The Necessity for Special Education in Electrotherapeutics* (undated pamphlet), xxv–xxxii; Senior, "Rationalizing Electrotherapy in Neurology," 56.
91 Wellington Adams, *Electricity: Its Application in Medicine and Surgery* (Detroit: George S. Davis, 1891), 2; see also Julius Althaus, *Report on Modern Medical Electric and Galvanic Instruments* (London: T. Richards, 1874); and Robert Amory, *A*

Treatise on Electrolysis and Its Applications to the Therapeutic and Surgical Treatment in Disease (New York: William Wood, 1886).
92 Jacalyn Duffin documents this worldwide trend in Chapter 6 of her *History of Medicine: A Scandalously Short Introduction,* 2nd ed. (Toronto: University of Toronto Press, 2010).
93 Ibid., 81–84.

4

Cosmic Moderns
Re-Enchanting the Body in Canada's Atomic Age, 1931–51

BETH A. ROBERTSON

If the modern could be geographically located in present-day Canada, then it just might be in Kitchener-Waterloo. The area, after all, is currently known as "Silicon Valley North" – one of many "globally competitive technology clusters" around the world that has produced numerous examples of cutting-edge innovation.[1] What might not be as widely known is that southern Ontario, including Kitchener-Waterloo, has historically been a hotbed of what Catherine Albanese has called "metaphysical religions."[2] A close neighbour to the "burnt-over district" south of the border, the region welcomed a range of beliefs and faith communities beginning in the early nineteenth century.[3] This range includes the movement of spiritualism – a religion that first emerged in the mid-nineteenth century and argued that the living could communicate with the dead. In fact, Leah, Margaret, and Kate Fox, the sisters at the centre of early spiritualism, had moved from southern Ontario before heading in the 1840s to New York, where they first heard the mysterious knocks that would make them famous. Their visit with family members in Canada during the 1850s once more solidified southern Ontario as a place where enthusiasm for the supernatural flourished.[4] Such a past might seem to be distant from the contemporary high-tech aspirations of the area, but it might not be entirely coincidental. As Jeffrey Sconce, Jill Galvan, and others have demonstrated, spiritualists historically made use of scientific and technological imaginaries to understand changes in the world around them and their place within it.[5] Spiritualism and its sustained engagement with new scientific ideas and technologies thus served as one more arena of many in which Canadians learned how to be "modern."[6]

This chapter aims to build upon such insights by investigating the discourse of a small spiritualist group in Kitchener-Waterloo from the 1930s to the 1950s. The medium of the group, Thomas Lacey, delivered a series of lectures in the 1930s and into the Cold War era while under the alleged influence of otherworldly intelligences. Drawing on contemporary, even cutting-edge, theories of the atom, he explained the inner workings of spirit, the universe, and in turn the human body. While advocating the belief that such scientific discoveries marked the dawn of a new age in which the body would play a key role, Lacey's teachings powerfully conveyed how the tenets of modern physics influenced not merely the centre but also the peripheries of postwar culture – each of which critically repositioned the body in what was heralded as the advent of an inspiring yet ominous atomic age.

Science and Technology at the Fringe

In 1993, Richard Jarrell wrote that, though the discipline of the history of science had for "so long wedded itself to great men and great ideas," it was finally "freeing itself, if slowly, from the thrall of a narrow vision of science that places a premium upon so-called pure science."[7] It is difficult to clarify exactly how far Jarrell envisioned that the discipline would go to "free itself" from a historically preconditioned idea of pure science. His words nevertheless point to the utility, perhaps even the necessity, of looking to the margins of the scientific endeavour. By examining such uses of scientific and technological representations, we can gain a deeper understanding of how such ideas were conceptualized historically in the broader culture as people attempted to come to terms with changing understandings of the natural world and, perhaps most importantly, its effects on the human body. Just as Dorotea Gucciardo explains how electricity's popular allure in earlier decades was directly applied to the body, so too this chapter focuses on a case study in Kitchener-Waterloo to examine how atomic energy reshaped occult discourse in Canada from the 1930s to the 1950s to become a powerful analogy of the possibilities of bodily rejuvenation.

In the 1930s, a small group in Kitchener-Waterloo considered the links between the revelations of a new atomic age and the esoteric world. Thomas Lacey led the group as the main medium for several years. An immigrant from Derbyshire, England, Lacey was recognized as a talented medium in well-known spiritualist publications such as *Two Worlds* and for a time acted as a medium at the Lily Dale spiritualist retreat in New York.[8] Little else is known of the man other than that he established a spiritualist group in Kitchener-Waterloo as early as 1931. Holding séances and communicating with spirits, sometimes through an aluminum conical device referred to as a séance trumpet, he gave a series of lectures while under the alleged influence of spirits throughout the 1930s and into the 1960s. To reduce any external tampering that might unduly shape the contents of the messages received, some of these lectures were recorded in a room with only Lacey present and referred to as "The Room Alone Lectures." The "lectures" presented an interesting blend of beliefs, tacking from spiritualism to theosophy to Gnosticism on any given night.[9]

Lacey's followers compiled, transcribed, and preserved recordings of these lectures for future use for themselves and other interested practitioners. The group itself was fairly eclectic, including members from both Kitchener and Waterloo. Despite their strong investment in exploring spiritualist and occult practices, most were devout Lutherans. Otto G. Smith, a local musician and the head choirmaster of St. Matthew's Lutheran Church in Kitchener, hosted many of the meetings in his home from the 1930s to the 1960s.[10] Smith performed a dual role in many of the séances as he often played the organ to produce the appropriate "vibrations" for communication to take place as well as frequently acted as the main note taker for the group.[11] Other members included Lutheran minister Garnet Schultz; a local bank manager and his wife; Mr. and Mrs. E.W. Sheldon, who also occasionally hosted meetings; as well as Decima Laing, the wife of a prominent Waterloo businessman, John M. Laing. Decima was also a long-standing member of St. John's Lutheran Church in Waterloo who gave generously of her time and capital to the Waterloo Lutheran Seminary.[12] Sidney Wright, a prominent local entrepreneur and office supplies businessman in Kitchener, also regularly attended the meetings. Wright shared record-keeping duties with Smith while maintaining connections with other occult groups in nearby St. Catharines.[13]

None of the regular participants was a professional scientist, yet the notes of the séance group nevertheless reveal a mutual interest in scientific and technological developments. Maintaining "careful records" over time, Lacey's group dutifully assembled "interesting and convincing data," as Smith described, alongside a variety of "unsolicited test messages and physical phenomena" that they believed proved the empirical and esoteric importance of Lacey's mediumship. Although his lectures ostensibly addressed "a wide range of topics," rapid shifts in scientific findings and technological advances became particularly prominent in terms of what information the group chose to preserve.¹⁴ Shifting conceptions of the behaviour of sound waves as it related to wireless communication, advances in transport, as well as changing notions of space and time loomed large in many of the recorded conversations between the living and the dead.[15]

Arguably, these group meetings could be disregarded as utterly peripheral and irrelevant to the history of science and technology in Canada or elsewhere. Yet these séances might in fact reveal a great deal about how science and technology, and indeed conceptions of the modern, were being redrawn during this period.[16] In the first half of the twentieth century and well into the Cold War, the sciences were a site of contestation and dispute. The emergence of what became known as "modern physics" introduced multiple new ideas about how the universe operated, revolutionizing how the scientist, and in turn the layperson, understood the world around them.[17] This climate left a great deal of room for individuals to dispute what could be considered unnecessary scientific dogma. As Michael Gordin and others have argued, eccentric theorists from J.B. Rhine to Immanuel Velikovsky attempted to lay claim to scientific authority from the 1930s to the 1970s, challenging the boundaries of empirical inquiry to give rise to "the birth of the modern fringe."[18]

Spiritualists gravitated toward this unpredictably persuasive fringe, and there were many historical reasons for just such an attraction. Since the nineteenth century, spiritualists had maintained a strong affinity with the burgeoning professional sciences, even as they stretched and reshaped newly introduced scientific ideas for their own ends. It can even be said that both spiritualism and science helped to define the other, since each was granted new meaning in what was heralded as the age of "the modern."[19] Emergent technologies and scientific theories frequently provided the basis for occult understandings, whether it be metaphors describing mediums as "spiritual telegraphs" or new discoveries such as the X-ray attaining an almost supernatural status.[20] References to "occult science" proliferated as individuals perceived their pursuits as critical links between what seemed like the increasingly disparate realms of science and religion.[21]

Prime examples of attempts to join science and spiritualist hypotheses were the debates that surrounded a strange paranormal condensation referred to as "teleplasm" or "ectoplasm." This substance was closely aligned with the possibilities inherent to the human body and mind – a fact only emphasized by its propensity to emerge from various orifices of the medium.[22] In the nineteenth century and well into the twentieth century, several observers explained the substance in light of biology. Charles Richet, a famous French physiologist, was reportedly the first to coin the term "ectoplasm" during a series of investigations in 1894 of the Italian medium Eusapia Palladino.[23] He argued that the material was, in fact, protoplasm – a significant parallel since protoplasm had come to occupy a special place in biological understandings around the turn of the century, becoming, as Robert Brain has described, "the crucial object for investigating some of the greatest mysteries of living substance," whether "automaticity and plasticity, memory and heredity, temporality, autonomy and immortality."[24]

The conclusions of Richet deeply influenced how ectoplasm was understood, and investigators continued to study ectoplasm in light of its seemingly biological characteristics well beyond the *fin de siècle*. Canadian medical doctor T. Glen Hamilton, in a book published posthumously in 1942, quoted Richet's theories of ectoplasm at length in an attempt to make sense of his own encounters with the condensation. Convinced of the significance of this substance, especially in terms of what it could reveal about normal and "supranormal" biology, Hamilton dedicated the remaining years of his life to studying the materializations after they began appearing in his psychical investigations from 1928 on. He himself described ectoplasmic objects as "pseudopods" – temporary projections of unusual cell membrane or amoeboid that would form different types of simulacra capable of performing an array of physical feats in the séance room. The "plasmic masses" appeared in a multitude of different shapes, sizes, and textures, transforming into diverse "fibrous structures" of varying density and cohesion, with each maintaining a specific and unique function.[25]

Hamilton and those like him continued to draw such parallels between accepted science and its marginal incarnations when examining paranormal phenomena. Rather than hearken to biology, however, others made use of shifting theories of physics, particularly those pertaining to atomic structure and energy. Dr. Harvey Agnew, Hamilton's colleague and then the secretary of the Canadian Medical Association, who would later be remembered as "the father of hospital administration," remained enthusiastic that burgeoning theories about the atom and radioactivity could help to explain the inexplicable.[26] Citing the work of Robert Millikan on cosmic rays, Agnew proposed that the paranormal condensations of "teleplasm" or "ectoplasm" could actually be the by-products of high-energy radiation. The general fragility of the material, especially when exposed to light, could similarly be compared with contemporary experiments that demonstrated how atomic orbitals widened and became unstable when exposed to a powerful light source. As he further explained, "when a strong beam of light is directed into the atom the orbit of electrons around the central proton is considerably enlarged. The atom expands and because of this enlarged orbit some of the electrons may escape. This is associated with the existence of isotopes wherein a number of the elements may have varying atomic weights."[27] Agnew paralleled such experiments to studies in ectoplasm, known to promptly disintegrate when exposed to white light. He argued that ectoplasm behaved this way because of the atoms that composed the unpredictable substance, which had not yet "become stabilized" to withstand such exposure.[28]

Agnew's insights reveal a great deal about how the world was beginning to be understood in the first half of the twentieth century. These significant changes marked the beginning of what Matthew Lavine has referred to as "the first atomic age," which predated by some decades detonation of the first atomic bomb in July 1945. As Lavine has argued, recognizing the significance of this early atomic history is vital to understanding the full range of knowledge about atomic energy that circulated at this time.[29] Although Trinity, the world's first nuclear test, and the subsequent bombings of Hiroshima and Nagasaki were literally earth shattering, atomic energy had long begun to transform how the natural world was conceived – and in turn could be destroyed. These shifts had wider repercussions than is often assumed. A significant amount of the scholarship on the atomic age, or "nuclear culture" as it has also been called, has focused heavily on the more official realms of the government, politics, and the military. Yet, as Jonathan Hogg, Christoph Lauct, and others have shown, the influence of atomic power extended much further, reaching from laboratories and lecture halls to newspapers and other public forums and even, as I argue, to darkened séance rooms.[30]

Atomizing the Occult in 1930s Canada

Harvey Agnew was hardly alone in linking occult energies with new theories arising from modern physics, which maintained a wide appeal among spiritualists and occultists.[31] Unfolding conceptions of the atom, atomic energy, radiation, and nuclear power, after all, served as examples of multiple, interrelated, invisible forces that spiritualists insisted proved the presence of the inexplicable and the unseen.[32] Similarly, the unintuitiveness of emerging theories such as relativity compelled many spiritualists and occultists. As Stephen G. Brush and Ariel Segal have pointed out, Einstein's theories of relativity seemed to be "contrary to common sense" when they were introduced.[33] Like "occult science," relativity challenged static notions of time and space, pointing to physical energies and notions of matter that seemed to be unheard of until that time. Its proposals were so far outside the realm of contemporary understandings, in fact, that some described it as a move toward metaphysics – a direction that not all viewed as problematic.[34] Indeed, many occultists believed that relativity confirmed their faith, particularly their belief in the presence of multiple dimensions of existence that rested beyond normal human perception.[35]

These links were not lost on Thomas Lacey and his followers. During their group meetings, the spirit of famous inventor Thomas Edison frequently made an appearance, predicting various advances in science, communication, and transportation. One such prediction in 1933 involved the automobile and how the internal combustion engine would soon be replaced with something much more efficient, presumably an electric battery that Edison himself had worked on in life.[36] During the same evening, Edison and other spirits elaborated on the shifting nature of matter, likening the transformation of water (or, more precisely, hydrogen and oxygen) from liquid to gas to a spiritual transfiguration.[37] In the years to follow, otherworldly entities similarly alluded to the scientific discovery of various new rays "that penetrate ... the solids of matter." Although undetected by human perception, Lacey's spirit guides argued that humanity would soon learn, through scientific as well as spiritual enlightenment, to "utilize these pulsations" in such a way as to alter radically the composition not only of the natural world around them but also of their physical bodies. Through the development of "instruments powerful enough" to "see into the microcosmic world," the linkage of the metaphysical and the empirical made manifest through "occult science" would only become more apparent and pave the way for a deeper understanding of multiple dimensions of reality and human materiality more broadly.[38]

Theories of matter and the nature of the material world continued to feature prominently in many of the group's lengthy discussions. Musings and predictions regarding atomic life played a significant role. Or, as one spirit guide allegedly explained through Lacey in 1934,

> there is no such thing as matter. Everything is vibrating with cosmic life. The chair upon which you are sitting is nothing but a series of ... atomic and electronic pulsations that ... constitute that which your limited senses perceive. To you it is solid matter, and in reality it is but a series of electronic pulsations, so is the human body, so is everything that manifests in the physical plane.[39]

As this quotation indicates, the occult fascination with the atom and its role in both natural and supernatural processes directly shaped discourses of the body. It might be assumed that those pursuing esoteric mysteries would be fascinated only by the affairs of the spirit. However, similar to what historian Marie Griffiths and others have argued, faith, whether of the occult variety or not, often compelled individuals to pay close attention to the inner workings and health of the body.[40] As many historians of the body would agree, this focus was not without significance.[41] For Lacey, and the occult group that met alongside him, the body played a significant role in what they foresaw as the future transfiguration of humankind. They discussed the body at length – dividing it into various material and spiritual components. The spirit guide "White Eagle" clarified in 1936 that "we have not only the physical body, but also an 'etheric' (according to some schools) or a 'vital' body according to others ... Next we come to the desire body. Some call it the astral body."[42]

As in other occult and spiritualist groups, nineteenth-century ideas of "ether" and the "etheric" persisted in the language of Lacey's spirit guides and fellow participants when discussing the body and other topics. Although this could be interpreted as an inability to engage with modern theories, a more complex picture emerges when analyzing how occultists used these terms. Undoubtedly, the idea of ether underwent some significant transformations in this period, but neither did it wholly disappear, and in fact many continued to argue that the tenets of modern physics could still be compatible with the presence of this now classical substance.[43] Rather than disregard or ignore such shifts, occultists such as Lacey and his followers made these transformations apparent as they employed ideas of ether in conjunction with emergent atomic theoretical models. Or, as "White Eagle" claimed in 1936, "the etheric body," through which the medium could project himself onto other dimensions, was composed of "denser atoms" than the physical body, a trait that indicated its more highly evolved state.[44]

New discoveries about atoms and atomic power held huge potential for the restoration and reconfiguration of the body, etheric or otherwise, Lacey argued, discoveries that would eventually propel humankind into a new era of heightened scientific and spiritual understanding. In 1936, he predicted the coming of spiritual masters who would soon "take their place and dwell in a physical body amongst men." Lacey claimed that their influence would lead to rapid advances in social and medical science, including the harnessing of "cosmic rays" to heal and perfect the human body through innovative surgical methods.[45] His reference to cosmic rays might be dismissed as hyperbole, yet his mention of them merits some attention when taking account of the scientific research being conducted at the time. Robert Millikan first introduced the term "cosmic rays" in 1925 after undertaking an intensive study of radiation that physicist Victor Hess first discovered in 1912. Through such studies, Millikan was able to confirm, as Hess postulated, that the rays did have an extraterrestrial origin. As a result of such discoveries, these rays captured the public imagination and remained "a staple of science journalism for decades," as Lavine has described. Demonstrating the mysterious "energies that ... lurked inside the atom," cosmic rays might not have been fully understood by the layperson. These newly identified rays nevertheless came to represent what seemed to be the increasing esoteric nature of the atom itself.[46]

The inclusion of radium within a host of allopathic medications and cosmetics promising health and long life had come to an end by the mid-1930s. Yet hope still existed that radiation could lead to new cures for a number of serious diseases.[47] Cosmic rays in particular garnered much fascination, and it was postulated that this truly alien radiation might not only contain healing properties but also drive the process of evolution to create more physically advanced human beings.[48] Considering such a context, Lacey's reference to cosmic rays and his claim that they could heal and eventually perfect the human body were not necessarily peculiar to the discourses of this occult group of Kitchener-Waterloo. Rather, his assertions demonstrate that the atom, and the energy that it produced, provided a language for occultists in the 1930s. This "atomic" discourse aligned what was perceived as scientific and medical advancement in the wake of new discoveries, with spiritual, even embodied, progress.

Embodying the Atom during the Cold War

Thomas Lacey's group continued to meet throughout the rest of the 1930s and as the Second World War raged, but the records of their group meetings decreased significantly in number compared with previous years. The precise reason for the lack of reporting is unclear; however, when rigorous record keeping was once more taken up shortly after the conflict had ceased, it was apparent that the group had not emerged unscathed. The spirits that allegedly appeared in the fall of 1946 warned that "suspicions are rife, [and] distrust is the prevailing spirit everywhere." Rapid technological change since the war had created a smaller, more interconnected world, yet this did not necessarily breed optimism, especially in the shadow of the atomic bomb: "The world is getting smaller, distances are less and time is shorter, and those things we have thought of as barriers no longer exist. If a man so willed now, the whole world could be destroyed."[49]

Yet atomic power retained the mystique that it held before the war despite, or perhaps because of, the destruction that it could wreak. The atom and atomic power were regularly equated with the force of mind, spirit, and in turn, body that highlighted both the spiritual and the violent potential of humankind. Or, as Lacey described,

> in the atom are inner worlds. A cohesiveness about those tiny worlds that scientists disturb or smash the atom to break down the power of the constituents of the atom ... How Mighty are the voltages of Atoms? ... Who exerts that power? ... At this point we speak of things that pertain to ourselves. That force that we have to build machines that the atoms might be broken apart.[50]

Newly discovered mysteries of the atom could also lead humans to prevent the annihilation of which they were newly capable. Now that the full destructive power of the atom had been unleashed, science could dedicate itself to "building instead of destroying." Or, as Lacey lectured in 1948, scientists who wielded "disintegrating forces that were so tremendous" have "come to the light of the reverse ... This is the beginning of the transmutation of substances, to its complete manipulation and projection."[51] The discovery of nuclear transmutation in which elements or isotopes could be transformed into another through the use of particle accelerators such as cyclotrons had placed "creation with[in] the hands of man."[52]

In emphasizing the beneficial aspects of atomic energy, Lacey and his fellow spiritualists joined others in the late 1940s in promoting what Scott Zeman has referred to as the narrative of the "bright atomic future." Emerging in the wake of the Hiroshima and Nagasaki bombings, this narrative put forward the idea that, though atomic forces could indeed be devastating, such energy also had the capacity to provide "limitless power" as well as "an end to war, disease and even poverty."[53] This language was built and maintained primarily upon the idea of the "peaceful atom" – which emphasized innovations such as the study of nuclear fission and the promise of "an inexhaustible energy source." Similarly, advances in nuclear medicine were perceived as potentially "humankind's final conquest of disease."[54]

This optimistic view of nuclear medicine seemed to be justified in the postwar years, especially upon the discovery that nuclear reactors could be used to produce radioactive isotopes. The use of radioisotopes for medical purposes was by no means new at the end of the Second World War, and medical scientists had already begun to manipulate them for the purpose of cancer therapy since the 1930s. The development of cyclotrons that Lacey referred to made such applications more feasible by making artificial radioactive isotopes available through a tightly constrained system of exchange among physicians, life scientists, physicists, and chemists. Their circulation expanded once nuclear reactors built for the war effort were repurposed to produce radioisotopes in the postwar years. These isotopes could serve a variety of medical purposes, including becoming effective molecular tracers that could be implanted in the body and followed as they progressed through the biological system, rendering once invisible molecular processes apparent to scientific scrutiny.[55]

Occultists grappled with this new reality, which saw both the great devastation and the promise of atomic energy, particularly as it was rendered upon the body. Perhaps appropriately, Lacey envisioned future scientific experiments in which the body would be destroyed and then rebuilt, its atomic material broken down by rays of radiation, only to be restructured once more. As Lacey described in 1948 while in a trance, scientists could "actually melt [the body] away, and wherever they may wish it to appear again, lo! It grows before their eyes ... almost as quick as the switch is turned that makes it possible."[56] This was merely an extension of what was already being revealed about the human body. Alluding to recent discoveries regarding the behaviour of radioactive isotopes, particularly their medical uses, and the regeneration of cells, Lacey claimed that "every minute all in our body has changed, and we can truly claim that there is not one iota of reality in our physical body that has not completely been transformed."[57] The links that he made between the body and atomic energy reflected and animated broader discourses of the atom and its physical effects. Newfound discoveries both legitimized and provided evidence for long-held occult beliefs in spiritual regeneration and transfiguration, discoveries that demonstrate how the world, and in turn the body, are "filled with these cosmic wonders."[58] Lacey argued that the body, especially when joined with other bodies, could wield greater power than "even the atomic bomb."[59] By "consolidating and conditioning this body," Lacey insisted, "it will have the power" to bring about "the end for which it is rightfully and righteously intended" – an end that would thrust humanity into "the mental age" in which matter and spirit could be both manipulated and renewed.[60] This new age could only be imagined through the atom and the power that it held, which provided a "deeper understanding of the essence of things."[61]

Lacey and his followers no doubt continued to express some anxiety about atomic power into the late 1940s, particularly if it was recklessly used by "man," who tended to "overstep his powers" especially now that something had been "placed in his hands ... by which he could unthoug[h]tfully very easily destroy himself."[62] Yet Lacey counterbalanced such caution with an almost unbridled optimism as they awaited a new age or, more precisely, an "atomic age." "You are coming into an atomic age, in fact the atomic age is here and all the wonders that will be derived from it. If man will accept these wonders in the manner they are intended to accept them, there will be no wars."[63] These "wonders" included limitless energy in which vehicles would be "entirely propelled by an atomic unit" that would radically alter the methods and speed of transportation. Meanwhile, heating, lighting, and a host of domestic technologies would be fitted with power cartridges fuelled by "atomic energy, something that would last a lifetime." Such developments in themselves could ensure that, "within a few years, the whole universe could be revolutionized."[64]

Even more than limitless power, however, the optimism that Lacey expressed regarding atomic energy was especially reserved for what it meant for the rejuvenation of the human body. Harnessing such energy would ultimately ensure that humanity no longer experienced "any sense of the age that creeps upon us," that "all the live cells of our body should be completely rejuvenated ... so that there should not be any deterioration of all those things that go up to make our physical life."[65] The "atomic content" of each individual, "those deep forces which lay hidden within us," would grant people "superhuman" powers that would far outpace what they currently possessed.[66] In the end, such discoveries would perfect the body and even distinguish humans on Earth from other beings who populated the universe. Speaking of inhabitants of other planets, Lacey contended that even these more enlightened creatures had not developed anything equivalent to atomic energy, a statement corroborated by an acclaimed extraterrestrial being speaking through him: "We haven't seen anything like that of course or atomic industrial power!"[67]

Atomic energy thus became for Lacey and his followers a means to envision a new future for the human body in which it would not deteriorate but reach its full potential of supernormal, even extraterrestrial, existence. And it is perhaps here, within these articulations, that the Lacey group revealed the main reason for their attraction to such energies. Not only did atomic energy represent one more example of unseen forces that they insisted confirmed long-held esoteric beliefs, but also the atom itself came to be viewed as a powerful analogy of the central tenet of spiritualist and occult understandings – the much-sought rejuvenation and immortality of humanity. The atom and the energy that it generated paralleled and granted new meaning to the human body in that each "released energies" that "go out, they multiply, they live, they do not die!" Even when pulled apart and apparently obliterated, the atom and the human body similarly continued to survive, according to Lacey and his followers, producing unimaginable and unseen energy that could heal as well as destroy.[68]

Conclusion

The twentieth-century incarnation of modern physics and particularly ideas of atomic energy played a critical role in how Canadian occultists in Kitchener-Waterloo and elsewhere re-envisioned the body. Drawing on contemporary understandings of the atom and how it could be used as an energy source as well as a means of restoring the body, Thomas Lacey, the main medium of the group, rearticulated occult understandings of materiality. Acquiring a distinctive "atomic" discourse to describe the inner workings of spirit and the physical world, Lacey insisted that the findings of modern physics offered spiritual insights into the corporeal and metaphysical complexities of the human body. Indeed, conceptions of the atom and the energy that it could produce provided a compelling analogy to describe how the body could reach its full immortal potential – which spiritualists and occultists had long clung to as evidence of their faith. These revelations could, in turn, beckon a new, modern "atomic age" in which the body, and the natural world more broadly, could be transfigured and rejuvenated.

This group was hardly alone in linking physics and atomic energy so closely to the experience of modernity and its future possibilities. Although Thomas Lacey and the occult group that he founded could be dismissed as far beyond any conception of "pure" science, their ideas provide a unique insight into the multiple uses of modern physics in comprehending the physical world as well as the more intimate aspects of the body. These uses vividly convey the unexpected diversions of scientific knowledge and technological representations within the prevailing culture. More so, they demonstrate how individuals came to understand themselves and the world around them in what they believed to be a radically new age. Employing an atomic discourse to rearticulate occult understandings in accordance with emergent scientific ideas and technologies, the occult group in Kitchener-Waterloo grappled with both the limits and the vast potential of their own embodiment in the era of the modern.

NOTES

1 Larisa V. Shavinina, "Silicon Phenomenon: Introduction to Some Important Issues," in *Silicon Valley North: A High-Tech Cluster of Innovation and Entrepreneurship*, ed. Larisa V. Shavinina (Amsterdam: Elsevier, 2004), 3.
2 Catherine L. Albanese, *A Republic of Mind and Spirit: A Cultural History of American Metaphysical Religion* (New Haven, CT: Yale University Press, 2007), 4–10.
3 Ibid., 180.
4 Ibid., 179–253; Molly McGarry, *Ghosts of Futures Past: Spiritualism and the Cultural Politics of Nineteenth-Century America* (Berkeley: University of California Press, 2005), 1–3; Stanley Edward McMullin, *Anatomy of a Séance: A History of Spirit Communication in Central Canada, 1850–1950* (Montreal/Kingston: McGill-Queen's University Press, 2004), 22–41.
5 Jeffrey Sconce, *Haunted Media: Electronic Presence from Telegraphy to Television* (Durham, NC: Duke University Press, 2000), 13–16; Jill Galvan, *The Sympathetic Medium: Feminine Channeling, the Occult, and Communication Technologies, 1859–1919* (Ithaca, NY: Cornell University Press, 2010), 1–22.
6 And in this I mean not to define modernity as much as to come to terms with "the experience of modern life" similar to Torbjörn Wandel's analysis in "Too Late for Modernity," *Journal of Historical Sociology* 18, 3 (2005): 255–68.
7 Richard Jarrell, "Measuring Scientific Activity in Canada and Australia before 1915: Exploring Some Possibilities," *Scientia Canadensis: Canadian Journal of the History of Science, Technology, and Medicine* 17, 1–2 (1993): 27.
8 This retreat originated from a series of "camp meetings" in the 1870s in western New York State that later developed into a permanent community that would become Lily Dale – an internationally recognized spiritualist resort. See *Seventy-Fifth Anniversary of the Lily Dale Assembly: A Condensed History* (New York: Lily Dale, 1952), 8–50.
9 McMullin, *Anatomy of a Séance*, 161, 172, 195.
10 W.V. Uttley, *A History of Kitchener, Ontario* (Waterloo: Wilfrid Laurier University Press, 1975), 352.
11 McMullin, *Anatomy of a Séance*, 162. The audio recordings and transcriptions of the lectures have been preserved in the Thomas Lacey Papers, Room Alone Lectures GA67, University of Waterloo Library.
12 Murray Davidson, "Waterloo Lutheran University Receives $42,000 from Decima Laing Estate," press release, Waterloo Lutheran University, April 2, 1963, Wilfrid Laurier News Releases 1960–69, 023–1967, Wilfrid Laurier University Library.
13 Sidney Wright maintained a long-standing friendship with Fred Maines, a minister of the Church of Divine Revelation in St. Catharines, which he founded alongside his wife, Minnie Maines, and his sister-in-law, Jenny O'Hara Pincock. For a more detailed discussion of this series of séances, see Beth A. Robertson, "Radiant Healing: Gender, Belief, and Alternative Medicine in St. Catharines, Ontario, Canada, 1927–1935," *Nova Religio: The Journal of Alternative and Emergent Religions* 18, 1 (2014): 16–36.
14 O.G. Smith, "Canadian's Remarkable Seances," *Two Worlds*, June 23, 1933, 485–86.
15 "White Eagle's Inner Teaching Group: The Three Fold Aura," May 20, 1936, Thomas Lacey Papers, GA67(1), University of Waterloo Library.

16 Bernhard Rieger, *Technology and the Culture of Modernity in Britain and Germany, 1890–1945* (New York: Cambridge University Press, 2005), 2–4.
17 Richard Stanley, *Einstein's Generation: The Origins of the Relativity Revolution* (Chicago: University of Chicago Press, 2008), 2–4; Helge Kragh, "Introduction," in *History of Modern Physics*, ed. Helge Kragh, Geert Vanpaemel, and Pierre Marage (Turnhout, Belgium: Brepols, 2002), 11–12.
18 Michael D. Gordin, *The Pseudoscience Wars: Immanuel Velikovsky and the Birth of the Modern Fringe* (Chicago: University of Chicago Press, 2012), 2–3; H.M. Collins and T.J. Pinch, "The Construction of the Paranormal: Nothing Unscientific Is Happening," in *On the Margins of Science: The Social Construction of Rejected Knowledge*, ed. Roy Wallis (Keele, UK: Keele University, 1979), 237–70; Milena Wazeck, *Einstein's Opponents: The Public Controversy about the Theory of Relativity in the 1920s*, trans. Geoffrey S. Koby (Cambridge, UK: Cambridge University Press, 2014), 66–76; Jeroen van Dongen, "On Einstein's Opponents and Other Crackpots," *Studies in History and Philosophy of Modern Physics* 41, 1 (2010): 78–80.
19 Alex Owen, *The Place of Enchantment: British Occultism and the Culture of the Modern* (Chicago: University of Chicago Press, 2004), 7–9.
20 Galvan, *The Sympathetic Medium*, 23; Matthew Lavine, *The First Atomic Age: Scientists, Radiations, and the American Public, 1895–1945* (Houndmills, UK: Palgrave Macmillan, 2013), 38–39; Sconce, *Haunted Media*, 21–58.
21 Diana Basham, *The Trial of Woman: Feminism and the Occult Sciences in Victorian Literature and Society* (Basingstoke, UK: Palgrave Macmillan, 1992), vii.
22 Beth A. Robertson, "Spirits of Transnationalism: Gender, Race, and Cross-Correspondence in Early Twentieth-Century North America," *Gender and History* 27, 1 (2015): 154; Beth A. Robertson, *Science of the Seance: Transnational Networks and Gendered Bodies in the Study of Psychic Phenomena, 1918–40* (Vancouver: UBC Press, 2016), 146–56, 163–68.
23 Courtenay Grean Raia, "From Ether Theory to Ether Theology: Oliver Lodge and the Physics of Immortality," *Journal of the History of the Behavioral Sciences* 43, 1 (2007): 20; Courtenay Grean Raia, "The Substance of Things Hoped For: Faith, Science, and Psychical Research in the Victorian *Fin de Siècle*" (PhD diss., University of California, Los Angeles, 2005), 46.
24 Robert Michael Brain, "Materializing the Medium: Ectoplasm and the Quest for Supra-Normal Biology in *Fin-de-Siècle* Science and Art," in *Vibratory Modernism*, ed. Anthony Enns and Shelley Trower (Houndmills, UK: Palgrave Macmillan, 2013), 116.
25 T. Glen Hamilton, *Intention and Survival: Psychical Research Studies and the Bearing of Intentional Actions by Trance Personalities on the Problem of Human Survival* (Toronto: Macmillan, 1942), 8–9, 51–53.
26 G. Harvey Agnew, honorary degree recipient, May 14, 1955, Honorary Degree Recipients, Campus History, University of Saskatchewan Archives and Special Collections, https://library.usask.ca/archives/campus-history/honorary-degrees.php?id=89&view=detail&keyword=&campuses=.
27 Harvey Agnew to T.G. Hamilton, April 7, 1931, Hamilton Family Fonds, MSS14 (A.79–41), box 4, folder 3, University of Manitoba Archives and Special Collections.

28 Ibid. For more on Agnew's and others' insights into ectoplasm or teleplasm, see Robertson, *Science of the Seance*, 238–39.
29 Lavine, *The First Atomic Age*, 2–8.
30 This exploration of the more social and cultural effects of atomic energy has led to the theorizing of "nuclear culture" more broadly, as can be seen in the 2012 special issue of the *British Journal of the History of Science*. See Jonathan Hogg and Christoph Laucht, "Introduction: British Nuclear Culture," *British Journal of the History of Science* 45, 4 (2012): 479–93.
31 Even Sir Oliver Lodge, perhaps one of the most well-known occultists who challenged Einstein's theories, by no means utterly rejected the theories of modern physics or new understandings of atomic energy. Rather, he tried to identify how atomic forces and older conceptions of "ether" and energy interrelated, especially in relation to uncanny, psychical processes. Sir Oliver Lodge, *Ether and Reality: A Series of Discourses on the Many Functions of the Ether of Space* (1925; reprinted, Cambridge, UK: Cambridge University Press, 2012), 26–30, 121–80; Sir Oliver Lodge, "Sir Oliver Lodge on the Possibilities of the Human Spirit," *Light: A Journal of Psychical, Occult, and Mystical Research* 47, 2414 (April 16, 1927): 182–85.
32 As argued, for instance, by Canadian spiritualist Jenny O'Hara Pincock in *Trails of Truth* (Los Angeles: Austin, 1930), 13, 15.
33 Stephen G. Brush and Ariel Segal, *Making Twentieth-Century Science: How Theories Became Knowledge* (Oxford: Oxford University Press, 2015), 334.
34 Stanley Goldberg, "Putting New Wine in Old Bottles: The Assimilation of Relativity in America," in *The Comparative Reception of Relativity*, ed. Thomas F. Glick (Dordrecht: D. Reidel, 1987), 10.
35 Wazeck, *Einstein's Opponents*, 24–25.
36 Smith, "Canadian's Remarkable Seances," 485–86.
37 Ibid.
38 "The Mind," November 22, 1934, Thomas Lacey Papers, GA67(1), University of Waterloo Library.
39 Ibid.
40 Marie Griffith, *Born Again Bodies: Flesh and Spirit in American Christianity* (Berkeley: University of California Press, 2004), 108.
41 For some foundational studies of the history and cultural role of the body, see Michel Foucault, *Discipline and Punish: The Birth of the Prison*, trans. Alan Sheridan (1977; reprinted, New York: Vintage Books, 1995), 136; and Judith Butler, *Bodies that Matter: On the Discursive Limits of Sex* (New York: Routledge, 1993), 30–31, 67–68, 94–95.
42 "White Eagle's Inner Teaching Group: The Three Fold Aura," May 20, 1936, Thomas Lacey Papers, GA67(1), University of Waterloo Library.
43 Ludwik Kostro, *Einstein and the Ether* (Montreal: Apeiron, 2000), 1–78. See also Andrew Warwick's "Cambridge Mathematics and Cavendish Physics: Cunningham, Campbell, and Einstein's Relativity 1905–1911, Part I: The Uses of Theory," *Studies in History and Philosophy of Science* 23, 4 (1992): 625–56; and Stephen G. Brush, "Why Was Relativity Accepted?," *Physics in Perspective* 1 (1999): 190–91, 194.

44 "White Eagle's Inner Teaching Group: The Three Fold Aura," Thomas Lacey Papers.
45 "White Eagle's Inner Teaching Group: The Works of the Masters," September 30, 1936, Thomas Lacey Papers, GA67(1), University of Waterloo Library.
46 Lavine, *The First Atomic Age*, 15–16.
47 Ibid., 158–66.
48 Ibid., 168–70.
49 Séance notes, September 22, 1946, Thomas Lacey Papers, GA67(1), University of Waterloo Library.
50 "Love Ye One Another," September 16, 1947, Thomas Lacey Papers, GA67(1), University of Waterloo Library.
51 Séance notes, April 27, 1948, Thomas Lacey Papers, GA67(1), University of Waterloo Library.
52 Ibid.
53 Scott C. Zeman, "'To See ... Things Dangerous to Come To': *Life* Magazine and the Atomic Age in the United States, 1945–1965," in *The Nuclear Age in Popular Media: A Transnational History, 1945–1965*, ed. Dick Van Lente (Houndmills, UK: Palgrave Macmillan, 2012), 61. This promise of atomic power or nuclear "things" more generally is also highlighted by Gabrielle Hecht, *Being Nuclear: Africans and the Global Uranium Trade* (Cambridge, MA: MIT Press, 2012), 6–8.
54 Zeman, "'To See ... Things,'" 63–67.
55 Angela N.H. Creager, *Life Atomic: A History of Radioisotopes in Science and Medicine* (Chicago: University of Chicago Press, 2013), 3–4, 62–63, 70–71, 124; Gerald Kutcher, *Contested Medicine: Cancer Research and the Military* (Chicago: University of Chicago Press, 2009), 1–20. One early foreign shipment was to the nuclear laboratories in Chalk River, Ontario. This shipment was made upon the request of its newly minted supervisor, Wilfred Bennett Lewis, a scientist who would have a profound influence on the development of nuclear energy in Canada for the following three decades. Ruth Fawcett, *Nuclear Pursuits: The Scientific Biography of Wilfred Bennett Lewis* (Montreal/Kingston: McGill-Queen's University Press, 1994), 33–64.
56 "Applying the Principle and Transmutation in Nature and Self," May 11, 1948, Thomas Lacey Papers, GA67(2), University of Waterloo Library.
57 "The Mysteries that Reveal the Nature of God," May 23, 1948, Thomas Lacey Papers, GA67(2), University of Waterloo Library. For context, see Paul C. Aebersold, "Radioisotopes – New Keys to Knowledge," in *Annual Report of the Board of Regents of the Smithsonian Institution* (Washington, DC: Government Printing Office, 1953), 219–40. Aebersold wrote about how the discovery of isotopes and the "tracer studies," especially since the 1930s, permitted medical scientists not only to see the movements of compounds within the body but also to reveal "that the atomic turnover in our bodies is quite rapid and quite complete ... Indeed, it has been shown that in a year approximately 98 percent of the atoms in us now will be replaced by other atoms that we take in" (232).
58 "The Oneness of All Things," June 6, 1948, Thomas Lacey Papers, GA67(2), University of Waterloo Library.
59 "The Vital Importance of Unity," August 17, 1948, Thomas Lacey Papers, GA67(2), University of Waterloo Library.

60 "Amarai's Lecture," August 22, 1948, Thomas Lacey Papers, GA67(2), University of Waterloo Library.
61 "The Application of the Principle," July 11, 1948, Thomas Lacey Papers, GA67(2), University of Waterloo Library.
62 "In the Room Alone: Thomas Edison and Luther Burbank Discuss the Value of Their Work," September 3, 1949, Thomas Lacey Papers, Room Alone Lectures, September 1948–July 1951, GA67, University of Waterloo Library.
63 "In the Room Alone," October 15, 1949, Thomas Lacey Papers, Room Alone Lectures, September 1948–July 1951, GA67, University of Waterloo Library.
64 "In the Room Alone: Travel, Light, and Power in the Spiritual Age," November 19, 1949, Thomas Lacey Papers, Room Alone Lectures, September 1948–July 1951, GA67, University of Waterloo Library.
65 "In the Room Alone: John and Tom Discuss the Growth of the Senses and They Join 'The Wise Men on Their Journey,'" December 3, 1949, Thomas Lacey Papers, Room Alone Lectures, September 1948–July 1951, GA67, University of Waterloo Library.
66 "Power Available," January 14, 1950, Thomas Lacey Papers, Room Alone Lectures, January–December 1950, GA67, University of Waterloo Library.
67 "In the Room Alone: John Speaks of the Experience Called Death, John and Tom Visit Another Planet and See the Life There," January 14, 1950, Thomas Lacey Papers, Room Alone Lectures, September 1948–July 1951, GA67, University of Waterloo Library.
68 "Power Available," Thomas Lacey Papers.

PART 2

TECHNOLOGIES

5

The Second Industrial Revolution in Canadian History

JAMES HULL

My intention in this chapter is to make a specific suggestion regarding the place of the history of science and technology in Canada's broader history. It echoes a suggestion made some years ago by the distinguished economic historian Peter Temin, who proposed that close attention to the Second Industrial Revolution as a "particular problem" would help to overcome a lack of focus in economic history scholarship.[1] I suggest that attention to the Second Industrial Revolution in a Canadian context can be a useful means by which historians of Canadian science and technology bring greater clarity to their work; I also suggest the significance of that work to other aspects of Canadian history and in helping to bring greater clarity to a concept in historical writing that, though it has been around for over one hundred years, remains ill defined.[2] Not only can historians not agree among themselves on some fundamental questions concerning the Second Industrial Revolution, but also sometimes they cannot individually make up their minds. Peter Stearns in his standard work *The Industrial Revolution in World History* says on one page that the "'second industrial revolution' in western Europe ... may have brought more changes to more people than the first revolution did," but about one hundred pages later he says that the term "is misleading – many fundamental trends simply intensified."[3] Clearly, some help is needed, but can the study of the Second Industrial Revolution in Canada help? I think that it can.

The late Ian Drummond, offering debate in the *Canadian Historical Review*, argued that the "industrial transformation that followed Canada's Confederation was in part a diffusion of technology and organization from the 'first industrial revolution' ... but chiefly a very rapid adoption and digestion of the changes that characterized the 'second,' along with the organizational forms ... that accompanied that second revolution."[4] Marvin McInnis has been so bold as to claim that "Canada was arguably the most successful exploiter of the new technology of the Second Industrial Revolution" and has offered an account of the Laurier boom years in those terms.[5] Like Yves Gingras in his study of physics and research in Canada, McInnis draws special attention to the role of the Canadian system of engineering education.[6]

If Drummond is right – and I think that he is – then the modernization of the Canadian economy and Canadian society that is sometimes conflated with the industrialization and urbanization of Canada is in fact principally a phenomenon of the Second Industrial Revolution. Indeed, this more specific focus would assist us in bringing greater utility to what is often rightly pilloried as an analytically flabby concept, "modernization." The modernization that I am talking about here is not some vague transition to an urban industrial society but one rooted in a particular system of production, based upon science, with dimensions reaching from the very points of production on factory floors and elsewhere out into the broader society. Yes, new industrial technologies transformed work, but they transformed as well or were part of transformations of human relations, material culture, the state, and the natural environment.

In the first part of this chapter, I offer a definition of the Second Industrial Revolution and instantiate that definition with reference to the Canadian historical experience. What was the Second Industrial Revolution, where did it happen, when did it happen, and where did Canada as, we might say, a rapid second industrializer fit in? Then I get at the nuts and bolts – or maybe the beakers and transformers – of what was in the first place a revolution in production. Changes occurred at several levels, including the structure of the economy, industries, firms, and shop floors. Finally, a revolution isn't a revolution if it is just a Kondratiev or a wave of gadgets; it must be transformative beyond the technological, narrowly conceived. Stearns was right, I think, with his first observation; in certain respects, the Second Industrial Revolution – perhaps because it linked an understanding of the natural world, science, with an understanding of the manipulation of the natural world, technology – was even more transformative, more responsible for making the modern world, than was the (First) Industrial Revolution. We see this looking at the rise of the state, understandings of the environment, and gender.

The Revolution

The Second Industrial Revolution was not simply the second phase of the (First) Industrial Revolution. Early industrialization has its own historiography, often concerned with different patterns of industrialization in the different British North American colonies.[7] And, in an ambitious recent work, Robert Sweeny reminds us that such paths are not inevitable outcomes of impersonal economic forces but the deep consequences of choices made in specific places by specific people with specific ideas about property, gender, and many other matters.[8] We can date the start of the Second Industrial Revolution to the 1860s and 1870s when the giant chemical firm BASF in Germany began placing more formally qualified chemists in positions of control over production in place of the skilled *Meister*, while in the United States the Pennsylvania Railroad hired Charles Dudley to investigate and standardize its materials.[9] This places us in the era of Confederation and the origin of the National Policy. It also takes us directly into the argument about industrialization in Canada. How much was modern – large in scale, corporate in organization, drawing on new bodies of knowledge in production – and how much was still traditional and artisanal?[10] Thus far, that debate has largely been about labour, but we might suppose that some attention to technology in various productive settings would be salutary. Rick Szostak, in a theoretically and empirically well-grounded argument, identifies the Great Depression as the disastrous end to the Second Industrial Revolution. A long period of process innovation without new product innovation to match left industry with gross overcapacity and the need to shed jobs.[11] This could fit well with and add a new dimension to the traditional story from staples theorists that Canada had exhausted immediate investment opportunities in the new staples of pulp and paper and nonferrous metals, plus hydroelectricity – all of them important Second Industrial Revolution sectors. But the new, new staples (petroleum, potash, uranium) had not yet come on stream, to be joined by a new round of investment in hydroelectric megaprojects.[12]

A separate problem – or really a pair of problems – is the internal periodization of the Second Industrial Revolution. The first problem is accounting for the long and somewhat sluggish period of world economic history in the latter part of the nineteenth century that gave way to spectacular growth around the turn of the century and lasted to just before the outbreak of the

First World War. Part of that is well understood, including new gold flooding into the world economy – giving liquidity and ending the price deflation that was having distorting effects, notably in the American political economy – and falling ocean steamship rates that encouraged international commerce. But part of it as well was the lag between the introduction of new Second Industrial Revolution technologies and their full impacts in terms of productivity. So investment is being soaked up, and then one gets the returns, new technologies are not immediately superior to the old ones, but the subsequent improvements are where the gains are mostly realized, and there is learning by doing as users get better at doing things with the new technologies. Jeremy Greenwood and others have explored this problem in other contexts, examining the role of technology in what we might call revolutionary lag.[13] For Canada, we would like to say more certainly what roles technologies and new Second Industrial Revolution industries played in explaining the so-called Laurier boom. We know that it was more than early ripening strains of wheat and harvest excursion trains that allowed a wheat boom to happen, but there is still much about which we can argue. In an unpublished paper, McInnis states flatly that the explanation for the rapid absolute and comparative growth of the Canadian economy, especially at the beginning of the boom, "might be summed up as saying that the Second Industrial Revolution arrived in Canada."[14]

The second problem is the First World War. To what extent was that political and social cataclysm an economic watershed? Was it the war that modernized Canada? We have long assumed that it was, and textbook accounts still present Canadian history that way. But I think that Douglas McCalla has argued effectively that "the war did not affect in any fundamental way trends in the structure of the economy."[15] For historians of Canadian science and technology, this problem is especially of historiographic interest because of the wartime origins of the National Research Council (NRC). Much has been written about, and almost certainly too much significance has been attached to, that one institutional development even for an understanding of federal science and its application to economic development. As Stéphane Castonguay, looking at a period *ending* at the start of the First World War, explains, "government leaders wanted to enlarge and diversify the state apparatus to intervene more directly in the nation's economic affairs. Scientists successfully expounded the idea that scientific research was the key to national prosperity by solving problems related to industrial production and resource conservation."[16] My own study of Canadian universities and research before and during the First World

War argues that the Second Industrial Revolution is a far more useful context in which to examine the development of scientific industrial research than is the Great War.[17] Canada's universities were committed to research, applied science, and industry cooperation virtually from the get go and did not need the stimulus of the war to direct them in such channels. As early as 1895, we find W.L. Goodwin, head of the Kingston School of Mining and Agriculture, effectively the applied science faculty of Queen's University, confidently declaring that "scientific discoveries [are] organized by the great universities, scientific schools, and industrial corporations. Science and industry are at last wed."[18]

That the United States, along with Germany, led the Second Industrial Revolution while the United Kingdom tried to fight the second with the weapons of the first is a well-established picture even if some of the details can be disputed.[19] It does pose, however, an interesting issue for a country that has had strong British and American influences. How much of Canada's success was a consequence of continental economic integration? How much of that involved a common technological practice with the United States? I put it that way rather than a transfer of technology, though we might also like Arnold Pacey's term "technological dialogue."[20] This is something that Bruce Sinclair addressed many years ago in his essay on "Canadian Technology: British Traditions and American Influences."[21] We have good studies from Dianne Newell and Gordon Winder for mining and agricultural equipment, respectively, of the continental nature of the pool of technology from which Canadians drew and to which they contributed.[22] W.A.E. McBryde has identified early work in the refining of crude petroleum in southwestern Ontario as a "pilot plant" for this new chemical process industry in North America.[23] My own work on the pulp and paper industry documents an explicit attempt by public and private actors on both sides of the border to codify jointly a common body of applicable technical and applied scientific information. I have also in a preliminary way documented the evolution of a common North American regime of technical standards used by industry and public agencies.[24] We might also try further to assess, both qualitatively and quantitatively, the striking claim by McInnis, already mentioned, concerning Canada's notable success in exploiting the technologies of the Second Industrial Revolution.

A Revolution in Production

Well has it been said that the history of invention is a poor substitute for the history of technology. Also, if there is one thing that professional historians of Canadian science and technology find cringeworthy, then it is surely lists of Canadian firsts and the like. However, it is entirely fair to note developments in which they are salient and to ask how it came about that a modest economic player might have acquired the capability to make internationally significant contributions to the emerging science-based technology of the Second Industrial Revolution. Electrification, with Ontario Hydro and the huge literature about it, might be the standout example. Here we are in the mainstream since it has equally attracted the attention of historians of politics and economics, from H.V. Nelles's classic *Politics of Development* to Jamie Swift and Keith Stewart's account of the decline and fall of Ontario Hydro.[25] Beyond this, there is an extensive literature on Hydro Quebec, hydroelectric development in British Columbia, and to a lesser extent such development in other provinces. Somewhat less well developed is the literature on the electrical industries beyond the actual generation of the power, though we must acknowledge the work of Norman Ball and John Vardalas on Ferranti-Packard as providing a valuable model.[26] Historians of electrification have drawn attention to the link between lighting and enlightenment in discussions of the broader role of this technology in the creation of the modern world. Indeed, several recent studies explicitly link electric power development in different regions of Canada with modernization.[27] For chemicals and the chemicalization of industry, we can point to the development of electrochemistry at Shawinigan Falls, cellulose chemistry at McGill University, the emergence of Polymer Corporation, and early work in the chemical problems of petroleum production.[28] As perhaps a small aside here, Canada's modest if delicious contribution to material culture was the making of vanillin, once a scarce natural product, significantly more available with the discovery and development of an economical method of extracting it from pulp mill waste.[29] Sectors that really were "world class" – hydroelectricity, pulp and paper – acted as not just investment frontiers but also high-tech industrial frontiers, among other things attracting and retaining world-class people and making nationally and internationally significant contributions to technology.

Although the principal locus of production was the firm, some of what could be gained through the internalization of transaction costs and other efficiencies by corporate organization and corporate concentration described by Alfred Chandler could also be gained in other ways, in particular through the cooperation of independent firms in patent pools, power grids, cross-licensing arrangements, freight car pools, trade associations, and the like.[30] There are many technical aspects of this, and we do not know enough about it in any national context. A good start might be to look more at industry-wide cooperative research activities that certainly went on in Canadian industries ranging from textiles to metals to pulp and paper. They are particularly interesting because so often they involved not just interfirm cooperation for scientific industrial research but also cooperation among industry, government, and academia.[31] One thing that the organizational synthesis that has so informed some of the best American historical writing on modern America has been intensely interested in is the political economy of business, an interest that we should share.[32] Andre Siegel and I have recently argued that Canadian business, acting through organizations such as the Canadian Manufacturers' Association, was not solely wedded to an unimaginative defence of the tariff and other anticompetitive strategies. Rather, strategies also included persuading Canadians of the merits of Canadian-made goods, and making those goods indeed more meritorious, strategies that formed a package with enlisting the participation of governments in technical education and support of industrial research.[33] Technology is of course more than hardware, and productive processes are also work processes. One way in which shop floor practices were transformed in the Second Industrial Revolution was via the redesign of work processes by engineers: that is, Taylorism or scientific management. The shops of the major Canadian railways, which manufactured many of their own cars and locomotives, had long emerged as the most sophisticated and among the largest sites of secondary manufacturing in the country.[34] It was there that Taylorist scientific management techniques crossed the border when Henry Gantt brought them to the Canadian Pacific Railway's Angus Shops in Montreal.[35] The other important way in which control over production at the factory floor level changed was the growing reliance not on individual human sensory judgment but on a panoply of test and control instruments. A variety of devices interrogated the physical states of materials being transformed in production and formed the basis for process control decisions.[36] These instruments, usually but not always, were manufactured by American companies. Westinghouse pressure gauges, Brown pyrometers and flow meters, and Tycos recording thermometers were found in plants and mills throughout North American industry. But so was the Canadian standard freeness tester, developed in the 1920s by the Canadian Forest Products Laboratory and adopted by mills in the pulp and paper industry across North America.[37]

The story of the development of elaborate hierarchies of skill in the factories of the First Industrial Revolution is a familiar one. But what of the

Second Industrial Revolution? We can talk about a couple of related matters. First is the growing significance of one part of the labour force, semiskilled machine tenders, in what Lindy Biggs has so wonderfully termed the "rational factories" of the Second Industrial Revolution.[38] While material handling devices, assembly lines, and continuous process plants reduced the demand for grunt labour of the fetch-and-carry sort, and Taylorism and new process control technology deskilled certain work processes, the need for those with skill sets to tend the new machinery grew. As Stearns notes, for this increasingly predominant part of the labour force, some of whom had already felt the impact of the (First) Industrial Revolution, they now faced "further shifts in methods – some more fundamental than anything that had come before."[39] This would have been felt with special force, as in Canada, where the Second Industrial Revolution had its impact quickly after the first one. It is easy to be dismissive of "semiskilled" as a category, and many labour historians have been. But in fact doing so neglects the scope for getting better at quickly learned tasks and neglects the significance of the diagnostic if not necessarily operative skills of such workers. Also, employers found it to their advantage to offer efficiency wages rather than a Ricardian wage model.[40] Why? Because the share of total costs represented by wages was small, and the costs of interrupted production were large. But paying the workers good wages also allowed them to buy in to aspects of a middle-class lifestyle and just maybe some of the middle-class values that went with it.

Beyond the Factory

Jennifer Karns Alexander has invited us to think more broadly about "efficiency" as a cultural construct in the industrial and postindustrial era.[41] It was an ideology of engineers, certainly, but one that could be celebrated or condemned and that, in certain important respects, could be seen as antithetical to human freedom. Conceived in this way, the scientific efficiency of new productive methods had implications that immediately take us into social and political realms of the function of the state, class and gender, and the environment.

The state in the Second Industrial Revolution had a number of functions. Yes, it was, if absolutely necessary, supposed to shoot a few workers now and then *pour encourager les autres*. (Or maybe *pour décourager les autres*.) But that is a pretty heavy-handed approach. Certainly, the state was looked to for the provision of scientific and technical services. Notably, this included capstone bodies of national research on the model of the *Physikalisch-Technische Reichsanstalt*. More proximately, the National Physical Laboratory and the National Bureau of Standards provided models for our own NRC, though it was not a copy of anything else. If there is one thing that might claim to have a well-developed historiography in our field, then it is the NRC.[42] We have some studies of other federal scientific agencies, though not nearly enough.[43] At the provincial level, we probably know more about Quebec than any other province, but here too, fitting the work of the provincial research organizations and ministerial-level agencies into the narrative of federalism and province building would be welcome.[44] Governments were also broadly permissive where they were not actively supportive of industries' own measures to support, collectively, industrial research. When a Canadian branch of the British Society of Chemical Industry was formed in 1901 at the initiative of the Canadian Manufacturers' Association, the Ontario government provided a grant to support publication of its papers.[45]

Next, as part of the state's role in the legitimation of capital and for other reasons, both the federal and the provincial states were drawn into the regulation of industry. This included some technical matters, providing a veneer of democratic oversight for highly capitalized utilities, something investigated by Christopher Armstrong and H.V. Nelles.[46] The Canadian system of technical standards, crucial in making regulations as well as business-to-business relations, features relationships among federal scientific research bodies, engineering societies regulated at the provincial level, and an independent and unique national standards association. Tom Traves's persistently overlooked *The State and Enterprise* called attention to the importance of science and technology to the story of the rise of the regulatory state in Canada.[47] We need to take up that challenge. This could also be pursued profitably at the municipal level with attention to building codes and zoning, as Amy Slaton has done for the United States and as Chris Armstrong touches on in his recent study of modern architecture in Toronto.[48]

Finally, the state was expected to distribute socially much of the overhead costs of training new workforces for the new science-based industries of the Second Industrial Revolution. Although individual provinces addressed this, at the secondary level it became a national issue with the 1910 Royal Commission on Industrial Training and Technical Education.[49] At a higher level, this involved education for engineers, chemists, metallurgists, and so forth. Here again the Canadian story is part of an international one with various models of engineering education, though we developed our own specific pattern with its own specific features – apparently a successful one.[50]

The state had its own experts, but it also got to say who was an expert. Middle-class experts, with their credentials validated and enforced by the state, if self-administered, increasingly defined the rules of the game not just for themselves but also for those below and above them.[51] This was not the "rising middle class" (a historical cliché) but an emergent, hegemonic middle class that increasingly came to occupy the decision-making points not just in production, as did engineers, but also in the organizations and institutions of society more broadly. And, as Burton Bledstein has pointed out, engineers – those paradigmatic new professionals – led the way.[52] This has all been well worked out in an American context but unfortunately little explored in a Canadian one.[53]

The growth of the new corporations and the expansion of the regulatory state's civil service created vast numbers of new low-paying clerical jobs occupied chiefly by women. The technical aspects of these pink-collar ghettos have been explored by Graham Lowe.[54] But what about the professions? The story of women's struggle for medical education and thence entry into the modern medical profession is well known, as is the absolute intransigence of the legal profession. But what about the technical professions? Dentistry, which in some countries in Northern Europe professionalized as a female profession, was open in Canada to women from the start, though few entered it.[55] More broadly, was it the case, as it was with mechanical engineers in the United States, that by the time women were asking to be admitted, men controlling professional societies were ready to say yes? It is good to celebrate the lives of the Elsie MacGills but more interesting to know the general story. Faculties of Household Science did cooperative research with the Biological Board of Canada on seafood.[56] Some of the young women trained by Clara Benson at the University of Toronto's Faculty of Household Science went on to work as technicians in Canadian munitions factories in the Great War, and Benson's own analytical techniques for food chemistry found application in the standardization of munitions production.[57] But we would like to know on a systematic basis what kind of science, with what aims, was taught in Faculties of Household Science and where the graduates went.

Further, I suspect that the Second Industrial Revolution saw an analogue of that industrious revolution that Jan de Vries has described for the (First) Industrial Revolution and that women's decisions as consumers had a great deal to do with it.[58] This came about through female influence on patterns of household consumption in response to branded mass marketing. Donica Belisle points out that "budgeting had become women's responsibility in many parts of Canada in the late nineteenth and early twentieth centuries."[59]

Finally, that quintessentially modern virtue "efficiency" is nowhere so clearly seen than in Second Industrial Revolution attitudes to the environment and natural resources. For the bringing of science to the crucial forest products sector, it was 1907–13 when most of the important things happened, including the creation of the first university-level programs in forestry, the founding of the Forest Products Laboratories of Canada, and the establishment of the federal Commission of Conservation. Interest in conservation must be understood in the context of a desire for more advantageous usage of natural resources. Uniting many conservationists, engineers, and other progressive reformers was an absolute moral abhorrence of waste. "Implementation of conservationist principles required a reformulation of the industry-government partnership by the inclusion of scientifically trained professionals."[60] Andrew Stuhl's chapter in this volume points out that the Canadian Arctic Expedition that began in 1913 addressed, *inter alia*, the use and conservation of natural resources in the North.[61]

Conclusion

I have used this chapter to ask questions but not answer them. What I hope I have shown is that questions about the Second Industrial Revolution are worth asking and that some worthwhile answers might be found in the history of Canadian science and technology. It might fairly be asked what my own, at least tentative, answers to some of the questions are. First, I think that McInnis, Drummond, and McCalla are right. Canada was an early and successful participant in the Second Industrial Revolution with most of the important changes accomplished before the First World War. I suspect that, in spite of anything said then or subsequently to the contrary, a remarkably flexible and responsive system of higher education and easy movement of people among industry, university, and government played an important role. And certainly the relationship with the United States, a relationship that in terms of applied science and technology is ill described in terms of dominance or dependence or derivativeness, was crucial.[62] Indeed, Canada's modernization was very much a North American one. Yes, we tooled up for industrial production on an American model, shared a common pool of technology, and coordinated our technical standards. But it was much more than that. The Second Industrial Revolution gave us a shared material culture, regulatory states whose differences are easier to exaggerate than are their similarities, and perhaps most strikingly, shared understandings of the natural environment. As perceptions of North America as an unlimited treasure trove themselves became unsustainable, they were superseded less by romanticism than by belief in the moral rightness of efficiency. Just as Lynn White found not just the roots of industrialism in medieval attitudes to nature but also the roots of contemporary ecological problems, so too might we find in the science-based wizardry that allowed a transformed material culture even deeper environmental challenges.[63]

NOTES

1 Peter Temin, "The Future of the New Economic History," *Journal of Interdisciplinary History* 12, 2 (1981): 179–97.
2 See James Hull, "The Second Industrial Revolution: The History of a Concept," *Storia della storiografia* 36 (1999): 81–90; and Ernst Homburg, "De 'Tweede Industriele Revolutie,' een problematisch historische concept," *Historisch tijdschrift* 8 (1986): 376–85. For an ambitious attempt to integrate the concept into a discussion of Russian economic history, see Tamás Szmrecsányi, "On the Historicity of the Second Industrial Revolution and the Applicability of Its Concept to the Russian Economy before 1917," *Economies et sociétés* 42, 3 (2008): 619–46.
3 Peter N. Stearns, *The Industrial Revolution in World History*, 4th ed. (Boulder, CO: Westview, 2013), 10, 114.
4 Ian Drummond, "Ontario's Industrial Revolution, 1867–1941," *Canadian Historical Review* 69, 3 (1988): 283–314.
5 Marvin McInnis, "Engineering Expertise and the Canadian Exploitation of the Technology of the Second Industrial Revolution," in *Technology and Human Capital in Historical Perspective*, ed. Jonas Ljungberg and Jan-Pieter Smits (Basingstoke, UK: Palgrave Macmillan, 2005), 49–78.
6 Yves Gingras, *Physics and the Rise of Scientific Research in Canada*, trans. Peter Keating (Montreal/Kingston: McGill-Queen's University Press, 1991). The Second Industrial Revolution is used as a key concept to explain what was happening by several of the contributors to *Class, Community, and the Labour Movement: Wales and Canada, 1850–1930*, ed. Deian R. Hopkin and Gregory S. Kealey (St. John's: Llafur/Canadian Committee on Labour History, 1989). A few other historians have already made the Second Industrial Revolution central to their presentations of aspects of Canadian history; see, for example, Gregory P. Marchildon, "Portland Cement and the Second Industrial Revolution in Canada, 1885–1909," paper presented at the Fifth Canadian Business History Conference, Hamilton, 1998; and Gordon Winder, "Following America into the Second Industrial Revolution: New Rules of Competition and Ontario's Farm Machinery Industry, 1850–1930," *Canadian Geographer* 46, 4 (2002): 292–309.
7 See, for instance, John McCallum, *Unequal Beginnings: Agriculture and Economic Development in Quebec and Ontario until 1870* (Toronto: University of Toronto Press, 1980), and Graeme Wynn, *Timber Colony: A Historical Geography of Early Nineteenth Century New Brunswick* (Toronto: University of Toronto Press, 1981),

explaining New Brunswick's nonrevolutionary road to industrial capitalism. Both explanations are staples based, wheat and timber, respectively.

8   Robert C.H. Sweeny, *Why Did We Choose to Industrialize? Montreal, 1819–1849* (Montreal/Kingston: McGill-Queen's University Press, 2015).

9   For BASF, see Werner Abelshauser, Wolfgang von Hippel, Jeffrey Allan Johnson, and Raymond G. Stokes, *German Industry and Global Enterprise – BASF: The History of a Company* (Cambridge, UK: Cambridge University Press, 2004). For standardization, see Janet T. Knoedler and Anne Mayhew, "The Engineers and Standardization," *Business and Economic History* 23, 1 (1994): 141–51.

10 See Drummond, "Ontario's Industrial Revolution," and the accompanying responses to his argument by Louis P. Cain and Marjorie Cohen. See the chapter on "Toronto's Industrial Revolution" in Gregory S. Kealey, *Toronto Workers Respond to Industrial Capitalism, 1867–1892* (Toronto: University of Toronto Press, 1980). Also see Robert B. Kristofferson, *Craft Capitalism: Craftsworkers and Early Industrialization in Hamilton, Ontario* (Toronto: University of Toronto Press, 2007); and Marvin McInnis, "Just How Industrialized Was the Canadian Economy in 1890?," unpublished paper, http://qed.econ.queensu.ca/faculty/mcinnis/HowIndustrialized.pdf. McInnis argues for an early and extensive industrialization of the Canadian economy.

11 Rick Szostak, *Technological Innovation and the Great Depression* (Boulder, CO: Westview, 1995).

12 The classic statement is A.D. Safarian, *The Canadian Economy in the Great Depression* (Toronto: McClelland and Stewart, 1970). For the new staples, see H.V. Nelles, *The Politics of Development* (Toronto: Macmillan, 1974). For the new, new staples, see John Richards and Larry Pratt, *Prairie Capitalism: Power and Influence in the New West* (Toronto: McClelland and Stewart, 1979); and Robert Bothwell, *Eldorado: Canada's National Uranium Company* (Toronto: University of Toronto Press, 1984). For postwar hydro, see (among many other studies) the chapter by Daniel Macfarlane in this volume; James L. Kenny and Andrew Secord, "Public Power for Industry: A Re-Examination of the New Brunswick Case, 1940–1960," *Acadiensis* 30, 2 (2001): 84–108; Jeremy Mouat, *The Business of Power* (Victoria: Sono Nis Press, 1997); and Yves Bélanger and Robert Comeau, eds., *Hydro-Québec: Autre temps, autres défis* (Montréal: Université du Québec à Montréal, 1995).

13 For example, Jeremy Greenwood, *The Third Industrial Revolution* (Washington, DC: AEI Press, 1997).

14 Marvin McInnis, "Canadian Economic Development in the Wheat Boom: A Reassessment," unpublished paper, http://qed.econ.queensu.ca/faculty/mcinnis/Cdadevelopment1.pdf.

15 Douglas McCalla, "The Economic Impact of the Great War," in *Canada and the First World War: Essays in Honour of Robert Craig Brown*, ed. David MacKenzie (Toronto: University of Toronto Press, 2005), 148.

16 Stéphane Castonguay, "Naturalizing Federalism: Insect Outbreaks and the Centralization of Entomological Research in Canada 1884–1914," *Canadian Historical Review* 85, 1 (2004): 1–34.

17 James Hull, "'A Stern Matron Who Stands beside the Chair in Every Council of War or Industry': The First World War and the Development of Scientific Research at Canadian Universities," in *Cultures, Communities, and Conflict: Histories of Canadian Universities and War*, ed. Paul Stortz and E. Lisa Panayotidis (Toronto: University of Toronto Press, 2012), 146–74.
18 W.L. Goodwin, "The Signs of the Times," *Queen's Journal* 23, 3 (1895): 42, as quoted in A.B. McKillop, *Matters of Mind: The University in Ontario, 1791–1951* (Toronto: University of Toronto Press, 1994), 167.
19 Thomas K. McCraw, ed., *Creating Modern Capitalism* (Cambridge, MA: Harvard University Press, 1997).
20 Arnold Pacey, *Technology in World Civilization* (Cambridge, MA: MIT Press, 1991).
21 Bruce Sinclair, "Canadian Technology: British Traditions and American Influences," *Technology and Culture* 20, 1 (1979): 108–23.
22 Dianne Newell, *Technology on the Frontier: Mining in Old Ontario* (Vancouver: UBC Press, 1986), 35–36; Gordon M. Winder, "Technology Transfer in the Ontario Harvester Industry 1830–1900," *Scientia Canadensis: Canadian Journal of the History of Science, Technology, and Medicine* 18, 1 (1994): 38–88.
23 W.A.E. McBryde, "Ontario: Early Pilot Plant for the Chemical Refining of Petroleum in North America," *Ontario History* 79, 3 (1987): 203–30.
24 James Hull, "Strictly by the Book: Textbooks and the Control of Production in the North American Pulp and Paper Industry," *History of Education* 27, 1 (1998): 85–95; James Hull, "Technical Standards and the Integration of the U.S. and Canadian Economies," *American Review of Canadian Studies* 32, 1 (2002): 123–42.
25 H.V. Nelles, *The Politics of Development* (Toronto: Macmillan, 1974); Jamie Swift and Keith Stewart, *Hydro: The Decline and Fall of Ontario's Electric Empire* (Toronto: Between the Lines, 2004). See also Daniel Macfarlane's chapter in this volume.
26 Norman Ball and John N. Vardalas, *Ferranti-Packard: Pioneers in Canadian Electrical Manufacturing* (Montreal/Kingston: McGill-Queen's University Press, 1994).
27 James L. Kenny and Andrew Secord, "Engineering Modernity: Hydroelectric Develoment in New Brunswick, 1945–1970," *Acadiensis* 39, 1 (2010): 3–26; Lionel Bradley King, "The Electrification of Nova Scotia, 1884–1973: Technological Modernization as a Response to Regional Disparity" (PhD diss., University of Toronto, 1999); Alexander Netherton, "From Rentiership to Continental Modernization: Shifting Paradigms of State Intervention in Hydro in Manitoba, 1922–1977" (PhD diss., Carleton University, 1993).
28 It would be useful to contrast the developments at Shawinigan Falls with those at Niagara Falls, New York. Martha W. Langford, "Shawinigan Chemicals Limited: History of a Canadian Scientific Innovator" (PhD diss., Université de Montréal, 1987); Claude Bellavance, *Shawinigan Water and Power, 1898–1963: Formation et déclin d'un groupe industriel au Québec* (Montréal: Boréal, 1994). Compare Martha Moore Trescott, *The Rise of the American Electrochemicals Industry, 1880–1910* (Westport, CT: Greenwood, 1981). For McGill University, see James Hull, "Federal Science and Education for Industry at McGill, 1913–38," *Historical Studies in Education* 13, 1 (2001): 1–18. For Polymer Corporation, see Matthew J. Bellamy, *Profiting the*

Crown: Canada's Polymer Corporation 1942–1990 (Montreal/Kingston: McGill-Queen's University Press, 2005). For petroleum, see McBryde, "Ontario." See also the chapter by Eda Kranakis in this volume.

29 The discovery was made by George Tomlinson at McGill University while working on his PhD dissertation under the supervision of Harold Hibbert. Hibbert held the Eddy Chair of Industrial and Cellulose Chemistry. For the process, see US Patent 2069185A, "Manufacture of Vanillin from Waste Sulphite Pulp Liquor," filed by Hibbert and Tomlinson in 1934 and granted in 1937. At the start of the 1980s, a single Canadian plant produced 60 percent of the world's supply of vanillin. Martin B. Hocking, "Vanillin: Synthetic Flavoring from Spent Sulfite Liquor," *Journal of Chemical Education* 74, 9 (1997): 1055.

30 The classic study is Alfred D. Chandler, *The Visible Hand: The Managerial Revolution in American Business* (Cambridge, MA: Harvard University Press, 1977). However, as William Lazonick points out, what is really important is not so much the firm but the "organization which links business firms engaged in interrelated productive activities." William Lazonick, *Business Organization and the Myth of the Market Economy* (Cambridge, MA: Harvard University Press, 1991), 8. This is perhaps especially relevant to Canada, which has had fewer giant "Chandlerian" corporations.

31 Some sense of the extent of this can be gleaned from the list of scientific and technical societies in Canada compiled by the Canadian National Research Council and included in the US *Bulletin of the National Research Council* 76 (1930). See also Philip C. Enros, "'The Onery Council of Scientific and Industrial Pretence': Universities in the Early NRC's Plans for Industrial Research," *Scientia Canadensis: Canadian Journal of the History of Science, Technology, and Medicine* 15, 2 (1991): 41–51.

32 Louis Galambos, "Technology, Political Economy, and Professionalization: Central Themes of the Organizational Synthesis," *Business History Review* 57, 4 (1983): 471–93.

33 Andre Siegel and James Hull, "Made in Canada! The Canadian Manufacturers' Association's Promotion of Canadian-Made Goods, 1911–1921," *Journal of the Canadian Historical Association* 25, 1 (2014): 1–32.

34 Paul Craven and Tom Traves, "Canadian Railways as Manufacturers," *Canadian Historical Association Historical Papers* (1983): 254–81.

35 Robert Nahuet, "Une expérience canadienne de Taylorisme: Le cas des usines Angus du Canadien Pacifique" (MA thesis, Université de Québec a Montréal, 1994). See also James W. Rinehart, *The Tyranny of Work*, 3rd ed. (Toronto: Harcourt, Brace, 1996); and Craig Heron and Bryan Palmer, "Through the Prism of the Strike," *Canadian Historical Review* 58, 4 (1977): 423–58.

36 This is discussed in Stuart Bennett, "'The Industrial Instrument – Master of Industry, Servant of Management': Automatic Control in the Process Industries, 1900–1940," *Technology and Culture* 32, 1 (1991): 69–81.

37 W.G. Mitchell, "Review History of Pulp and Paper Research Institute of Canada 1925–1937," McGill University Archives, RG2 c66, 25–28; Allen Abrams, "Report on

Testing Freeness of Pulp," *Paper Trade Journal* 84 (1927): TAPPI Section, 110; James d'A. Clark, *Pulp Technology and Treatment for Paper* (San Francisco: Freeman, 1978), 511. Roughly, "freeness" is a measure of the speed at which water drains away from pulp as it is formed up into paper on the wire of the papermaking machine.

38   Lindy Biggs, *The Rational Factory: Architecture, Technology, and Work in America's Age of Mass Production* (Baltimore: Johns Hopkins University Press, 1996).

39   Stearns, *Industrial Revolution*, 162.

40   For a general treatment of this concept, see George Akerlof and Janet Yellen, *Efficiency Wage Models of the Labor Market* (Cambridge, UK: Cambridge University Press, 1986). The classic case is of course Henry Ford's five-dollar wage. See Daniel Raff and Lawrence Summers, "Did Henry Ford Pay Efficiency Wages?," *Journal of Labor Economics* 5, 4 (1987): 57–86.

41   Jennifer Karns Alexander, *The Mantra of Efficiency: From Waterwheel to Social Control* (Baltimore: Johns Hopkins University Press, 2008).

42   See Richard A. Jarrell and Yves Gingras, eds., *Building Canadian Science*, special issue of *Scientia Canadensis: Canadian Journal of the History of Science, Technology, and Medicine* 15, 2 (1991).

43   Morris Zaslow, *Reading the Rocks: The Story of the Geological Survey of Canada* (Toronto: Macmillan, 1975); T.H. Anstey, *One Hundred Harvests* (Ottawa: Agriculture Canada, 1986); Kenneth Johnstone, *The Aquatic Explorers: A History of the Fisheries Research Board of Canada* (Toronto: University of Toronto Press, 1977); Anthony N. Stranges, "Canada's Mines Branch and Its Synthetic Fuel Program for Energy Independence," *Technology and Culture* 32, 3 (1991): 521–54.

44   Frances Anderson, Olga Berseneff-Ferry, and Paul Dufour, "Le développement des conseils de recherche provinciaux: Quelques problematiques historiographiques," *HSTC Bulletin: Journal of the History of Canadian Science, Technology, and Medicine* 7, 1 (1983): 27–44.

45   "Chemistry in Its Relation to the Arts and Manufactures," *Industrial Canada* 1 (1901): 253–58; "The Society of Chemical Industry," *Industrial Canada* 2 (1902): 195–96. For a retrospective look, see *Journal of the Society of Chemical Industry* 50 (July 1931): 27. See also Colin A. Russell, with Noel George Coley and Gerrylynn K. Roberts, *Chemists by Profession: The Origins and Rise of the Royal Institute of Chemistry* (Milton Keynes, UK: Open University Press, 1977).

46   Christopher Armstrong and H.V. Nelles, *Monopoly's Moment: The Organization and Regulation of Canadian Utilities, 1830–1930* (Philadelphia: Temple University Press, 1986).

47   Tom Traves, *The State and Enterprise* (Toronto: University of Toronto Press, 1979).

48   Amy E. Slaton, *Reinforced Concrete and the Modernization of American Building, 1900–1930* (Baltimore: Johns Hopkins University Press, 2001); Christopher Armstrong, *Making Toronto Modern: Architecture and Design 1895–1975* (Montreal/Kingston: McGill-Queen's University Press, 2014).

49   See Robert M. Stamp, "Technical Education, the National Policy, and Federal-Provincial Relations in Canadian Education, 1899–1919," *Canadian Historical Review* 52, 4 (1971): 404–23; Oisin Patrick Rafferty, "Apprenticeship's Legacy: The Social and Educational Goals of Technical Education in Ontario, 1860–1911" (PhD

diss., McMaster University, 1995); and *Report of the Royal Commission on Industrial Training and Technical Education* (Ottawa: King's Printer, 1910).

50 The best single study is Richard White, *The Skule Story: The University of Toronto Faculty of Applied Science and Engineering, 1873–2000* (Toronto: Faculty of Applied Science and Engineering and University of Toronto Press, 2000).

51 For efficiency, reform, expertise, and the Canadian state in this era, see Douglas Owram, *The Government Generation: Canadian Intellectuals and the State, 1900–1945* (Toronto: University of Toronto Press, 1986).

52 Burton J. Bledstein, *The Culture of Professionalism: The Middle Class and the Development of Higher Education in America* (New York: Norton, 1976).

53 J. Rodney Millard, *The Master Spirit of the Age: Canadian Engineers and the Politics of Professionalism* (Toronto: University of Toronto Press, 1988); R.D. Gidney and W.P.J. Miller, *Professional Gentlemen: The Professions in Nineteenth Century Ontario* (Toronto: University of Toronto Press, 1994).

54 Graham S. Lowe, *Women in the Administrative Revolution* (Toronto: University of Toronto Press, 1987).

55 Tracey L. Adams, *A Dentist and a Gentleman: Gender and the Rise of Dentistry in Ontario* (Toronto: University of Toronto Press, 2000). Contrast Richard White, *Gentlemen Engineers: The Working Lives of Frank and Walter Shanly* (Toronto: University of Toronto Press, 1999).

56 Johnstone, *The Aquatic Explorers*, 86.

57 Ralph A. Bradshaw, "Clara Cynthia Benson," *ASBMB Today*, March 2006, 17; Susan Bustos, "The Joy of Cooking, with Gunpowder," *Inkling Magazine*, February 21, 2007.

58 Jan de Vries, "The Industrial Revolution and the Industrious Revolution," *Journal of Economic History* 54, 2 (1994): 249–70. See also Donica Belisle, "Toward a Canadian Consumer History," *Labour/Le travail* 52 (2003): 181–206.

59 Donica Belisle, *Retail Nation: Department Stores and the Making of Modern Canada* (Vancouver: UBC Press, 2011), 134. See also Cynthia R. Comacchio, *The Infinite Bonds of Family: Domesticity in Canada, 1850–1940* (Toronto: University of Toronto Press, 1999); Bettina Bradbury, *Working Families: Age, Gender, and Daily Survival in Industrializing Montreal* (Toronto: McClelland and Stewart, 1993); and Cynthia Wright, "'Feminine Trifles of Vast Importance': Writing Gender into the History of Consumption," in *Gender Conflicts: New Essays in Women's History*, ed. Franca Iacovetta and Mariana Valverde (Toronto: University of Toronto Press, 1992), 230. For women's agency in household consumption and production, see Joy Parr, "What Makes Washday Less Blue? Gender, Nation, and Technology Choice in Postwar Canada," *Technology and Culture* 38, 1 (1997): 153–86. Compare Ruth Schwarz Cowan, "How the Refrigerator Got Its Hum," in *The Social Shaping of Technology*, 2nd ed., ed. Donald MacKenzie and Judy Wajcman (Buckingham, UK: Open University Press, 1999), 208–18.

60 Ken Drushka, *Canada's Forests: A History* (Durham, NC: Forest History Society, 2003), 48.

61 See also Graeme Wynn, *Canada and Arctic North America: An Environmental History* (Santa Barbara, CA: ABC-CLIO, 2007).

62 James Hull, "Watts across the Border: Technology and the Integration of the North American Economy in the Second Industrial Revolution," *Left History* 19, 2 (2015–16): 13–32.
63 Lynn White Jr., "The Historical Roots of Our Ecologic Crisis," *Science* 155, 2767 (1967): 1203–7. Environmental issues are explored in a separate section of this volume. For a general introduction to Canadian environmental history, see Laurel Sefton MacDowell, *An Environmental History of Canada* (Vancouver: UBC Press, 2012).

6

Mysteries of the New Phone Explained
Introducing Dial Telephones and Automatic Service to Bell Canada Subscribers in the 1920s

JAN HADLAW

From the perspective of most North American telephone users, the change from manual, operator-assisted telephone service to dial or automatic telephone service was the biggest change in telephony from its early dissemination in the 1880s until the rise of the mobile phone more than a century later. The magnitude of this sociotechnical change was clear at the time and the subject of much discussion, some anxiety, and even a little controversy. During the mid-1920s, executives with the Bell Canada Telephone Company (hereafter Bell) predicted that managing what they called the "psychological changes" associated with the introduction of dial service to central Canadian cities would be just as important as managing the technical and organizational aspects of automating telephony. However, the replacement of manual telephony by dial telephony has generally been overlooked by historians of both technology and communications.[1] Perhaps that is because the conversion proceeded unevenly, occurring at different times in different places: some North Americans experienced the changeover during the interwar years, but others had already been using dial service for many years, while still others continued using operator service for decades to come. Nevertheless, the introduction of dial telephone service is an important part of a larger history of how North Americans came to see themselves as "modern" – and in fact did "become modern" – by becoming competent users of the very technologies – such as electrical appliances, automobiles, radios, and telephones – by which the modern age was commonly defined.

The strangeness of the dial telephone in much of central Canada during the interwar years was evocatively captured in a publicity movie produced in 1929 to commemorate the opening of the first automatic exchange in Hamilton, Ontario.[2] It shows local telephone pioneer Hugh Cossart Baker Jr. and his wife seated in a garden. Between them on a table sits a new dial handset telephone, which Baker invites his wife to use. She obligingly does so – by raising the phone handset to her mouth and asking for the operator. This was precisely the right way to use the telephone most commonly in service in Canada's major cities during this period, a Northern Electric candlestick telephone connected to a manual exchange staffed by friendly, helpful operators. It was not the right way to use a dial handset telephone that relied on mechanical switches to connect callers without any human assistance or intervention. What makes the "misuse" of the dial telephone in this short promotional film so intriguing is that Mrs. Baker was familiar with telephones and telephony. Her husband had been responsible for bringing the first commercial telephone service to Canada in 1877. In 1880, he had been granted the charter to build a national telephone company in Hamilton, a charter that had enabled the formation of Bell, which soon became Canada's largest and most influential telephone company. Hamilton's new automatic telephone exchange, which the film was promoting, was in fact the "Baker" exchange, named in honour of Hugh Baker.

Regardless of how much or how little interest Mrs. Baker had in her husband's business affairs, her confusion about how to use the new dial telephone offers a neat demonstration of how there was nothing particularly "natural" or "obvious" about the operations of the dial telephone or automatic telephony more generally, and it raises the question that drives this chapter: how did people learn to dial? The eventual ubiquity and the contemporary taken-for-grantedness achieved by automatic telephone service and similar technologies too often discourage historians from thinking about the fact that at one time they were strange, mysterious, unfamiliar things, which people had to learn about or be taught to use. But there are inevitably moments in the life cycles of technologies that allow us to see them with fresh eyes. Historian Carolyn Marvin has identified one of them as the moment "when old technologies were new" and ordinary people had to negotiate their relationships with them.[3] This is a moment when technical competence has not yet been acquired and naturalized or when there is ambiguity over the practical or symbolic uses of a technological device –

when technical and social knowledges and practices are indeterminate or in a process of negotiation. It is one of the moments, as literary theorist Bill Brown has argued, when historians can more easily perceive how technological "things" have mediated human relations in addition to how human relations have mediated technological "things."[4]

Like other chapters in this volume – such as Stein's work on the use of advertising campaigns to teach Canadians that it was safe to fly in winter, Gucciardo's description of Canadians' early encounters with electricity through electrotherapy, and Macfarlane's account of the campaign undertaken to "manufacture consent" among residents living in the path of the proposed St. Lawrence Seaway route in the 1950s – this chapter is concerned with how Canadians were introduced to and "sold" on new modern technologies and new modern social relations and practices. Rather than tracing the development of the dial telephone and its associated technological systems, I consider the pragmatic aspects of its adoption by telephone users in the major cities of central Canada. Drawing on newspaper accounts and corporate records, I demonstrate that telephone users' anxieties about and antipathy toward the introduction of automatic service were due as much to their reluctance to abandon familiar sociable practices of "telephoning" as to suspicions about Bell's motives for doing away with operator-assisted calling. I then examine Bell's wide-reaching program of subscriber education that recognized and sought to accommodate the public's expectations of personal service, which manual telephone service and Bell itself had done much to engender. Bell's educational campaign was developed to ensure that telephone users in major cities such as Toronto, Montreal, and Hamilton acquired the technical competence with the new dial telephone that was required for smooth operation of the larger telephone system. Bell described the dial system as a "modern development in the art of communications" and in keeping with the period's "tendency to do things by machine that yesterday were done by hand."[5] By replacing human operators with automatic switches and integrating the telephone user into the operation of the larger telephone system, dial telephony aligned with and enacted those modern values associated with mechanization and standardization: speed, efficiency, anonymity, predictability. Historian James Vernon has argued that one of the key features of modernity, facilitated in part by technological transformations in communications and transportation, was the creation of a "society of strangers." The conversion of telephony from manual to dial, with its more impersonal interactions, can be seen as a manifestation and an enactment of these new forms of social relations in early-twentieth-century Canada.[6] Thus, Bell's extensive subscriber education campaign should be interpreted as a kind of pedagogy of modernity. Looking closely at how Bell sought to educate its subscribers (and telephone users more generally) provides valuable insights into the roles that technology and technology education played in Canada's biggest cities during the interwar years. It also demonstrates the second "origin" of the telephone, whereby it became a more impersonal technological device – a way of perceiving and using the telephone that remains familiar to most Canadians today.

From "Calling" to "Dialling"

Dial service, or automatic switching, was not a new technology when Bell introduced automatic exchanges to the big cities of central Canada in the mid-1920s. The first successful automatic system had been developed decades earlier, in 1889, by Almon B. Strowger, a Kansas City undertaker who suspected a local telephone operator of diverting his clients' calls (and thus patronage) to a competitor.[7] Automatic switching was the technological solution that Strowger devised to do away with an operator's ability to route and screen telephone calls. Dial service was initially attractive to small, independent, and undercapitalized telephone entrepreneurs, less because of a concern over operator malfeasance than because of its promise to minimize labour costs. In Canada, most early automatic exchanges were set up to service northern and prairie boomtowns and resource camps – places that typically had populations largely made up of transient rural workers and where finding operators could be difficult. Canada's first automatic exchange was set up in Whitehorse, Yukon, in 1901 during the Klondike Gold Rush.[8] Early adopters of dial service included towns that served as railway hubs – such as Woodstock, New Brunswick, in 1903, and West Toronto Junction, Ontario, in 1905 – and coal-mining towns, such as Sidney Mines, Nova Scotia, in 1904.[9]

While many small, independent telephone companies set up automatic exchanges during the early years of the twentieth century, major North American telephone systems continued to rely on manual exchanges, including those of the dominant Bell System companies. Bell Canada operated as a de facto subsidiary of the American Telephone and Telegraph Company (AT&T), and its networks in urban central Canada were the most established and extensive in the country.[10] This meant that Bell incurred substantial labour costs associated with employing the many operators who routed calls across its network. Nevertheless, despite the incentive of lowering labour costs, there were several good reasons for the company to be cautious about converting its systems from manual to dial. Although creating an automatic exchange where no telephone network existed was, comparatively speaking, the cost-effective choice, by 1920 the prospect of converting large, established networks from manual to automatic switching posed its own particular challenges.

Perhaps surprisingly, technical challenges associated with "cutting over" from manual to automatic service were the lesser of Bell's concerns. The conversion could cause significant service disruptions if not carefully planned and executed, but by the early 1920s a technical protocol for converting from manual to automatic switching without interruption of service had been developed.[11] As the number of North American telephone subscriptions grew exponentially during the early 1920s, and the costs and technical limitations of established manual telephony became increasingly obvious, many Bell System companies began the process of converting to automatic service. A 1923 agreement signed with AT&T gave Bell Canada access to the latest research on switching transmission as well as to the practical experiences of other Bell System companies.[12] It was the financial and logistical challenges associated with converting to dial service that caused Bell to delay implementing this change in major urban centres as long as possible. Purchasing new switching equipment and constructing new central exchange buildings to house it would require significant capital investment. These costs and the size of Bell's network required that conversion to dial service proceed incrementally. In 1923, Bell manager Frank M. Kennedy estimated that conversion of Toronto's telephones from manual to automatic systems would cost $4 million and take ten years to complete.[13]

Another significant factor in Bell's reluctance to convert its systems to automatic switching was the good quality of its existing equipment. After the period of intense competition between telephone companies that followed the expiration of Bell's patents in 1885, a parliamentary inquiry into the telephone industry in 1905 had imposed a regulatory regime that allowed Bell to operate as a regulated monopoly but required it to maintain high standards of service.[14] As a consequence, many of Bell's exchanges, as its Montreal division manager Frank G. Webber noted, were "of the most modern and expensive type," and Bell considered it a "false economy to scrap them for the automatics."[15] Bell's policy of leasing – rather than selling – telephones to its subscribers was another way that it controlled the quality of its service and another reason to delay the introduction of dial service. Because Bell owned the telephones connected to its system, any conversion from manual to dial service would require Bell to purchase thousands of new dial telephones to replace subscribers' still-functional manual telephones and then absorb the cost of the redundant devices.[16] For all of these reasons, it made business sense for Bell to hold off the introduction of a new switching system until demand was high enough that it could recuperate its investment in a timely manner. That point was reached in large central Canadian cities in the early 1920s, when a spike in demand for residential telephone service resulted in significantly increased numbers of middle-class subscribers, a heavier volume of use, and a sharp rise in Bell's revenues.[17] Even under these promising conditions, Bell's plans for conversion to automatic service called for incremental implementation, with the company acting only when and where the number of subscriptions and connections outgrew the capacity of existing exchanges to handle the traffic.

The influx of new subscribers during the early 1920s provided Bell with higher revenues that helped to justify conversion to dial service. It also exacerbated what was known as the "subscriber problem" – how subscribers imagined and used the telephone. To place a call with manual service, a subscriber simply picked up the receiver, thereby signalling the operator, who responded according to Bell's established protocol by asking for the desired number. The caller would then state the number – or sometimes, in smaller communities, just the name – of the person whom he or she wished to speak to, and the operator would repeat the number to verify its accuracy before making the connection. As a Bell advertisement proclaimed, in the experience of manual or operator-assisted telephony, all successful telephone calls involved *two* conversations among *three* people. This was a widely shared experience and what Bell subscribers in central Canadian cities were accustomed to in the early 1920s.

Although the first telephone operators were young men trained in technical aspects of electricity, by the early 1880s Bell began to hire "lady operators" because they were soft spoken and courteous with subscribers and proved to be more adaptable and disciplined employees.[18] Well aware that its operators represented the company to its subscribers, Bell sought young, educated, lower-middle-class and "respectable" working-class women and trained them thoroughly on every aspect of their job, especially on "the right way of speaking," which included learning standard responses to subscriber queries and delivering them using tone and inflection that conveyed sincerity without wasting time. Bell liked to insist that the interactions between telephone operators and users remain business-like, but these exchanges were ripe with opportunities for subscribers to indulge in chit-chat, make

Mysteries of the New Phone Explained 149

Figure 6.1 Early Bell advertising (June 1911) portraying the telephone as a dependable employee in the office and a helpful servant in the home. | BCHC, 37633–1. Courtesy of the Bell Canada Historical Collection.

requests for information, and, in too many instances, vent frustration with telephone service. Bell had originally encouraged the idea of telephony as a "personalized service." Subscribers' views of operators had been shaped by early Bell advertising, which had portrayed the telephone as a dependable employee in the office and a helpful servant in the home (Figure 6.1), and subscribers expected operators to behave in ways that substantiated those representations.

Stories abounded of operators complying with subscribers' requests for recipes for tea biscuits or advice on relationships or to keep an ear on a sleeping baby while the mother ran an errand. These stories might have been slightly exaggerated, but subscribers did regularly call on operators to answer questions and provide assistance on matters that had little or nothing to do with telephone service.[19]

If the qualities associated with women's gender and upbringing made them perfect operators, the same qualities made them targets of the offensive behaviour of some subscribers.[20] As reliant as subscribers were on telephone operators, they were often more generous with complaints than with

Figure 6.2 Cartoons often made fun of both subscribers and operators. | *Star Weekly* (Toronto), February 24, 1923. BCHC Newspaper Clippings. Courtesy of the Bell Canada Historical Collection.

compliments – blaming operators for wrong numbers, disconnected calls, and busy signals. These complaints were so common that they became the stuff of jokes and cartoons (Figure 6.2). An operator working at an Ottawa exchange noted that "operating" "requires a lot of devotion and brings very little gratitude from the public."[21] Many subscribers felt proprietorial and possessive about their right to the services that they imagined the telephone operator should provide, and they felt justified in taking operators to task for real, or more likely perceived, lapses in service. It was not uncommon for subscribers to be rude and abusive with operators, who were trained not to respond in kind and to focus instead on keeping the number of connections high. Telephone companies, notably Bell, tried to insist that subscribers adhere to proper telephone etiquette, especially regarding the treatment of operators, but so ingrained was the idea that operators were literally their servants that in one instance a subscriber who was denied service because he swore at an operator took his complaint to the courts to get his service reinstated.[22]

Once it had made the decision to introduce automatic switching in its major urban exchanges, Bell concluded that it had to wean subscribers off their expectation of personalized operator service while still promoting a perception of automated service as preferable and progressive. In 1923, a full year before it introduced its first automatic exchange in Toronto, Bell began streamlining its system-wide operator protocol for connecting calls. Rather than repeating the telephone number back to the caller, operators were instructed henceforth simply to say "thank you" and connect the call.[23] Shortly afterward, Bell issued a directive to subscribers that, beginning February 20, 1923, its operators would no longer respond to callers' requests for the time, by which many set their watches and clocks.[24] Bell hoped that this would have the immediate benefit of decreasing some of the growing volume of telephone traffic and allow operators to route more calls. The larger, long-term intention, however, was to make subscribers' experiences of telephone service more routine and mechanical.

Bell's new directives were met with equal parts humour and irritation. Most subscribers soon accommodated themselves to more perfunctory interactions with telephone operators, but some thought that Bell was simply trying to increase profits by making them do work that had always been done by operators. Others resented what they perceived as Bell's reduction of service without a commensurate reduction of rate. One Toronto subscriber wrote to the *Globe* to express his dismay over "this silly action of the Bell Tel Co.," reminding Bell that "their patrons pay for service ... If said patron wants to know the time, courtesy demands that the young lady who is paid by us should glance at the clock and give the correct time."[25]

Responses to Bell's imminent introduction of dial service indicate that subscribers recognized these ostensibly technical changes as being cultural changes that would affect their expectations and experiences of telephone sociability. As Michèle Martin writes, "the fact that the telephone allowed oral communication without visual contact created a kind of intimacy which people previously had not experienced."[26] She suggests that the "contradictory effect created by the telephone of feeling nearby and far away at the same time seemed to embolden some people," but I argue that the feeling of distant-intimacy associated with the experience of telephoning also likely created empathy for the operators. Many subscribers were genuinely worried about operators losing their livelihoods with the introduction of automatic switching. This concern can be seen as an expression of the era's growing anxiety about automation and the threat of "technological unemployment," but it also points to the complicated relationship between subscribers and operators.[27] Although some subscribers complained about operators, many felt a kind of acquaintanceship with or friendliness toward the "hello girls" and looked forward to the sociability that "calling" afforded.

With the introduction of automatic service, the experience of telephony changed from "calling" to "dialling."[28] In addition to losing the personal service provided by the operator and having to learn to operate the dial telephone, automatic service would require telephone users to accommodate themselves to a new, standardized system of phone numbers. Telephone numbers had been assigned to subscribers in the order in which they had joined the network, so that in Hamilton, for example, the number of the seventh subscriber in the Regent exchange would be Regent 7, and so on. This worked perfectly when calls were switched manually, and callers had only to tell the operator the name of the exchange and the number to which they wished to be connected. However, for a caller to place a call using an electromechanical automatic switching system, all telephone numbers had to have the same number of digits or "pulls" on the dial.[29] Hamilton's five-pull system meant that all subscribers' telephone numbers consisting of fewer than four digits needed to be reconfigured as part of the conversion from manual to automatic service, being prefixed by zeros in addition to the first letter of the exchange name. Thus, Regent 7 became R 0007, and Regent 32 became R 0032. It was essential for the smooth operation of the new automatic system that callers use these new numbers. Some subscribers, especially business owners, resented the inconveniences that would accompany the conversion to dial telephones, especially the need to change telephone numbers, and therefore stationery, signs, and promotional materials. Both the irritation and the excitement that subscribers felt were captured in a Toronto theatre's cheeky notice announcing its new telephone number while chiding Bell for doing "the meanest thing [and replacing] the old 'phone with one of those 'swing-me-around clock thing-a-me-giggs.'"[30]

Bell understood how subscribers' experiences of telephony would be transformed by the replacement of operator service with a mechanized system, and it strategically introduced procedures intended to reduce opportunities for sociable interactions between callers and operators, such as replacing operators' traditionally sociable greeting with the more impersonal "number, please?"[31] Initiatives to streamline interactions between operators and callers were estimated to have saved the company $40,000. Although not always well received, they did serve as an important step in reforming subscribers' expectations.[32] Automatic switching required the subscriber to take over what had been the work of the operator, and Bell had real concerns about the public's willingness and ability to perform the necessary steps correctly and consistently.[33] As one Bell executive explained, "the chief difference between the service results in dial and manual offices is the much greater extent to which the subscriber's service is affected by his use of the equipment." For this reason, the executive argued, "much effort must be applied to the improvement of the subscribers' use of the equipment."[34] This sentiment succinctly described both the motivation and the goal of the "how-to-dial" subscriber education program that Bell introduced in central Canadian cities during the mid-1920s.

Improving Subscribers' Use

Historian of technology Joseph Corn has drawn attention to the emergence during the late nineteenth century of instructional texts and users' manuals and their role as a new and important means by which ordinary North Americans learned how to use and maintain technological devices. He suggests that by "textualizing technics," the manufacturers that included these texts with their products sought to "modify behaviour and ... control the way consumers used products." Corn proposes that, just as manufacturers sought to standardize their products through design and production, so too they employed highly didactic instructional literature in an effort to increase the chances that users would "think and act in uniform ways toward already standardized machines [and] make their mechanical possessions perform as the designers and manufacturers had intended."[35] Like the manufacturers described by Corn, telephone companies relied on printed materials to instruct subscribers on the correct use of the telephone from the early 1880s on. Bell Canada followed the example of its parent company, AT&T, and took every opportunity to provide subscribers with technical information on how to use the telephone as well as to promote what it deemed to be proper telephone etiquette. Unlike manufacturers who sold their products to consumers, Bell's telephone-leasing arrangement provided the company with the opportunity to send subscribers printed information on an ongoing basis. New subscribers received booklets with detailed illustrated instructions for use of the dial telephones, and they were subsequently sent circulars offering advice and telephone news with each monthly bill. In addition to listing telephone numbers, the directories provided to all subscribers featured instructions and advice.

When Bell decided to introduce automatic service in central Canada's major cities during the mid-1920s, it developed a remarkably comprehensive approach that employed diverse strategies of communication to teach the public about the dial telephone. In addition to its instructional booklets, circulars, and newspaper advertisements, Bell used window displays and, perhaps most significantly, group and one-on-one demonstrations that allowed subscribers as well as the general public to observe the operation of the dial mechanism and to practise using the dial telephone.[36] Although Bell referred to its campaign as a "subscriber education" program, it was in fact intended to educate the wider public about the new dial system and the new technique of dialling. Bell was able to communicate directly with its subscribers through its mailings and telephone directories. Demonstrations, displays, advertisements, and newspaper features allowed it to teach telephone users who were nonsubscribers how to use the new phone.

Bell drew on AT&T's advice to guide its conversion to dial service in Toronto (1924) and Montreal (1925). It also kept detailed records of the advertising and promotional materials that it produced and the events and activities that it sponsored for these launches. When Bell decided to introduce dial service to Hamilton in September 1929, it compiled these materials into a comprehensive handbook that set out the year-long implementation plan for Hamilton's Baker Exchange.[37] In addition to organizational and technical information, it outlined the subscriber education program, identifying strategies and providing templates for communicating with subscribers, the press, service organizations and clubs, and the public. Introduced to Bell supervisory staff at a four-day educational conference in July 1928, the handbook provided a look at the diverse strategies that Bell undertook to educate and reassure telephone users. In contrast to the manufacturers described in Corn's case studies of "textualizing technics," Bell's approach to subscriber education made extensive use of three distinct modes of communication or pedagogy: printed materials, displays, and demonstrations.

Printed Materials

In addition to its use of how-to-dial booklets, blotters, and advertisements, Bell's Commercial Department relied on personal and professional contacts with the press to have stories about the dial telephone and the coming conversion to dial service placed in the news. Prior to the Toronto and Montreal conversions, Bell invited newspaper editors and reporters to visit its telephone exchanges for a tour of the facilities and a detailed explanation of the new automatic switching equipment. To ensure that reporters understood - and could explain - the new system, Bell's Publicity Department distributed booklets that provided detailed instructions on how to dial and information about the technical equipment. Acutely aware of subscribers' concerns about operators' job security, Bell representatives also took these opportunities to deny vigorously that dial service would cause operators to lose their jobs.[38] Bell found the time spent on visits with "newspaper men" to be particularly worthwhile: one executive reported that "some very good newspaper articles with excellent publicity value have been obtained in this way, sometimes with dialling instructions worked into the write-up."[39]

In the lead-up to the Toronto, Montreal, and Hamilton conversions, local newspapers featured large, often full-page articles that described dial service and provided step-by-step instructions on how to dial. Articles announcing Toronto's conversion to dial telephony began appearing in newspapers in January 1923, over a year and a half before Toronto's first automatic exchange, the Grover Exchange, began operation.[40] Similarly, coverage in Montreal newspapers began in June 1923, almost two years in advance of the opening of the Lancaster Exchange in April 1925.[41] One Toronto *Star* feature article – "When You Get a Wrong Number – Whose Fault Is It?" – explained the many ways in which subscribers were responsible for the majority of service complaints levelled against operators.[42] It was shortly followed by another article – "Soon Be Your Own Central" – that presented dial service as the solution to the problem of wrong numbers and offered readers a primer on automatic telephony, including instructions on how to operate a dial telephone. The author's step-by-step description of the process of dialling was introduced in a manner that alluded to Bell's view of the task that confronted the company: "Mr. Frank Kennedy, the manager of [Bell's Toronto office], says he admires the nerve of the *Star Weekly* in attempting the task [of explaining automatic telephony] and he wishes us well and gives us all possible data, figures, statistics and specifications. But as for getting it across ... Well, then, here goes." The step-by-step description explained not only the dialling process – "you remove the receiver off the hook and listen" – but also reminded callers to "[look] in the directory for the number."[43]

To communicate with subscribers, Bell made extensive use of its directories and account mailers. A year in advance of the opening of the Baker Exchange in Hamilton, Bell began using the "account stuffers" that normally accompanied subscribers' bills to report on the construction of the exchange, plans for the cutover, and information about direct dial service. Telephone directories distributed to subscribers on December 1, 1928, listed the new standardized phone numbers, explained the reason for the initial zeros, and provided detailed dialling instructions.

In Hamilton, subscribers being "switched over" also received numerous personalized letters from the district manager reminding them of the change and of their new telephone number. Large business subscribers were targeted

Figure 6.3 Bell produced a wide range of promotional materials with "how-to-dial" instructions, including advertisements, booklets, and blotters, such as this one, distributed at the Made-in-Canada Exhibition, June 3–8, 1929. | BCHC 31097–2, exhibit 34. Courtesy of the Bell Canada Historical Collection.

first. In September 1928, a year in advance of the opening of the Baker Exchange, representatives from Bell's Commercial Department visited its large business subscribers to explain why their telephone numbers were changing and to advise them to amend stationery or advertising materials accordingly. Shortly after these visits, each business was sent a personalized letter featuring its new telephone number and reiterating the importance of using it exclusively. Smaller business subscribers were not visited but received similar personalized letters. Finally, all residential subscribers with "short" telephone numbers were notified of the impending change and provided with their new numbers by personal letters in the first week of November 1928. Numerous "reminder" letters were sent to all Bell subscribers in the months that followed.

Displays and Demonstrations

Bell used displays in the storefronts of its business offices and local stores as a way of communicating news about the dial telephone to the general public. Seen by hundreds or thousands of passersby every day, these displays reached subscribers and nonsubscribers alike, and they did so by combining elements of both instructional texts and demonstrations. An article in the telephone industry trade journal *Telephony* noted that storefront

Mysteries of the New Phone Explained 157

Figure 6.4 Window display in the G.F. Glassco Fur Store, King Street East, Hamilton, 1928. | BCHC 36126–2. Courtesy of the Bell Canada Historical Collection.

displays were an effective way of presenting factual information to people who had "so much reading matter constantly thrust at them [that they] read nothing more than a line or two."[44] Bell's dial telephone display in Hamilton's G.F. Glassco Fur Store's window featured an oversized dial candlestick telephone and a large-scale model of the dial itself, with a real dial handset telephone positioned closest to the viewer. Placards placed in a circular arrangement described the sequence of steps involved in dialling. Illuminated at night, storefront displays had the advantage of promoting dial service twenty-four hours a day. The window display also featured a placard with the locations of Bell's demonstration offices inviting people to visit and receive a demonstration of the dial telephone.[45] Such eye-catching and informative storefront displays were effective means of reaching large audiences during the interwar years – much more so than today.[46]

Demonstrations were another key element of Bell's subscriber education program. They provided an opportunity for Bell representatives to demonstrate how to use a dial telephone, explain how dial service worked, respond to people's questions, and address concerns about the loss of operator service. Demonstrations were not restricted to subscribers, and they were held in many locations in order to reach the largest possible number of telephone users. Typically, demonstrations had as their centrepiece a large wooden demonstration dial and two working dial telephones installed for instruction purposes. Sometimes a film was shown. In almost all instances, Bell employed its telephone operators to demonstrate how the dial telephone worked.[47] The qualities that made these young women "perfect operators" also made them ideally suited to the job of teaching the public how to dial. Their youth, gender, and friendly demeanour put audiences at ease and made the dial telephone appear to be an easy technology to master. Furthermore, the employment of operators to demonstrate dial technology was a pragmatic acknowledgment of the intimacy that Martin suggested the telephone engendered.[48] The familiar practice of calling brought telephone users into intimate conversations with operators and with each other: when callers requested a connection, asked a question, or complained about service, it was a person who responded to them. Demonstrations and one-on-one training led by operators offered a form of interaction reassuringly similar to central Canadian urbanites' experiences and expectations of telephone service. Perhaps more cynically, operator-led demonstrations might have also promoted the impression that operators themselves looked forward to the new dial system and thus dampened the public's concern about their replacement by automation.

Demonstration centres or "dial service schools" for the public were set up in Bell's business offices approximately two months before a planned date of conversion.[49] This allowed those in attendance to see how to operate the telephone and become familiar with the new automatic system. The "schools" were open to the public six days a week from Monday to Saturday between 10 a.m. and 10 p.m., making it convenient for people from all walks of life to visit at times that suited them.[50] In some cases, demonstrators invited willing participants to practise using the dial telephone, thereby proving to audiences that anyone could learn to use the new technology.

Figure 6.5 How-to-dial demonstration in Bell office, Montreal, 1925. | BCHC 5887. Courtesy of the Bell Canada Historical Collection.

A month before a planned cutover, demonstrations were given to targeted groups of telephone users. Bell Commercial Department representatives organized talks and demonstrations for service clubs, such as the Kiwanis and Rotary Clubs, and sent representatives to workplaces, particularly those increasingly reliant on telephone service, such as newspaper offices and police and fire departments. Demonstrations were also held for students from the fifth grade and up. Bell considered schools to be an "excellent field for talks on dialling instructions [because] children, after instruction, go home and show their parents how it's done."[51]

For subscribers located in a new automatic exchange area, Bell arranged one-on-one demonstrations or "interviews" consisting of a visit from a member of what Bell called its "flying squadron" to explain the new dial telephone and its place in the larger automatic system and to allow the subscriber to practise dialling. Bell started exchanging new dial telephones for subscribers' manual telephones about two weeks ahead of the conversion day. The visits during which these new dial telephones were installed offered yet another opportunity to educate subscribers on how to dial and to assess the telephone's location within the home or business. Ensuring

Figure 6.6 A Bell employee using an automatic switching "dial train" to demonstrate the operation of the new dial system. Bell Canada Business Office, 8 Main Street East, Hamilton, 1928. | BCHC 36126–1. Courtesy of the Bell Canada Historical Collection.

that the telephone was installed in a well-lit location was deemed important so that subscribers could avoid making dialling errors resulting in wrong numbers or uncompleted calls or having to call central for assistance.

Although the introduction of automatic switching in Toronto and Montreal was preceded and accompanied by the extensive use of newspaper coverage – how-to-dial stories and reports of dial-related events, displays, and demonstrations – this communication strategy was less evident in the conversion in Hamilton in 1929. The cutover there was reported in both of the city's daily newspapers, but feature stories appeared only in the month preceding it. This more modest lead-in to the conversion might have occurred because the Baker Exchange was relatively small, and conversion affected fewer subscribers than in Grover in Toronto or Lancaster in Montreal. Or Bell might have believed that, given the scope of the coverage of the introduction of dial service in Toronto, subscribers in nearby Hamilton were likely to be somewhat familiar with the operation of the dial telephone. Bell did not, however, reduce its use of displays, demonstrations, or interviews in Hamilton, suggesting that the company found public demonstrations and personal in-home interviews – face-to-face and not textual communications – to be the most effective ways of educating the public to use the new dial telephone technology.

Conclusion

Following the introduction of dial service in the major cities of central Canada, Bell tracked and documented faulty uses of the new dial telephones in order to inform its future subscriber education campaigns. Bell's "service observations" identified common errors that telephone users made, such as beginning to dial before getting a ring tone, dialling extra or insufficient digits, forcing the movement of the dial, and confusing letters and numbers when dialling. These observations show that users adjusted slowly and unevenly to the new requirements of dial telephony. Like the film clip of Mrs. Baker described in the introduction to this chapter, users' errors remind us that the process of dialling a telephone – which most urban Canadians soon came to do without thinking – was the result of complex negotiations of technical and nontechnical factors, none of which was inherently or obviously logical from the user's perspective. So, though it might seem to be obvious to state that people had to learn how to use the modern technologies that entered their lives, paying attention to how they learned or were taught to use them helps to explain how those new but soon to be taken-for-granted knowledges and practices shaped human and technological relationships. It is rare to find evidence that helps to explain how people were taught about technologies in the past. User manuals and instruction booklets are often perceived as the primary means for educating users and owners of new technologies during the early twentieth century, and because such textual evidence is likely to have survived and be accessible to historians, it is easy to overlook the varied ways that people learned to use new technologies and integrate technological knowledge into everyday practices. The comprehensive documentation of Bell's subscriber education program during the mid-1920s offers a rare and valuable overview of how technological education was a complex, coordinated, and crucial undertaking. It also demonstrates Bell's awareness that the introduction of the new system required not only that subscribers be trained to use dial telephones properly but also that they learn to imagine themselves as a new modern telephone user – one disciplined in the use of the telephone and a fully integrated component of the operation of the larger telephone system.

NOTES

Many thanks to Lise Nöel and Janie Théorêt of the Bell Canada Historical Collection for their generous assistance with the research for this chapter. Thanks as well to the anonymous reviewers and the editors of this volume for their helpful feedback and suggestions.

1 This chapter draws insights from a number of important sociocultural histories of Canadian telephony, including Christopher Armstrong and H.V. Nelles, *Monopoly's Moment: The Organization and Regulation of Canadian Utilities, 1830–1930* (Philadelphia: Temple University Press, 1986); Robert MacDougall, *The People's Network: The Political Economy of the Telephone in the Gilded Age* (Philadelphia: University of Pennsylvania Press, 2014); Michèle Martin, *"Hello Central?": Gender, Technology, and Culture and the Formation of Telephone Systems* (Montreal/Kingston: McGill-Queen's University Press, 1991); and Jean-Guy Rens, *The Invisible Empire: A History of the Telecommunications Industry in Canada, 1846–1956* (Montreal/Kingston: McGill-Queen's University Press, 2001). Claude Fischer's *America Calling: A Social History of the Telephone to 1940* (Berkeley: University of California Press, 1992) includes a chapter on "Educating the Public" about the telephone in the late nineteenth and early twentieth centuries.
2 Bell, *Bell Telephone at Hamilton*, black-and-white silent film, 1929, Bell Canada Historical Collection (hereafter BCHC).
3 Carolyn Marvin, *When Old Technologies Were New: Thinking about Electric Communication in the Late Nineteenth Century* (New York: Oxford University Press, 1988).
4 Bill Brown, "Thing Theory," *Critical Inquiry* 28, 1 (2001): 4. Brown identifies the moment when technologies are new and the moment when they break down as particularly valuable for the insights that they allow. Joseph J. Corn offers another perspective by documenting consumer frustration with technological goods in *User Unfriendly: Consumer Struggles with Personal Technologies* (Baltimore: Johns Hopkins University Press, 2011).
5 Bell, "Introduction of Dial Service in Hamilton, Ont. Educational Conference for Supervisory Employees, July 9–13, 1928," BCHC 31097-1, section A, sheet 6.
6 James Vernon, *Distant Strangers: How Britain Became Modern* (Berkeley: University of California Press, 2014). Vernon locates the earliest materialization of this "profoundly new and modern social condition" in Britain in the late eighteenth and early nineteenth centuries but argues that the model spread elsewhere and continued well into the twentieth century (xi).
7 R.B. Hill, "The Early Years of the Strowger System," *Bell Laboratories Record* 31, 1 (1953): 95–103.
8 John Wiley, a former employee of Strowger's Automatic Electric Company, established the automatic exchange in Whitehorse. Bell, "No. 119 Sat. Jan. 27, 1940 – Automatic Telephones," BCHC, Central Office Equipment: Step-by-Step.
9 Bell, "Historical Sketch of Automatic Telephony (1951)," BCHC, Central Office Equipment: Step-by-Step.

10 In 1907, Bell gave up its claim to national networks in order to secure and expand its profitable central Canadian market. Armstrong and Nelles, *Monopoly's Moment*, 184–85.
11 In the 1920s, *Telephony*, the trade journal of the telephone industry, featured numerous articles on automatic dial service, including articles on "tying" automatic switching to manual systems. See, for example, A.B. Smith, "Automatic Telephone Switching," *Telephony*, January 10, 1920, 14–17; Northwestern Bell, "Tying Automatic to Manual Systems," *Telephony*, April 17, 1920, 16–20; and F.L. Baer, "From Manual to Automatic Switching," *Telephony*, July 10, 1920, 16–19.
12 Laurence B. Mussio, *Becoming Bell: The Remarkable Story of a Canadian Enterprise* (Montreal: Bell Canada, 2005), 36.
13 "Phone Efficiency to Cost Company about $4,000,000," *Globe* (Toronto), January 22, 1923.
14 Robert E. Babe, *Telecommunications in Canada: Technology, Industry, and Government* (Toronto: University of Toronto Press, 1990), 97–101.
15 "Automatic Phone at Rotary Club," *Gazette* (Montreal), February 11, 1925.
16 In addition to the cost to Bell of purchasing new dial telephones, the strain that production of the new telephones would place on Bell's manufacturing arm, Northern Electric, was a consideration. Typically, the manual subscriber sets that Bell replaced with dial telephones would be refurbished and converted to dial sets.
17 Mussio, *Becoming Bell*, 39–40.
18 Historians of communications Michèle Martin and Venus Green discuss the importance of women's gendered roles in establishing protocols for telephone use and in developind the telephone industry more broadly. See Martin, *"Hello Central?,"* 56–67; and Venus Green, *Race on the Line: Gender, Labor, and Technology in the Bell System, 1880–1980* (Durham, NC: Duke University Press, 2001), 53–69. Green points out that "the feminization of telephone operating during this period [was a rare] example of management exercising control over the workplace to provide a better service rather than an attempt to reduce wages" (57).
19 A newspaper article published in 1923 recounted the story of a Saskatoon man who turned to the operator for baking advice while his wife was away: "Say, Information, please how do you make tea biscuits?" The article went on to explain that, with the introduction of dial telephony, this "service" would no longer be available. The humorous account of the exchange between a telephone subscriber and the operator is notable for a description of manual telephony that highlighted its social nature. After passing on her recipe, the operator checked with the caller: "Have you got it down?" "Yes. Thanks very much, Information. I'll send you some of the biscuits." "Can't Abuse Central with Dial Service," *Star* (Toronto), January 20, 1923.
20 Green writes that Bell managers used both racial and gender exclusion to create their "image of the ideal telephone operator as a woman who conformed to nineteenth-century notions of virtue and piety: a 'lady.'" African American women were not included in this conception of womanhood, and they were hired as operators only during the Second World War because of labour market pressure and the efforts of the civil rights movement. See Green, *Race on the Line*, 53, 195.

21 She went on to observe that Bell operators "put their femininity to the service of the community. Very few men would be patient enough to perform the duties of an operator." Bell, "Life Story: B. Lalonde, Operator" 1906, cited in Martin, *"Hello Central?,"* 59.

22 M.N. Campbell, letter to the editor, *Herald* (Montreal), June 25, 1923.

23 This was first put into effect in Toronto's downtown exchanges. In late January 1923, Frank M. Kennedy, Bell's resident manager, reported to the North Toronto Ratepayers' Association that "the system was working out satisfactorily," noting that "the 'wrong numbers' were, if anything, a little less numerous than under the old system." "Phone Efficiency to Cost Company about $4,000,000," *Globe* (Toronto), January 22, 1923.

24 This change was expected to result in savings of $40,000 a year for Bell. "Must Consult Clocks Now," *Toronto Telegram* , February 20, 1923.

25 R.R. Hopkins, letter to the editor, *Globe* (Toronto), February 24, 1923.

26 Michèle Martin, "Gender and Early Telephone Culture," in *Sound Studies Reader*, ed. Jonathan Sterne (London: Routledge, 2012), 343. See also R.T. Barrett, "The Telephone as a Social Force," *Bell Telephone Quarterly* 19 (1940): 129–38.

27 Amy Sue Bix, *Inventing Ourselves Out of Jobs: America's Debate over Technological Unemployment, 1929–1981* (Baltimore: Johns Hopkins University Press, 2000).

28 The distinction between these two conceptions of telephony was pointed out in the article "Can't Abuse Central with Dial Service," *Star* (Toronto), January 20, 1923, which noted that "residents no longer call each other on the 'phone.' They dial each other instead."

29 Whereas Hamilton had a five-pull system, Toronto had a six-pull system. It is unclear why Bell chose to implement different systems, especially given the geographic proximity between the two cities.

30 Regent Theatre (Toronto) program, April 6, 1925, BCHC, Newspaper Clippings.

31 "Number, Please?," *Ottawa Citizen,* November 10, 1925.

32 "Must Consult Clocks Now," *Toronto Telegram,* February 20, 1923.

33 An article in the Montreal *Gazette* reporting on Bell's demonstration of the dial telephone at the Rotary Club noted that the "manager of BTCo expresses anxiety to instruct [the] public on latest methods," and the author similarly questioned how long it might take for subscribers "to become sufficiently expert" to use the service competently. "Automatic Phone at Rotary Club," *Gazette* (Montreal), February 11, 1925.

34 Bell, "Introduction of Dial Service in Hamilton," BCHC 31097–1, section U, sheet 1.

35 Joseph J. Corn, "'Textualizing Technics': Owner's Manuals and the Reading of Objects," in *American Material Culture: The Shape of the Field,* ed. Ann Smart Martin and J. Ritchie Garrison (Winterthur, DE: Winterthur Museum, 1997), 170, 191, 194.

36 The use of demonstrations can be seen as a return to a strategy that electrical inventors and manufacturers – including Alexander Graham Bell – used with success to promote their technologies in the late nineteenth century. Armstrong and Nelles, *Monopoly's Moment,* 63–65, describe the combined effect of dramatic demonstrations and "skillful public relations" as a "theatre of science" that was part "public entertainment" and part "moral pageant."

37 Bell, "Introduction of Dial Service in Hamilton," BCHC 31097–1.
38 Bell included scripted responses to inquiries from the press about the effect of dial service on operators' employment in its Hamilton handbook. Bell, "Introduction of Dial Service in Hamilton," BCHC 31097–1, section I-1, part 2, exhibit 12A, "Girls to Retain Jobs."
39 Bell, "Introduction of Dial Service in Hamilton," BCHC 31097–1, section I-1, sheet 18.
40 "Can't Abuse Central with Dial Service," *Star* (Toronto), January 20, 1923; "300 Waiting for Phones," *Toronto Telegram*, January 22, 1923; "York County News," *Mail* (Toronto), January 22, 1923; "Phone Efficiency to Cost Company about $4,000,000," *Globe* (Toronto), January 22, 1923.
41 "Automatic Systems for Telephones," *Gazette* (Montreal), June 26, 1923.
42 "When You Get a Wrong Number – Whose Fault Is It?," *Star* (Toronto), February 17, 1923.
43 "Soon Be Your Own Central," *Star Weekly* (Toronto), May 17, 1923.
44 F.H. Williams, "Effective Window Display Publicity," *Telephony*, May 29, 1920, 13–14.
45 Suggestions on how to arrange window displays were described in Bell, "Introduction of Dial Service in Hamilton," BCHC 31097–1, section I-1, sheet 18.
46 On this topic, see especially Keith Walden, "Speaking Modern: Language, Culture, and Hegemony in Grocery Window Displays, 1870–1920," *Canadian Historical Review* 70, 3 (1989): 285–310.
47 Press coverage indicates that demonstrations held for police and fire departments were conducted by male Bell representatives, typically district managers or Commercial Department representatives.
48 Martin, "Gender and Early Telephone Culture."
49 Sometimes Bell also rented vacant stores in high-traffic areas for use as demonstration centres. Bell, "Introduction of Dial Service in Hamilton," BCHC 31097–1, section I-1, sheet 17, item 17.
50 Ibid., section I-1, sheet 16, item 14.
51 Ibid., section I-1, sheet 17, item 16.

Small Science
Trained Acquaintance and the One-Man Research Team

DAVID THEODORE

In 1970, a young scientist named Christopher Thompson started a new job at the Montreal Neurological Institute (MNI). Founded in 1934 and affiliated with the Faculty of Medicine of McGill University, the MNI was both an active hospital and a scientific research centre. It had an illustrious history of combining basic research in neuroscience and neurological illnesses (e.g., studies in brain imaging and memory) with clinical neurology and neurosurgery (e.g., surgical treatments for epilepsy).[1] Thanks to a Major Equipment Grant from the Medical Research Council, a team of MNI researchers led by neurosurgeon Pierre Gloor had just acquired a "laboratory computing system," namely a Digital Equipment Corporation (DEC) PDP-12 minicomputer.[2] In the grant application, the team had outlined "several lines of research" in which "versatile data analysis by a digital computer would be required," including investigations of thermoregulatory mechanisms, electroencephalography (EEG), and stereotaxic surgery.[3] William Feindel, the MNI's illustrious and energetic director, couldn't wait. In a letter to Gloor, he wrote that "we have been ready now for a year and a half to utilise computer techniques, had the equipment been available."[4] In other words, rather than use the computer for one big research project coordinating a large number of brain researchers, the MNI team planned to undertake smaller, individualized investigations. Thompson's task, then, was to program and operate the computer for a variety of small research projects. Over the next decade, Thompson and the PDP-12 together formed a sort of

Small Science

Figure 7.1 Digital Equipment Corporation PDP-12 minicomputer installed in Christopher Thompson's laboratory at the Montreal Neurological Institute, c. 1970. | Courtesy of Christopher Thompson.

one-man interdisciplinary research team, a central node in the MNI's ambitious and influential program to incorporate computing into brain studies.

On the surface, Thompson was an unusual choice for the job, for he had no medical training. He held only a master's degree in physics from the University of Otago in New Zealand. But he was fresh from a stint working with a minicomputer on geological gamma ray spectroscopy for Atomic Energy of Canada, so he had acquaintance with two areas crucial to the MNI researchers: namely, computing and radioisotopes.[5] Thompson also had experience with rejigging computer equipment, for he had helped to make a mobile computerized analysis unit. Members of the survey team were attempting to automate analysis with computers to make geological surveying cheaper and faster. They loaded a Hewlett Packard 2115 computer inside aircraft so that they could carry out the analysis in the field.[6] The fact that Thompson had learned these skills on the job also showed that he had both the ability and the disposition to learn about areas of expertise outside his own.

At the MNI, Thompson quickly demonstrated his aptitude and flexibility. Take, for instance, the use of computing in neurosurgery. The grant application identified the need for a computer to perform the real-time location of

Figure 7.2 Photograph of the display on a Tektronix 4002 terminal taken during an operation at the Montreal Neurological Institute. The neurosurgeon could change the display by operating the joystick shown at bottom right. | Courtesy of Christopher Thompson.

a probe during stereotaxic surgery. The MNI had used this surgical technique on about 115 patients, usually Parkinson's patients, between 1963 and 1968.[7] It involved putting a probe into the patient's brain through a borehole in the skull. An electrode at the tip of the probe was used to find the target, and then the surgeon replaced the electrode with a cutting tool in order to remove brain tissue. The patient was awake during the procedure.[8] Acquaintance with putting computers into aircraft came in handy for Thompson, who had to move the computer to just outside the operating room and bring the computer screen into the operating room near the patient.

More importantly, Thompson also had to learn just enough neuroanatomy to understand how the image of the location of the probe moved in relation to anatomical structures. He did not become a neurosurgeon, but he had to acquire skills and knowledge to work closely in the operating room with one.

In this chapter, I explore the idea that Thompson's scientific work at the MNI manifests *small science*. Small science, as I characterize it, is an identifiable kind of scientific work carried out in certain architectural settings, with specific modes of collaboration and scales of activity. My ambition here is to show that small science is a useful concept in studying scientific practices. I argue, first, that it identifies how and where science gets done and, second, that it can sharpen our thinking about how history of science gets done. It forms a potent model for thinking about the scale and nature of postwar workplaces and practices more generally. As well, small science might help to complicate our story of what modern science and technology in Canada involve. Small science precisely combines high technology and advanced practices in a kind of artisanal scale of scientific production that many scholars have suggested died out after the Second World War with the rise of so-called big science.[9] In this way, a scholarly focus on small science can challenge some of the teleologies of our discipline.

I want to focus on two interlocked characteristic elements of small science, the "one-man research team" and the idea of "trained acquaintance." Mathematician Norbert Wiener introduced the latter term in 1948. He wanted to stress that, to participate actively in a multidisciplinary research team (he had especially in mind his own field of cybernetics), a scientist did not need to have full mastery of a discipline, only to be *just enough* of an expert to understand and criticize a true expert's intellectual contribution.[10] The one-man research team is simply the idea that an individual scientist can embody the relevant multidisciplinary training normally identified with a multidisciplinary research team. The one-man research team, though, is not a scientist with fully fledged expertise in each of the necessary disciplines but one who has a *trained acquaintance* with key concepts and practices in another discipline. Still, the one-man research team is not necessarily an autonomous unit; such a scientist can collaborate within a larger research team, but evidently, when a scientist with trained acquaintance is included, the overall team size can be smaller.

My case study follows Thompson's involvement in the program of computer-based clinical biomedical research that started at the Montreal Neurological Institute around 1970. Rather than a multiple-man team of scientists, one of whom had expertise in computer programming, another familiar with computer engineering and hardware, a third trained in neurosurgery and neuroanatomy, and yet another able to work with radioactive isotopes, Thompson himself had just enough trained acquaintance in each area to constitute a one-man team. This case study is an example but not an epitome. It is meant as a helpful illustration of a possible instance of small science, but it might turn out that small science is a useful concept, even if my case study is not generalizable. In the last part of the chapter, then, I will evaluate my use of this case study and speculate about further uses for the concept of small science in new scholarship on history, science, and technology in Canada.

Big Science, Little Science

In the literature, scholars define small science by first assessing the features of big science. Big science refers to the unprecedented transformation in scope and scale that scientific practices underwent in the twentieth century, especially in physics research. Big science experiments require national and multinational funding, large teams of interdisciplinary collaborators, and large laboratory compounds containing relatively large pieces of specialized equipment, as well as the complex logistical considerations and coordination that go with them.[11]

In his classic 1963 book *Little Science, Big Science*, Derek John de Solla Price made some additional enduring distinctions.[12] He insisted on two criteria not specifically related to an increase in quantity and scale, one temporal, one qualitative. First, Price used statistics to argue that big science has a direct impact on society and politics. Whereas little science can have a broad impact through technology transfer, big science itself is a sociopolitical phenomenon. In other words, because of the scale of funding, people, and material involved, big science has political, economic, and social significance independent of the scientific knowledge produced. If a big science project such as the Large Hadron Collider is carried out, then it will affect governments, careers, architecture, and cities regardless of any resulting social changes wrought by new ideas within physics.[13] Second, Price argued that big science was something new. Before the twentieth century, all science in Europe was little science.[14] If Price was right, then any little science that still exists is a holdover from former practices; the change to big science is inexorable.[15] But there is another way of distinguishing between scientific practices that makes little science and big science complementary rather than oppositional practices. H.M. Collins, for instance, asks whether innovation or creativity necessarily characterizes little science: "Could there be, as many scientists believe, a kind of science – let's call it 'developing science' – that can be done best by autonomous individuals or small teams using intuition and craft practices, and another kind – 'mature science' – that can be done best by coordinated large teams routinizing their science as far as possible?"[16] For Collins, little science might only be a stage (developing science) on an inexorable path to big or mature science.

I wish to distinguish small science from all three of these categories: big science, little science, and Collins's relational notion of development. Big science denotes a shift from individual researchers or laboratories to large collaborative teams. If that shift is historical, then small science is the persistence of "craft practices" alongside routinized science; if that shift is developmental, then small science denotes craft practices that simply do not develop into routinized science. If big science requires large teams working in large laboratories using multiple integrated arrays of equipment, small science denotes research that an individual can oversee in a laboratory not large enough to be an institution. Finally, scholars sometimes argue for a version of big science versus little science in which the former is pluridisciplinary and the latter is focused within one discipline.[17] Little science, as Price and Weinberg conceived it, is discipline specific. It denotes individual researchers who experiment within their own disciplines.[18] Big science attacks complex problems and uses teams of researchers from multiple disciplines to address them.[19] In this fashion, Ernest Rutherford (another New Zealand–born physicist) working in his laboratory at McGill University exemplifies little science; the construction and operation of the Underground Area 1 particle detector experiment at the European Council for Nuclear Research exemplify big science. In contrast, small science is a modern phenomenon that addresses research in a multidisciplinary fashion yet involves only a few scientists at a time.

Trained Acquaintance

How can multidisciplinary research be carried out without a large multidisciplinary research team? In those cases, the instances of small science, the bridging between disciplines is done through "trained acquaintance," the term coined by Norbert Wiener in 1948.[20] He emphasized trained acquaintance as a crucial component of postwar scientific life. Wiener recognized that, for a number of scientific issues, scientific specialization was a barrier. He saw the need for specialists to traverse the "no-man's land between the various established fields," arguing that a "proper exploration of these blank spaces on the map of science could only be made by a team of scientists, each a specialist in his own field but each possessing a thoroughly sound and trained acquaintance with the fields of his neighbours."[21] What was required was not expertise or even mastery. His discussion was normative but also categorical; Wiener both testified, albeit anecdotally, that trained acquaintance existed and argued that there should be more of it.

He discussed trained acquaintance by example. His first example parallels my medical case study. Christopher Thompson was a trained physicist who had to become acquainted with neurophysiology; Wiener looked at the links between a mathematician (himself) and a physiologist (his collaborator, Arturo Rosenblueth):

> If a physiologist, who knows no mathematics, works together with a mathematician who knows no physiology, the one will be unable to state his problem in terms that the other can manipulate, and the second will be unable to put the answers in any form that the first can understand ...
>
> The mathematician need not have the skill to conduct a physiological experiment, but he must have the skill to understand one, to criticize one, and to suggest one. The physiologist need not be able to prove a certain mathematical theorem, but he must be able to grasp its physiological significance and tell the mathematician for what he should look.[22]

Wiener's narrative is meant to show that trained acquaintance is acquired on a need-to-know basis. Other examples in the book describe his encounters with engineering design, communications engineering, neurology, psychology, sociology, and anthropology. Where Thompson's case diverges from Wiener's examples is not in the details of these collaborations but in Wiener's motivation. That is, his goal was to institute the "new science of cybernetics," a unified theory of communication and control in both machines and living tissue.[23] Wiener's story about trained acquaintance comes at the beginning of his discussion of the path to a generalized mathematical statement of communication and control. Thompson had no such goal. He simply participated in the ongoing research activity at the MNI.

Through his examples of interdisciplinary collaboration, Wiener implicitly argued that training scientists for work that could bridge specialized fields requires something other than formal education – and something not quite apprenticeship or on-the-job training. The examples show (but do not specify) that trained acquaintance is more than just a disposition to understand both conceptual and pragmatic issues in another scientist's disciplinary modes. The main difference between Wiener's use of the term and mine is that Wiener consistently referred to a team of collaborators; he never saw interdisciplinary research as embodied in one scientist. I want to say instead that Wiener himself, because of his trained acquaintance, should be thought of as an interdisciplinary researcher – a one-man research team. Thompson too.

Wiener was not concerned with small science. Moreover, his stress on interdisciplinary science can be seen as a justification of big science, of the necessity for collaboration. However, his examples distance his work from large-scale technoscience. Instead, he focused on promoting and explaining cybernetics, which would give scientists currently separated in silos a vocabulary and mathematics that would enable collaboration. Wiener anticipated that trained acquaintance would help to deal with nomenclature, specifically with the problem that, for scientific work, "every single notion receives a separate name from each group."[24] Trained acquaintance might be an important precondition that allows the linguistic development of the pidgins and creoles that activate historian Peter Galison's "trading zones."[25]

Architecture and Space

There are two ways in which architecture and space are integral to the concept of small science, the first to do with metaphor and the second to do with the physical environments of scientific practices.

First, metaphor. Wiener talked about trained acquaintance through spatial and architectural metaphors. He wrote that a scientist "will regard the next subject as something belonging to his colleague three doors down the corridor," an architectural metaphor; that "specialized fields are continually growing and invading new territory," a geographical metaphor; and that the "boundary regions of science ... offer the richest opportunities," a cartographical metaphor.[26] Wiener was not alone. Spatial metaphors have been common in the historiographical concepts that scholars use to study postwar interdisciplinary science: trading zones, archaeology, structure, foundations, perspective, and so on.[27] How useful is the metaphor of small and big? In *Image and Logic,* his history of physics, Galison grapples with the oddity of judging scientific practices by scale: "'Big physics' is about as helpful to the historian of science as 'big building' would be to the historian of architecture."[28] However, at almost the same time that Galison was writing, Dutch architect Rem Koolhaas and Toronto-based designer Bruce Mau published one of the most influential books in architecture of the past fifty years, precisely called *S,M,L,XL*.[29] The big building is indeed a helpful – even unavoidable – concept for historians of postwar architects. The point, even if Galison was right to be suspicious of big science (or big physics), is that the underlying spatial metaphors are not easily dismissed. Space matters, and scale matters, even at the level of metaphor.

There are other links between science and architecture that small science brings to light. One of the key postwar spatial metaphors is the idea of the network.[30] Wiener's search for like-minded colleagues famously engendered the Macy Cybernetics Conferences, celebrated as an important institution in the development of his interdisciplinary network.[31] In order for Wiener to interact at the conferences with both neurophysiologist Warren McCullough and anthropologist Margaret Mead, he needed to understand the limits of their expertise. But the equivalent networking was going on across postwar intellectual life. In architecture, for instance, Greek architect and planner Constantinos Doxiadis held a celebrated series of meetings specifically designed to work in the same interdisciplinary no-man's land that Wiener identified. These meetings deliberately brought together people already skilled in trained acquaintance. For instance, the first meeting of Marshall McLuhan and Buckminster Fuller took place on a boat trip around the Greek Islands organized by Doxiadis.[32] In both science and architecture, then, the search for metaphorical interdisciplinary space is directly correlated to physical spatial organization. Space is a metaphor for discussing interdisciplinarity, but architecture and territory are key to making interdisciplinarity work.[33]

The projects that Thompson undertook at the Montreal Neurological Institute also show the importance of metaphorical and physical spatial organization. My goal here is to suggest three supplementary characteristics of small science related to the notion that small science takes place in bounded place and time: small equipment, borrowed buildings, and ephemeral science.

The first two characteristics are intertwined. Small science involves small equipment and takes place in pre-existing sites. If big science involves large machines in purpose-built architecture – Livermore or the Bell Labs or the synchrotron in Saskatoon – then small science takes place with small equipment in "borrowed buildings."[34] This is clearly the case with early computing-based research at the MNI. An interwar hospital, eclectic on the outside, modern on the inside, it was designed by the prominent Montreal-based firm of Ross and Macdonald.[35] It opened in 1934 on the south slope of Mount Royal next to McGill University's football stadium, indicating its close physical and institutional ties with the Faculty of Medicine. It was linked by a bridge to the wards of the Royal Victoria Hospital, a pavilion-plan hospital that had opened in 1893.

Although it had been added to in 1952, it was basically unchanged when Thompson arrived. Bringing a computer in 1970 to a building designed and built in 1934 was an ad hoc use of existing environments. Note that the MNI did not open a *purpose*-built room for computers for another fourteen years – the McConnell Brain Imaging Centre, which opened in 1984.

Small Science 175

Figure 7.3 A contemporary view of the Montreal Neurological Institute (1934) with its prominent bridge to the Royal Victoria Hospital. | Courtesy of Don Toromanoff.

Whether large-scale digital imaging should count as big science is beyond the scope of this chapter, but the point remains: small science takes place in borrowed architectural settings.[36]

Small science also uses small equipment. The MNI researchers looked for a computer that would be "easily used by a variety of non-specialist experimenters."[37] In the summer of 1969, they set their eyes on the newly released DEC PDP-12 minicomputer. They argued that it was suitable for biomedical research because users had begun to share (and market) software and hardware that made it easy to link peripheral devices.[38] The minicomputer itself is perhaps characteristic of small science, at least in this period (i.e., still in the early stages of computer networking), because it was targeted at individual research laboratories.[39]

At the MNI, the minicomputer functioned as a boundary object, not for a discipline, but for the one-man research team. Thompson always worked as part of a team at the MNI. His specific expertise was in computing: even if he had no formal training in the discipline, the other researchers did not know how to use a computer at all. Their computing knowledge was below the threshold even of "non-specialist experimenters." They needed not only the machine but also someone to run it. His knowledge of computers came through improvisational flexibility. Thompson learned these fields

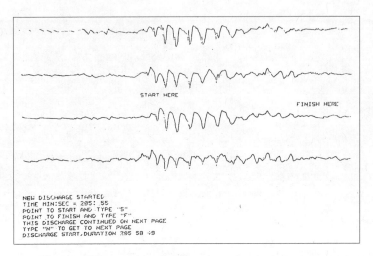

Figure 7.4 Polaroid image of computer display extracting features from EEG. Thompson later programmed the PDP-12 to monitor epileptic seizures. | Courtesy of Christopher Thompson.

on a need-to-know basis rather than through disciplinary training. Indeed, there was as yet no systematic education for programming and operating the minicomputer.[40] In other words, his practice satisfied Wiener's criteria for trained acquaintance. Like Wiener, through trained acquaintance, Thompson came to embody the traits of multidisciplinary research. Unlike Wiener, however, his ability to bridge disciplines was linked to the specific computer that he programmed and maintained in a borrowed building.

A third characteristic of small science is that the research program, however successful, does not endure. One fruitful project, identified in the original research grant, used the PDP-12 to visualize epileptiform discharges in the exposed human cortex. Epileptic seizures are infrequent, so researchers try to measure events between strikes. Once again the clinical researchers had long been ready to try this out, but Thompson and his computer made it happen.[41] He programmed the PDP-12 to map potential surface spatial and temporal representations of the discharging process using EEG.

Within a decade, other kinds of digital imaging techniques, namely computerized tomography (CT) and magnetic resonance imaging (MRI), took over this mapping, so the scientific work that the EEG team pioneered did not last. The label small science, then, can help us to privilege and investigate such temporally bounded techniques and practices, revealing not enduring concepts but provisional propositions.

Conclusion

What can we take from Christopher Thompson's career at the MNI about the *concept* of small science? What value might it have for historians? Is it only a heuristic, or might it actually pick out characteristic historical practices?

The first value of small science that I have proposed is that it allows historians to make use of Norbert Wiener's notion of trained acquaintance. Historians tend to discuss the highly specialized scientific training of postwar scientists, but for small science – and for interdisciplinary work more broadly – only acquaintance was needed, because the goal was not mastery but a circumscribed though accurate understanding of the operative concepts, instruments, and procedures. Small science, then, provides a framework to help us take stock of scientists with a different kind or degree of training than those with the highly specialized expertise typically thought to characterize postwar science.[42]

To study big science, that is, historians have been led to rearrange the contours and contents of scholarly work. Complex, multiauthored studies, it seems, might be required for the issues of big science in all of their complexity.[43] Price argued that for the historian big science is partly an information retrieval problem, so that historians now need training in bibliometrics.[44] And, more recently, historians have argued that interdisciplinary topics require interdisciplinary teams of specialist historians.[45] In parallel, small science activities such as Thompson's raise the question of whether historians need trained acquaintance in order to deal with postwar topics. What training do historians need to deal with small science? Certain topics or fields of study might require historians to develop trained acquaintance with multiple scientific disciplines in order to have intelligent discussions with scientists, understand the data that they produce, and incorporate those data into historical analyses of change over time.[46]

Overall, then, I have focused on reviving the notion of trained acquaintance. I have also discussed the importance of settings (borrowed buildings) and the idea that paying attention to small science allows us to privilege ephemeral scientific practices. I would like to end by speculating about five other possibilities for historians.

First, small science might be a Canadian phenomenon. Perhaps small science is science at a Canadian scale, done without the resources of big science. If true, then we might say that Canadian researchers now and then participate in big science projects, but normal science in Canada is small science. Small science might limn a nationalistic boundary, pointing to one characteristic manner of scientific education and research. One could argue that normal engineering emerged as big engineering as a form of "nation building" in Canada, as Daniel Macfarlane points out using the example of the St. Lawrence Seaway in his chapter in this volume.[47] I do not need to claim that the term applies only to Canada, only that it might have explanatory power when studying modern science in Canada. I am assuming that there is such a thing as "Canada," not investigating the conditions of its construction.[48]

The challenge might first be to approach this speculation empirically. Did Canada after 1945 have a significant number of scientists working as one-man research teams? Did scientists acquire skills through trained acquaintance? For instance, I have not stressed here the fact that Thompson was an immigrant trained elsewhere. It might be that immigrants to Canada took up jobs and careers associated with small science research as opposed to immigrants to the United States or Europe who became involved in multinational endeavours. Immigrants took up the characteristically Canadian way of doing (small) science.[49]

Second, small science might be a practice specific to clinical medical research. Thompson's case might be generalizable to all clinical research in neurosurgery, simply because the scope of practice is limited. For instance, the number of patients undergoing stereotaxic surgery at the MNI around 1970 was only about twenty a year. This kind of smallness is specific to a number of research areas; not all scientific practices gain in concept or application by becoming big.[50] In other words, the historical trend toward an increase in the scale of scientific practice does not mean that an increased scale is always beneficial or even possible. This category of small science might be especially useful for scholars in science and technology studies whose work is taken up by policy makers and regulators. Such scholars might use the concept of small science to critique unproductive research programs in medicine and elsewhere: that is, those in which the brute force, big science approach has so far paid few dividends.

Third, historians can engage small science by beginning to think about scientific activity circumscribed by and practised within small architecture. Operating rooms are small. Hospital encounters, even between a surgical team and a patient, are intimate. This spatial smallness is operative even if there exists a big number of similar small architectures. For instance, scholarship on laboratories tends to emphasize the singular laboratory building (e.g., the Salk Institute) or the aggregate of all laboratory buildings. Small science can point us to fragments of and partial zones in large buildings, decoupling us from the building or building type as a unit of analysis.[51] Here, too, we might look at the borrowing or repurposing of existing architectural settings for understanding the role of spatial environments in knowledge production. Focusing on small science can help us to interpret architecture in use rather than attend only to the prescriptive symbolic and functional intentions legible in the design and construction of purpose-built structures.

Fourth, small science can point to significant biographical or sociological characteristics of scientists. Just as scientific genius or creativity seems to be a property of persons, so too might a disposition to small science. (And here I would note that, despite Wiener's optimism, trained acquaintance might cause anxiety for trained scientists who expect to encounter expertise in a research team. I am still not sure that I would like to be a patient in an operating room with someone who is not an expert!) Similarly, certain groups of scientists might only be able to undertake research at a small scale (e.g., medical researchers discussed above). Small science can prosper in an era of big science because it has sociological advantages and niches.

Fifth, and finally, the most parsimonious explanation might lie in obscurity. Thompson might simply be a little-known scientist whose work has not been widely acknowledged. In other words, the story of early computer research at the MNI might not be the paradigmatic case study of small science that I have made it out to be but should serve only as a small indication of the immense amount of history of science in Canada yet to be done.

All of these speculations are meant to ward off, on the one hand, the old problem of using case studies to support a generalization about science and, on the other, the new problem of the heteroclite microhistory.[52] Overgeneralization and overspecificity are dangers because Thompson's evocative case is as yet the only empirical study that I have to offer. Still, it would be absurd to insist that a notional concept such as small science must pick out a family of identifiable objects in the world. It is surely so that notional concepts take on a life of their own in scholarly work precisely because they can be used in ways only genealogically related to the initial case study that engendered them. Big science, trained acquaintance, boundary objects, trading zones – all find their utility as scholarly concepts and not as facts about the world. Rather, they are modes for writing history. Likewise, small science is a matter of historiographical emphasis and orientation. By using it, scholars can emphasize space (borrowed buildings) rather than time, practice (trained acquaintance) rather than language, and objects (the PDP-12) rather than ideas. Recalling Wiener's claims about territory, we might say that small science both signals an overlooked landscape of scientific practices and provides the map to navigate it.

NOTES

Research for this chapter was supported by the Social Sciences and Humanities Research Council of Canada and the Pierre Elliott Trudeau Foundation. I would like to thank Annmarie Adams, Pierre-Gerlier Forest, Morgan Matheson, Christopher Thompson, and the editors of this volume for their comments, suggestions, help, and support.

1 See William Feindel and Richard Leblanc, eds., *The Wounded Brain Healed: The Golden Age of the Montreal Neurological Institute, 1934–1984* (Montreal/Kingston: McGill-Queen's University Press, 2016); and William Feindel, "Historical Vignette: The Montreal Neurological Institute," *Journal of Neurosurgery* 75, 5 (1991): 821–22.
2 A copy of the Medical Research Council of Canada (MRC) Computer Grant (ME-3822, typescript, 1969) is in the private collection of Christopher Thompson. The PDP computer line has a close connection to the history of postwar science in Canada; see Gordon Bell, "STARS: Rise and Fall of Minicomputers (Scanning Our Past)," *Proceedings of the IEEE* 102, 4 (2014): 629–38, http://ethw.org/index.php?-title=Rise_and_Fall_of_Minicomputers&oldid=112936: "In 1963, DEC built the PDP-5 to interface with a nuclear reactor at Canada's Chalk River Laboratories. The design came from a requirement to connect sensors and control registers to stabilize the reactor, whose main control computer was a PDP-1."
3 MRC Computer Grant, 1969.
4 Undated memo from William Feindel to P. Gloor, photostat, private collection of Christopher Thompson.
5 C.J. Thompson, "Activation Analysis with an On-Line PDP-9 Computer," *Nuclear Applications* 6, 6 (1969): 559–66.
6 R.W. Tolmie and C.J. Thompson, "Mobile Equipment for Combined Neutron Activation and X-Ray Fluorescence Analysis," in *Proceedings of the IAEA Symposium on Nuclear Techniques for Mineral Exploration and Exploitation*, 1969 (Krakow, Poland: IAEA Vienna, 1971), 7–26.
7 C.J. Thompson and G. Bertrand, "A Computer Program to Aid the Neurosurgeon to Locate Probes Used during Stereotaxic Surgery on Deep Cerebral Structures," *Computer Programs in Biomedicine* 2, 4 (1972): 265–76; G. Bertrand, A. Olivier, and C.J. Thompson, "Computer Display of Stereotaxic Brain Maps and Probe Tracts," *Acta Neurochirurgica: Supplementum* 21 (1974): 235–43.
8 A 1967 film showing Gilles Bertrand performing stereotaxic surgery has recently been rediscovered at Library and Archives Canada by Steven Palmer. The film was shown in the Meditheatre at the Man and His Health pavilion during Expo 67. Robert Cordier, dir., *Miracles in Modern Medicine/Miracles de la médecine moderne*, 1967 (eighteen minutes).
9 An early use of the term "big science" is in Alvin M. Weinberg, "Impact of Large-Scale Science on the United States," *Science* 134, 3473 (1961): 161–64. For historical

studies of big science, see J.H. Capshew and K.A. Rader, "Big Science: Price to the Present," *Osiris* 7, 1 (1992): 3–25; J.L. Heilbron, "Creativity and Big Science," *Physics Today* 45, 11 (1992): 42–47; and Peter Galison and Bruce Hevly, eds., *Big Science: The Growth of Large-Scale Research* (Stanford, CA: Stanford University Press, 1992), especially Hevly's afterword, "Reflections on Big Science and Big History," 355–63.
10   Norbert Wiener, *Cybernetics: Or, Control and Communication in the Animal and Machine* (Cambridge, MA: MIT Press, 1948), 25.
11   So-called big physics is a canonical example of big science; see Peter Galison, *Image and Logic: A Material Culture of Microphysics* (Chicago: University of Chicago Press, 1997).
12   Derek John de Solla Price, *Little Science, Big Science* (New York: Columbia University Press, 1963).
13   For instance, early on, the social, ethical, and legal issues raised by genome research became part of the research itself; see Daniel J. Kevles and Leroy Hood, eds., *The Code of Codes: Scientific and Social Issues in the Genome Project* (Cambridge, MA: Harvard University Press, 1992).
14   But compare John Robert Christianson, *On Tycho's Island: Tycho Brahe and His Assistants, 1570–1601* (Cambridge, UK: Cambridge University Press, 2000), who argues that big science was a feature of early modernity.
15   The impending disappearance of artisanal, small science is a trope of modern debates about scientific funding; see, for example, the editorial by Bruce Alberts, "The End of 'Small Science'?," *Science* 337, 6102 (2012): 1583.
16   H.M. Collins, "LIGO Becomes Big Science," *Historical Studies in the Physical and Biological Sciences* 33, 2 (2003): 264.
17   Bart Penders, Niki Vermeulen, and John N. Parker, eds., *Collaboration in the New Life Sciences* (London: Ashgate, 2012).
18   The example is drawn from Jeff Hughes, *The Manhattan Project: Big Science and the Atom Bomb* (New York: Columbia University Press, 2002), 1–7.
19   See, for example, National Research Council, *A Space Physics Paradox: Why Has Increased Funding Been Accompanied by Decreased Effectiveness in the Conduct of Space Physics Research?* (Washington, DC: National Academies Press, 1994), 13–14.
20   Wiener, *Cybernetics*. Wiener published an expanded second edition in 1961.
21   Ibid., 9.
22   Ibid., 8.
23   Ibid., 38.
24   Ibid., 8.
25   See Peter Galison, "Trading Zones: Coordinating Action and Belief," in *The Science Studies Reader*, ed. Mario Biagioli (London: Routledge, 1999), 137–60. Thank you to Edward Jones-Imhotep for this suggestion.
26   Wiener, *Cybernetics*, 8–9.
27   Interdisciplinary configurations occur in Galison, *Image and Logic*, 803–84; Michael E. Gorman, *Trading Zones and Interactional Expertise: Creating New Kinds of Collaboration* (Cambridge, MA: MIT Press, 2010); Michael E. Gorman, "Levels of Expertise and Trading Zones: A Framework for Multidisciplinary Collaboration," *Social Studies of Science* 32, 5–6 (2002): 933–42; H.M. Collins and R.J. Evans, "The

Third Wave of Science Studies: Studies of Expertise and Experience," *Social Studies of Sciences* 32, 2 (2002): 235–96; and S.L. Star and J.R. Griesemer, "Institutional Ecology, 'Translations,' and Boundary Objects: Amateurs and Professionals in Berkeley's Museum of Vertebrate Zoology, 1907–39," *Social Studies of Science* 19, 3 (1989): 387–420. Ilana Loewy has written extensively on boundary-work in medical sciences in "The Strength of Loose Concepts: Boundary Concepts, Federative Experimental Strategies, and Disciplinary Growth: The Case of Immunology," *History of Science* 30, 4 (1992): 371–439, and in "Historiography of Biomedicine: 'Bio,' 'Medicine,' and in Between," *Isis* 102, 1 (2011): 116–22.

28   Galison, *Image and Logic*, 553.

29   Rem Koolhaas and Bruce Mau, *S,M,L,XL*, ed. Jennifer Sigler (New York: Monacelli Press, 1995). On the book's importance, see Alicia Imperiale and Enrique Ramirez, eds., *#SMLXL*, special issue of *Journal of Architectural Education* 69, 2 (2015).

30   Wiener, *Cybernetics*, 31–32, noted the important work of Walter McCulloch and Walter Pitts on neural networks in the development of cybernetics.

31   See ibid., 26–27; and Steve J. Heims, *Constructing a Social Science for Postwar America: The Cybernetics Group, 1946–1953* (Cambridge, MA: MIT Press, 1993).

32   Urban planner Jacqueline Tyrwhitt, who helped to set up the graduate program at the University of Toronto, was a key node in the network. See Mark Wigley, "Network Fever," *Grey Room* 4 (2001): 82–122. Nam June Paik remarked on the confluence of ideas between Wiener and McLuhan in his acceptance speech for the 1998 Kyoto Prize in Arts and Philosophy entitled "Norbert Wiener and Marshall McLuhan: Communication Revolution."

33   Throughout *Image and Logic*, for instance, Galison argues that physics laboratories learned from the integrated factory techniques of industrial production. For data on interdisciplinarity in twentieth-century science, see Vincent Larivière and Yves Gingras, "Measuring Interdisciplinarity," in *Beyond Bibliometrics: Harnessing Multidimensional Indicators of Scholarly Impact*, ed. Blaise Cronin and Cassidy R. Sugimoto (Cambridge, MA: MIT Press, 2014), 187–200.

34   I borrow the term from Annmarie Adams, "Borrowed Buildings: Canada's Temporary Hospitals during World War I," *Canadian Bulletin of Medical History* 16, 1 (1999): 25–48.

35   On the architecture of the MNI, see William Feindel and Annmarie Adams, "Building the Institute," in Feindel and Leblanc, *The Wounded Brain Healed*, 441–58. See also "Montreal Neurological Institute," *Royal Architectural Institute of Canada Journal* 11, 10 (1934): 140–45. On the firm's prolific production, see Jacques Lachapelle, *Le fantasme métropolitain: L'architecture de Ross et Macdonald: Bureaux, magasins, et hôtels, 1905–1942* (Montréal: Presses de l'Université de Montréal, 2001).

36   Digital imaging spawned a new industry of medical equipment manufacturing and sales, and digital imaging requires the coordinated teamwork of hospital specialists, which, taken together, might constitute an instance of big science. See Joseph Dumit, *Picturing Personhood: Brain Scans and Biomedical Identity* (Princeton, NJ: Princeton University Press, 2004).

37   MRC Computer Grant, 1969.

38 Scholars have shown that the arrival of Internet-based communication tools and repositories warps the category of big science (i.e., when does an interconnected network of small scientists amount to big science?). See Leah A. Lievrouw, "Social Media and the Production of Knowledge: A Return to Little Science?," *Social Epistemology* 24, 3 (2010): 219–37. However, in these early years at the MNI, the minicomputer was a stand-alone installation, unconnected to any computer network.

39 For concise histories of computing and the minicomputer, see Martin Campbell-Kelly and William Aspray, *Computer: A History of the Information Machine* (New York: Basic Books, 1996); and Paul Ceruzzi, *A History of Modern Computing*, 2nd ed. (Cambridge, MA: MIT Press, 2003).

40 In Canada, universities first established computer science departments around 1964. For at least another decade, however, it was unclear whether computer science referred to designing computers, writing programs, or simply using computational methods. See Zbigniew Stachniak and Scott M. Campbell, *Computing in Canada: Building a Digital Future* (Ottawa: Canada Science and Technology Museum, 2009), 40–43.

41 J. Gotman, D.R. Skuce, C.J. Thompson, P. Gloor, J.R. Ives, and W.F. Ray, "Clinical Applications of Spectral Analysis and Extractions of Features from EEGs with Slow Waves in Adult Patients," *Electroencephalography and Clinical Neurophysiology* 35, 3 (1973): 225–35.

42 The notion of increasing specialization is common enough that one genre of scientific writing is the screed against specialization. See, for example, Rogers Hollingsworth, "The Snare of Specialization," *Bulletin of the Atomic Scientists* 40, 6 (June–July 1984): 34–37.

43 See Babak Ashrafi, "Big History?," in *Positioning the History of Science*, ed. Kostas Gavroglu and Jürgen Renn (Dordrecht: Springer, 2007), 7–11.

44 See Jonathan Furner, "Little Book, Big Book: Before and after *Little Science, Big Science*: A Review Article," *Journal of Librarianship and Information Science* 35, 2 (2003): 115–25.

45 See, for example, Paul Erickson, Judy L. Klein, Lorraine Daston, Rebecca Lemov, Thomas Sturm, and Michael D. Gordin, *How Reason Almost Lost Its Mind: The Strange Career of Cold War Rationality* (Chicago: University of Chicago Press, 2013).

46 See Yves Gingras, "Existe-t-il des chercheurs multidisciplinés?," in *Par-delà les frontières disciplinaires: L'interdisciplinarité: Actes de colloques* (Montréal: n.p., 1998), 65–73. I thank Tina Adcock for the suggestion that this might be the case for environmental historians.

47 Marco Adria makes a concomitant argument that the growth of industrialism (including big engineering projects) was a condition for the rise of nationalism. Marco Adria, *Technology and Nationalism* (Montreal/Kingston: McGill-Queen's University Press, 2010).

48 I thank an anonymous reviewer for insisting that this speculation is empty unless I can articulate my understanding of "Canada." I am looking for historiographical

utility, however, so I see the question of "whose Canada?" as an outcome and not a precondition.

49 For studies of immigration and science in Canada, see Sasha Mullally and David Wright, "La grande séduction? The Immigration of Foreign-Trained Physicians to Canada, c. 1954–76," *Journal of Canadian Studies* 41, 3 (2007): 67–89; and Laurence Monnais and David Wright, eds., *Doctors beyond Borders: The Transnational Migration of Physicians in the Twentieth Century* (Toronto: University of Toronto Press, 2016).

50 Karin Knorr Cetina has argued that biology is an individual, noncollaborative science, tied to the lab bench; see *Epistemic Cultures: How the Sciences Make Knowledge* (Cambridge, MA: Harvard University Press, 1999), 216–40. But others chart a movement from *in vitro* to *in silico* biology, in which research is structured as computationally mediated collaboration. See Hallam Stevens, *Life Out of Sequence: A Data-Driven History of Bioinformatics* (Chicago: University of Chicago Press, 2013); and Alberto Cambrosio and Peter Keating, *Biomedical Platforms: Realigning the Normal and the Pathological in Late-Twentieth-Century Medicine* (Cambridge, MA: MIT Press, 2003).

51 Historians of science looking at architecture tend toward prescriptivism or exceptionalism. See, respectively, William J. Rankin, "The Epistemology of the Suburbs: Knowledge, Production, and Corporate Laboratory Design," *Critical Inquiry* 36, 4 (2010): 771–806; and Scott G. Knowles and Stuart W. Leslie, "'Industrial Versailles': Eero Saarinen's Corporate Campuses for GM, IBM, and AT&T," *Isis* 92, 1 (2001): 1–33.

52 See Lorraine Daston, "Science Studies and the History of Science," *Critical Inquiry* 35, 4 (2009): 798–813.

8

Paris–Montreal–Babylon
The Modernist Genealogies of Gerald Bull

EDWARD JONES-IMHOTEP

Gerald Bull spent much of his life trying to collapse the ambiguity of an object. Not long after his release from a US prison in late 1985, the former McGill University engineering professor turned international arms dealer sat down in a small Spanish villa to write two redemptive histories: a thousand-page autobiography designed to exonerate him of his alleged crimes (illegal weapons sales to apartheid South Africa) and a technical genealogy that linked his controversial ballistics research to an icon of early twentieth-century modernity, the Paris Guns.[1] The German long-range guns that shelled Paris in the spring of 1918 were marvels of engineering and icons of scientific and industrial power. Deployed as weapons of psychological warfare against "the capital of modernity," their shells were the first human-made objects to reach outer space.[2] For five months, they bombarded the French capital from over 120 kilometres away before disappearing in August, almost without a trace.[3] Bull saw them as antecedents of his own ambiguous inventions – gargantuan cannons that straddled the line between scientific instrumentation and illicit weaponry. Built in the early 1960s and operated on the island of Barbados as part of McGill University's High Altitude Research Project (HARP), Bull's inventions were considered an impossibility by many: guns so powerful that they could fire instrumented projectiles into the stratosphere from the surface of the Earth.[4] All too conscious of his own weapons-dealing past, Bull ironically sought to connect HARP to the Paris Guns in order to *sever* their military connotations. He presented his devices as innovative and imaginative research instruments, unique (he claimed) precisely because they turned the technical sophistication and supreme violence of modern artillery toward peaceful, scientific ends.[5] Government teams had descended on the cannons in 1967, intent on dismantling and destroying them, erasing any evidence of their existence, but Bull had managed to save them as artifacts. Almost twenty years later, he still worried deeply about their place in a historical consciousness that would struggle to make sense of them. State of the art for their time, the huge cannons seemed to have come from nowhere. And so he sought to give them a genealogy. "Otherwise," he explained, "these enormous guns ... stand in isolation, dangling in history."[6]

Bull was an anomaly within Cold War Canada's interlocking worlds of academic research, government patronage, and military ambition. His fall from grace in 1980 was nearly as spectacular as his meteoric rise. The youngest doctorate earner in Canadian history in 1951, he would become the only foreigner after the Marquis de Lafayette and Winston Churchill to receive US citizenship through a special act of Congress. Hailed as a boy genius, courted by foreign governments, fawned over by ballistics experts and weapons engineers, he was a constant thorn in the side of Canadian government officials and the "cocktail scientists" he saw as their lackeys.[7] For all his anomalous qualities, though, Bull imbued his most spectacular projects – including his dream of a satellite-launching supergun – with typically modern concerns about progress, time, and memory. His attempts to redeem his inventions and himself while living as an expat in Belgium mobilized a broadly modern anxiety about loss and erasure.[8] And his attempt to create launch technologies for nonsuperpower nations, including Canada and Iraq, confronted widespread, late-modern anxieties about falling behind. Bull's story illustrates the crucial historical and geographical connections between Canadian science and technology and the broader, nation-based global anxieties of the modern age.

Bull not only cast his inventions, retrospectively, as part of a lost genealogy.[9] Throughout his career, he repeatedly presented them as part of a compelling *modernist* vision – a technological expression, built around space activities and weapons infrastructures, of what it meant to be an innovative, forward-thinking, contemporary nation in the middle decades of the twentieth century.[10] HARP was part of an increasingly stylized scientific and technological response in Canada to how middle-power nations could remain relevant and influential in an age of superpower rivalries. Technologically, that response took the form of a focus on niche programs (often focused on Canada's northern geography) and on mid-scale technoscientific projects that would give the country technological platforms to keep up with the superpowers.[11] In Bull's case in particular, that response represented an alternative to the high modernist orthodoxies of space activities and other high-profile megaprojects in the mid-twentieth century.[12] In place of expensive, large-scale, technologically complex systems, Bull sought inexpensive, flexible, small-scale, and comparatively simple infrastructures that would keep the "other" nations of the world from falling behind.[13] Resonating with parallel "middle-power" initiatives in Canada, it was an expression, through engineering technologies and their political representations, of how nations could remain (or become) modern.[14] And it was in pursuit of that vision that, on the eve of the Gulf War, Bull would repackage his Canadian-conceived vision of gun-launched satellites and space probes as part of a very different modernization project: Saddam Hussein's Project Babylon – a pact that would ultimately lead to Bull's assassination outside his Brussels apartment at the hands of foreign agents in March 1990.[15]

This chapter explores the key claim that underlay Bull's modernist vision: his contention that his most powerful cannons were not weapons at all, but scientific instruments. I argue that this claim was made both possible and implausible by a set of specific historical conditions, namely, the intersection of three broader histories that came together for a limited time and in particular ways from about 1950 to 1990: first, a conceptual ambiguity about projectiles and what kinds of objects they represented; second, a legal, political, and cultural uncertainty about islands and the activities that take place there; and third, a philosophical ambiguity about rationality and reason that achieved a particularly tortured state during the Cold War. The specifics of those histories, rather than the abstraction of Bull's linear genealogy – from the Paris Guns, to HARP, to Project Babylon – are what I examine here. Through them, I argue that the historical conditions that allowed Bull to plausibly claim that his superguns were peaceful instruments in the 1960s, at the height of the Cold War, also facilitated weapons designs and arms sales that undermined those claims in the 1970s and 1980s and that complicated his attempts to separate his technical consulting from outright weapons dealing. It was his continued espousal of a highly specific Cold War rationality, long after it had collapsed, that ultimately made his declarations both implausible and deadly. If the title of this chapter reads like the itinerary for an abstract idea – Paris, Montreal, Babylon – then this alternative trajectory – projectiles, islands, reason – will help to make sense of the complex historical conditions that made it possible.

Projectiles

Gerald Bull was born in North Bay, Ontario, in 1928, the ninth of ten children. His mother, Gertrude, died from complications following the birth of her tenth child in 1931. Her death, and the events it precipitated, would be a source of deep tension and resentment within Bull's immediate and extended family.[16] Shortly afterward, his father, George, suffered a mental breakdown and moved the remaining family to Trenton, where George's sister, Laura, took primary care of the children. After she died of cancer the following year, Bull's father remarried and largely abandoned his children to return to Toronto. Gerald and his siblings were raised together by their eldest sister, Bernice, in Sharbot Lake (north of Kingston) until the summer of 1935, when Gerald spent the summer with his maternal aunt and uncle, Edith and Philip LaBrosse, on their farm near Kingston. Gerald pleaded to stay permanently. Having won a small fortune in the Irish Sweepstakes, the LaBrosses enrolled Gerald in the private Jesuit boarding school, Regiopolis College, in Kingston. Upon his graduation, the only school that would immediately enrol the sixteen year old was the University of Toronto's Department of Aeronautical Engineering. After graduating and starting his doctoral studies in 1948, Bull began designing, building, and operating experimental wind tunnels, including the state-of-the-art facility at Downsview Airport, north of Toronto, in 1950. He received his PhD the following year in a special "fast-track" deal that sent him to work for the Canadian Armament Research Establishment (CARDE) in Valcartier, Quebec.

As his doctoral studies suggest, Bull came to his large guns not through artillery or ballistics but through aerodynamics. CARDE had recruited him to test the aerodynamic properties of a supersonic missile, the Velvet Glove, being designed by Canada as a defence against Soviet bombers.[17] Since Canada lacked a wind tunnel large enough for the missile, Bull sought to invert the testing problem. Rather than accelerating air to supersonic speeds and having it flow over a static missile body (as a wind tunnel would), he decided to accelerate the missile by firing it from a cannon.[18] Paper screens and high-speed photography registered its in-flight behaviour. A research problem in aerodynamics (the physics of air moving over a solid body) therefore became a problem of ballistics (the science and technology of projectiles). In this first instance, Bull saw paper screens, high-speed cameras, and the guns that would become his livelihood as the comparatively low-cost instruments of advanced engineering science. The rest of his career would trade on a characteristically modern ambivalence between guns and projectiles as research tools, on one hand, and their potential as weapons, on the other.

That ambivalence had not always existed. For centuries stretching back to the Scientific Revolution, the status of weapons as simultaneously theoretical and experimental objects had been both intellectually compelling and morally unproblematic.[19] Since the sixteenth century, gun-launched projectiles formed one of a trinity of natural philosophical objects that included pendula and heavenly bodies.[20] When Galileo (who taught ballistics at Padua) worked out his Law of Fall – the first law of classical physics – at the turn of the seventeenth century, he made the motions of both real and imagined cannonballs a centrepiece of his work.[21] Newton, while pondering (between 1684 and 1686) how to present the theory of gravitation in the *Principia Mathematica*, considered a lead shot propelled by larger and larger charges of gunpowder until it circled the Earth.[22] Ernst Mach, the famous Austrian positivist philosopher and physicist, and a major influence on the young Albert Einstein, worked out his influential ideas on supersonic flight in 1886 by photographing the shock wave produced by a bullet.[23] Even explorations of the upper atmosphere (of the kind that Bull would conduct) prominently featured firearms. In 1934, Dorothy Fisk's book *Exploring the Upper Atmosphere* sketched the regions of the atmosphere and the devices used to explore them. Alongside Mount Everest, weather balloons, and noctilucent clouds, Fisk located a shell from the Paris Guns, the same gargantuan cannons that Bull would weave into his fantastical genealogy of upper atmospheric research (see Figure 8.1).

Fisk was already writing in a period that felt compelled to draw sharper boundaries between scientific objects and military weapons. She questioned the place of weapons such as the Paris Guns in the history of science, even as she built enthusiastically on ballistics as a thought-thing. With the artillery barrages of the First World War still echoing culturally and politically, she hedged that, "if *for scientific purposes* we wanted to bombard the moon, the first thing to do would be to build a gun with a muzzle velocity of 7 miles per second," the escape velocity of the Earth. She then went on to calculate the velocity at which a shell fired from that gun would land on the lunar surface; for the sake of symmetry, she turned to the case of "a strange professor" who decided, again for scientific purposes, to bombard the Earth in return, saying that the Paris Guns would in fact produce the necessary muzzle velocity. Extending the example to mutual bombardment between

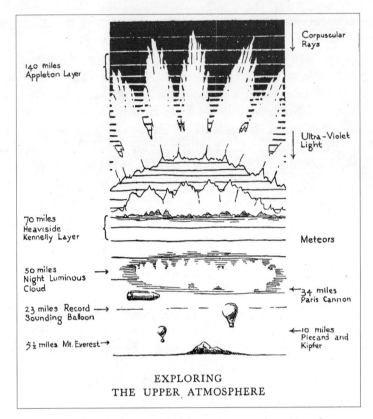

Figure 8.1 Dorothy Fisk's portrayal of the upper atmosphere. Note the shell from the Paris Guns at thirty-four miles. | Dorothy Fisk, *Exploring the Upper Atmosphere* (London: Faber and Faber, 1934), frontispiece. Illustrated by Leonard Starbuck.

the Earth and Mars, she explained how Martian engineers would have an easier task in constructing their gun because of the planet's lower escape velocity.[24]

Fisk's discussion traded both on the centrality of war to the modern imaginary and on the increasingly anxious attempts to "purify" science in the twentieth century, to distinguish it from technological applications and to cast both its ethos and its endeavours as something ethically opposed to military operations and warfare.[25] Even as her discussion questioned the scientific status of the Paris Guns, it placed them within a longer trajectory of firearms and projectiles as crucial thought-things for the fabric of modern science: Galileo working on questions of ballistics in the newly fortified

Padua; Newton using cannons to develop gravitational theory with the English Civil War still in living memory; Mach being drawn into a debate over whether the French had used illegal explosive bullets during the Franco-Prussian War; and Fisk invoking the Paris Guns barely a decade after the cataclysm of the First World War, its science-based warfare of machine guns and poison gas, and its inauguration of modern shell shock.[26]

Bull's work came to exemplify both the tension and the perceived potential of those complex associations. His disdain for government scientists (whom Bull saw as unimaginative yes-men) and his disregard for authority angered and antagonized officials in Ottawa. That disdain would only grow after Canadian officials killed off his dangerously ambiguous cannons. But his polyvalent ideas and skills helped him to rise quickly through the ranks at CARDE, where he became a section head in the Aeroballistics Section in July 1953.[27] In that rise, Bull saw his large-calibre guns as tools to achieve scientific and defence objectives simultaneously. Their ambiguity was in keeping with the studied ambiguity of the organization that employed him. The Defence Research Board (DRB), which oversaw CARDE and a number of other defence research establishments specializing in everything from cold weather science to upper atmospheric research, had been created in 1947 precisely to straddle the line between pure and applied research, conducting "basic" science in anticipation of military needs.[28] Under its aegis, Bull's aerodynamics testing soon morphed into a project (disguised as micrometeorite research) to use cannons as giant shotguns, skeet-shooting intercontinental ballistic missiles (ICBMs) as they bore down on North American cities.[29] In an article published in the *Montreal Star*, Bull discussed research on instrumented projectiles ("satellites") capable of radioing back telemetry and scientific readings from the upper atmosphere and near space; more clandestinely, the projectiles would study infrared radiation at the edge of the atmosphere and reveal information about warhead heating during re-entry.[30] But it was the vision of firing projectiles into orbit from the Earth's surface, one of the iconic thought experiments of classical physics, that became Bull's obsession. Disillusioned by the freeze in defence spending announced in 1960 and courted by McGill University for several years, Bull left CARDE to join McGill's Department of Mechanical Engineering in 1961 as a full professor. There Dean of Engineering Don Mordell encouraged Bull to develop large guns that might shoot objects into space.[31]

Bull's idea of developing cannons for space research was born partly from an archetypal modern anxiety about being left behind – one that Bull would exploit in pitching his projects first to the Canadian government and later to

Iraq. He saw cannon-fired space instruments as a way to bring space technology to nonsuperpower nations, such as Canada, that lacked ICBM programs and were anxious about being excluded from space activities. In the late 1950s, officials in the DRB had stressed that Canada could quickly be outpaced by the technological achievements of the superpowers, and they went so far as to advocate the construction of an international launch facility in Churchill, Manitoba.[32] Bull's initial idea for his gun-launched probes was to mount a powerful cannon in the nose of a rocket and have it fire a payload at a critical moment in its trajectory. Over the course of 1961–62, he instead came to see the gun barrel as a reusable first-stage booster for the rocket as well as a guidance and control system.[33] One potential benefit of the new system would be its low cost. But guns were attractive for two other qualities that stemmed from how they inverted the physics of liftoff. Unlike conventional launches from platforms, gun-launched rockets were accelerated under incredible forces, leaving the barrel at high speeds. As a result, they were not affected by surface winds that could throw conventional rockets off course as they slowly lifted off the launchpad. The dispersion area of a series of projectiles – their spatial distribution around a given target – was therefore comparatively small. This meant, first, that objects could be "placed" with great accuracy in the upper atmosphere and, second, that the "fallout" area of the launch debris – the wooden *sabots* that surrounded the projectiles, maximizing barrel pressures, as well as the various stages of the projectiles themselves – was restricted, making launches safer and security breaches less likely. As Bull explained it, "vehicle dispersion (in the case where there is no in-flight rocket-boost) can be closely controlled to both a predicted point in space, as well as impact into a relatively confined area." In this way, the technique resembled anti-aircraft fire much more closely than conventional rocket launches or artillery bombardments.[34] Its main drawback was that it generated enormous accelerations (requiring testing up to 10,000 g's) that threatened the delicate scientific instrumentation of the launched probes.[35]

In representing those qualities, Bull turned precisely to the kinds of abstractions that characterized the long history of projectiles in physics: disembodied parabolic curves and ranging graphs (see Figure 8.2). They make it clear that his interest in the 1960s was not in questions about shelling or bombardment but in questions about the detritus of launch, whose trajectories were also described by the laws of ballistics. The ambiguity suggested by these diagrams, though, would eventually attract fierce criticisms. Already in the mid-1960s, the diagrams suggested the long game: the

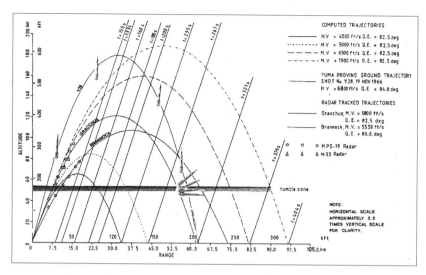

Figure 8.2 Flight envelopes for Bull's gun-launched Martlet "missiles." The potential ground impacts (shown most clearly for the curve Y28) were meant to illustrate the "fallout" from a launch. Bull's critics would later use such images to argue for the military character of HARP. | © 1988 *Paris Kanonen – the Paris Guns (Wilhelmgeschütze) and Project HARP* – Gerald V. Bull.

parabola extended ever outward until the thought experiments of Newton and the modernist fantasies of Fisk became concrete possibilities. For Bull, who acknowledged Jules Verne as a childhood inspiration and who once envisioned firing a man from a cannon and having him land on the moon, the cannons themselves were neutral; their nature was in their use.[36] As much as Bull dealt in pedestrian applications, even as he sold smaller long-range artillery to South Africa, Israel, and Iraq, these large guns remained for him scientific objects. His depictions of them would take on the brutalist, monumental, monolithic qualities of modern architecture. And it was the dream of finally building them that inspired him, with the support of the US military and later the Canadian defence establishment, to set up operations on the island of Barbados.

Islands

Like projectiles, islands have figured prominently in the intertwined histories of imperialism, technology, and science.[37] Starting in the mid-nineteenth century, their geographic and political status underwent a series of transformations that gave them an ambiguity central to Bull's story. British global power made islands central to naval strategy in the nineteenth century, in which they played at least two vital roles. The first was as coaling stations for refuelling, repairs, and trade route protection. The second was as points of landfall for an increasingly global network of underwater telegraph cables designed to avoid cable cutting and unreliable regimes. From the 1860s on, the United States adopted this model, acquiring a series of island territories that remain in its possession to this day – Samoa, Pearl Harbor, Guantanamo Bay, Puerto Rico, and the Virgin Islands.[38] As the United States expanded its possessions during the Second World War, acquiring islands and entire archipelagos from Japan and Germany, it transformed them into "stepping stones," sophisticated staging areas and forward-operating bases for military operations.[39] Immediately following the war, they were gathered into vast zones, such as the Pacific Proving Grounds, helping to shape a new militarized geography of oceans in which technologies played crucial roles.[40] US nuclear testing was used to justify the forced removal and relocation of Pacific Islanders in order to create laboratories out of islands such as the Johnston, Kwajalein, and Bikini Atolls. Britain depopulated Diego Garcia to transform it into part of a global geodetic network, a project that would underwrite future ground stations for the Echelon spy network (in which Canada participates), NASA's Mercury program, and the current GPS satellite system. Through official policies including the "Strategic Island Concept," which sought to acquire islands, like Diego Garcia, with specific qualities (strategic locations, small populations, isolated, and controlled by friendly powers), the United States moved to stockpile base rights before independence movements took hold in the 1960s.[41] Already within a decade of the Second World War, as Ruth Oldenziel has noted, islands had become central for the projection of America as an empire without colonies, able to exert its power globally while denying any formal foreign possessions on the model of the older European powers.[42]

Although the Pacific was the most prominent site for these postwar developments, the Atlantic and the Caribbean provided influential precedents and parallels. The contentious 1889 legal case involving Navassa, one of the famed Caribbean guano islands, helped the United States to develop the legal framework that cast its island possessions as liminal entities "belonging to" but not "part of" the United States.[43] Agreements during the Second World War granted the United States ninety-nine-year, rent-free leases of British ports in the North Atlantic and Caribbean, giving it control over British global communication lines.[44] As in the Pacific, these communication systems were instrumental in weaving together local geographies of Atlantic and Caribbean islands around the advanced technological

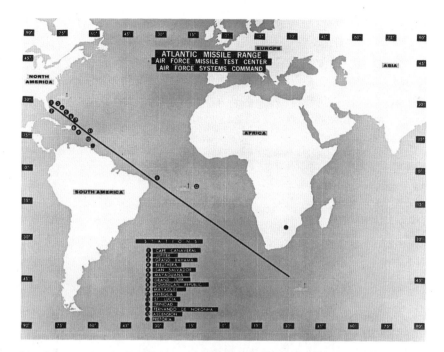

Figure 8.3 The Atlantic Missile Range. Created in 1949 and extended in 1952, the range formed a 5,000-mile test corridor for missiles launched from Cape Canaveral (1). Lying just southeast of St. Lucia (10), Barbados and the HARP site were integrated into the range's radar network and tracking stations. | State Library and Archives of Florida, Shelf Number 16024, Image Number RC04859.

and weapon systems of the Cold War. When the United States under Harry Truman was looking for a place to flight-test its missiles, it initially considered firing them along the Aleutian Islands, but the trajectory was considered too cold and remote. Officials thought about a range along the Baja Peninsula but scrapped the plan when a stray missile from White Sands landed in a cemetery in Juarez, Mexico. They finally settled on Cape Canaveral as a launch head for missiles overflying the Bahamas and out to Ascension Island. Knitting together a radar net strung across auxiliary Air Force bases that included former British possessions such as Grand Turk, St. Lucia, and Ascension, Truman signed a law in 1949 creating the Atlantic Missile Range – a 5,000-mile missile corridor stretching from the launch head at Cape Canaveral over the Bahamas, across the Lesser Antilles, and into the South Atlantic, with its web of radar stations and ship-based tracking equipment straddling Barbados (see Figure 8.3).

Bull and McGill University chose Barbados precisely because of its place within both the larger technopolitical geography of the Cold War and the broader historical geography of islands. They had originally considered a site in northern Quebec, but Bull feared the location would put the project within reach of his political enemies in the defence establishment.[45] The Barbados launch site was the latest in a series of research outposts that McGill had established on the island starting in 1954, when a British naval commander, Carlyon Bellairs, bequeathed his Barbados estate to the university as a centre for the scholarly study of the Tropics, creating the Bellairs Research Institute.[46] McGill's geography department used the island facilities for climatology and geomorphology and later set up a Tropical Research Laboratory run by the geographer Theo Hill. Along with "field stations" in the Canadian North, particularly those set up and run by Kenneth Hare under the Arctic Meteorology group with DRB funding, McGill's stations in Barbados connected research on the Tropics and the Arctic.[47] Early on, its researchers recapitulated a century-old maxim that stressed the privileged epistemic status of the Tropics, which (they claimed) illustrated certain natural processes, such as competition, more clearly than temperate areas.[48] The High Altitude Research Project was set up in 1962; its sixteen-inch, fifty-calibre guns, taken from US Navy surplus, were positioned on the beach below Seawell Airport, on the southeastern coast of the island. As Bull's collaborator, Charles Murphy, explained, the site was selected partly because of its near-equatorial location, where hurricanes, tropical winds, and ionospheric effects made it an interesting site for broad-scale atmospheric research. But the site's proximity to "the surrounding Eastern Test Range Facilities (Antigua, Trinidad, Ascension Island, etc.)" also integrated it into a radar and telemetry infrastructure that provided tracking for HARP's firings (see Figure 8.4).[49] As a joint program with the US military, the site operated through the Army Field Office under close coordination with Cape Canaveral.[50] Bull's research helped to further integrate Barbados into this Cold War geography of islands. HARP's activities were technically spread over three sites: Bull's ingenious Highwater compound, which straddled Quebec and Vermont, allowing Bull to strategically bypass specific weapons export and security restrictions; the HARP installation in Barbados; and an essentially identical facility at the Yuma Proving Grounds in the Sonoran Desert of Arizona. This last site was one of the largest military installations in the world and shared a number of features with militarized tropical islands – "emptiness," good weather, a large fallout area – that made it an ideal training ground for South Pacific operations during the Second

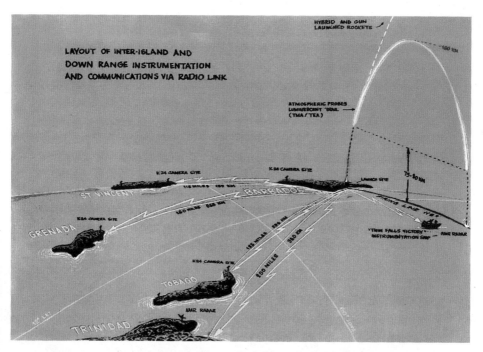

Figure 8.4 Layout of inter-island and down-range instrumentation and communications via radio link. The illustration depicts the strategic location of Barbados within a larger network of surrounding camera sites, radar installations, and telemetry ships. | © 1988 *Paris Kanonen – the Paris Guns (Wilhelmgeschütze) and Project HARP* – Gerald V. Bull.

World War.[51] In discussing launch results, Bull often fused the Barbados and Yuma sites, treating them interchangeably.[52] As part of the US Army's larger program in geophysical research, and McGill University's own research projects in the Arctic and Subarctic, Bull's launch site completed the triangle of Cold War hostile regions – Arctic, desert, and Tropics – linking it to both the existing Cold War geography of islands and the emerging hostile region of near space.[53]

The strategic appeal of islands in the second half of the twentieth century depended partly on the perception of them as politically docile territories, a portrayal aided by often weak political systems, small colonial or (recently) postcolonial populations, low environmental standards, and improvised labour laws.[54] Technical considerations aside, Bull was especially drawn to Barbados because of what he saw as the nation's deferential support for his research, an island of calm set against the political battles and predatory lobbying that, he felt, marked the research landscapes of Canada

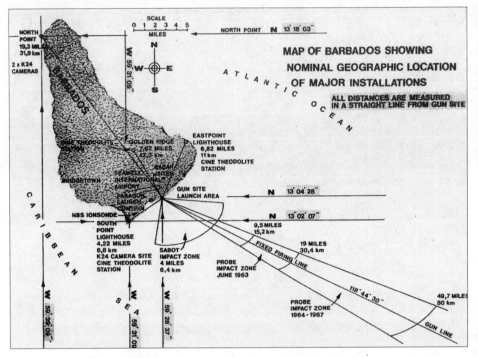

Figure 8.5 Map of Barbados showing nominal geographic location of major installations. Note how the only population centre shown is the capital, Bridgetown, and all distances are measured from the zero point of the HARP launch site. | © 1988 *Paris Kanonen – the Paris Guns (Wilhelmgeschütze) and Project HARP* – Gerald V. Bull.

and the United States. Throughout its history, HARP was highly valued by the Barbadian government, which saw it as an important employer of highly skilled domestic workers. In 1968, the country would issue commemorative postage stamps depicting the guns' arrival on the island and their first firing. The absence of any major domestic research activity on Barbados reinforced Bull's view of the island as essentially empty of the bureaucratic obstacles and petty disputes that Bull viewed as personal attacks in Canada and that he struggled to escape. HARP's official reports mirrored this view of Barbados as a blank canvas for his technological visions.[55] Echoing the physical displacement of people by complex technological systems in places such as Diego Garcia, the reports positioned HARP's research within a politically and demographically thin, but technologically thick, network of atmospheric probes, radar installations, and theodolites that characterized the military-scientific geography of the Caribbean and, more broadly, helped to transform coral atolls and Pacific lagoons into "natural" targets for guided missiles and nuclear detonations (see Figure 8.5).⁵⁶

Bull exploited that tense and ambiguous political geography of islands – part of their long history as havens for activities too clandestine or illicit or dangerous for mainlands.⁵⁷ In addition to the HARP launch site in Barbados, Bull would locate a test facility in Antigua. Both islands' nominally sovereign but pliable regimes allowed him to use them as way stations for later arms-smuggling operations with Israel and South Africa. But even before his journey into weapons sales and infamy, the specific context of HARP's Caribbean operations and the proximity of the region's different island nations worked continually to revive the ambiguous status of his cannons. HARP carried out some of its most intensive gun launches in late October 1960, during the Cuban Missile Crisis. Although Bull's enormous guns might have passed for expedient research tools in northern Quebec or at Highwater, where sparse populations and friendly borders made them politically inert, in Barbados their versatility and placement transformed them into potential Western weapons against Cuba and therefore potential targets for Cuban Air Force strikes. Bull would revive those tensions in June 1963 when, in a characteristically impolitic move, he glossed HARP's new altitude record of ninety-two kilometres by explaining that the guns were now capable of bombarding Havana.⁵⁸ The horrified response of Canadian officials mystified Bull. What rational person could see the huge and immobile HARP cannons – easy pickings for an airstrike – as weapons? But his dismissal and disbelief pointed to a final conflict at the centre of his work, this time between rationality and reason. The historical specificities of that conflict played out in his favour in the years surrounding the Cuban Missile Crisis, when Cold War rationality and global-scale nuclear strategy made his cannons plausibly innocuous. But his later misjudgment of that conflict, long after it had evolved into something more inimical to his interests, would prove to be fatal.

Rationality and Reason

The intellectual culture of the early Cold War, the period roughly between the division of Germany in late 1949 and the Tet offensive in 1968, traded on an important conflict between rationality and reason.⁵⁹ The history of those terms and of their underlying concepts is long and complex, but three elements of their larger history are important for Bull's story. First, though we often use the two interchangeably, *reason* and *rationality* are not identical. They have different histories and subtly different meanings. Historically, reason has encompassed both complexity and contingency and, unlike calculation for example, has proven difficult to automate.[60] Rationality has traditionally been directed at reducing complexity through techniques such as mathematical modelling or controlled experimentation. It rests on a kind of instrumental reason most often associated with economics and engineering.[61] Second, reason historically represented the highest of the mental faculties, drawing on all of the others (understanding, memory, judgment, imagination) to do its work. Because of that, it was seen as more general than, and superior to, rationality – *except* during the Cold War. And third, because of that complex history, it has been possible throughout history to be reasonable but not rational and vice versa.[62]

The Cold War made that historical possibility particularly stark. In the two decades following the Second World War, rationality expanded its domain from economics and engineering to political decision making, often in opposition to common sense. Practised by an array of civilian defence "think tanks" led by the RAND Corporation, that new rationality was formal, algorithm-based, acontextual, and predicated on rules that could be mechanically applied and therefore automated.[63] There it helped to generate the technically rational but morally absurd scenarios of the late 1950s and early 1960s, typified by the hypothetical schemes of Herman Kahn's doomsday machine and the analyses of mathematician and nuclear strategist Albert Wohlstetter (one of the inspirations for the character of Dr. Strangelove).[64] As Andrew Richter notes, Wohlstetter's 1954 RAND study on vulnerability, filled with elaborate charts and calculations, represented an archetype for rational analysis in the nuclear age.[65] It represented the most important example of an emerging type of strategic analysis based upon behavioural considerations and "hard" facts, rather than political calculations, and undergirded by a conviction that solutions to complex problems were simply a matter of numerical analysis. In doing so, the study manifestly ignored whether the threats in its scenarios even existed, focusing instead on imagined worlds of interesting and proliferating hypothetical vulnerabilities.[66]

In Canada, one of the places this kind of analysis flourished was among defence strategists and operational researchers at the Defence Research Board, the overseers of CARDE and the sponsors of Bull's early research on ballistics and anti–ballistic missile defence.[67] In 1951, strategists at the DRB had been the first in Canada to introduce the idea of "assured destruction" not only as a technological consequence of the Cold War that would materialize, in their view, around 1960, but also as a guarantor of geopolitical stability.⁶⁸ Whereas American civilian analysts often focused their strategic analyses on military superiority and first use of nuclear weapons, Canadian defence officials concentrated on the need for balanced strategic forces and the dangers of nuclear use.⁶⁹ As R.J. Sutherland, the most imaginative member of the early group, would go on to explain in 1960, stability lay (rationally but counterintuitively) in increasing the retaliatory force of *both* sides so that, within the new strategic calculations, "neither side can *rationally* strike first."⁷⁰ In this view, technological developments in long-range ballistic missiles generated a mutually assured destruction (MAD) – a stabilizing force, since no rational leader would authorize offensive use.⁷¹ The same calculations about strategic stability drove Sutherland, who had trained in operational research with the British and had conducted DRB analyses on artillery barrages and incendiary warfare during the Korean War, to conclude that targeting cities, rather than enemy nuclear forces, was preferable even if morally questionable.⁷² Although the strategy seemed absurd (and suicidal), since it would encourage the Soviets to target North American cities in return, Sutherland believed it made the stakes of nuclear exchange too dire and illustrated a key precept, the "rationality of irrationality": the desirability, within the seemingly twisted, mazelike calculus of Cold War strategizing, of making your opponent think that you were capable of irrational behaviour.

Bull, whose most trusted engineers called him "Dr. Strangelove," played precisely on this mixture of logic and absurdity, plainness and inversion, in positioning his project for a supergun. His early work on missiles and projectiles was framed by strategic discussions at the DRB and RAND. It was made possible by the reigning rationality-based calculus and the powerful imagined world it generated and relied upon. His research on antiballistic missile systems for the DRB, conducted while finishing his PhD and informed by the counter-city strategies of Sutherland and others, immersed Bull in the chess-game discussions on continental defence.⁷³ In particular, his analyses mirrored the acontextual character of Wohlstetter's work, driving home the idea that rationality could cut through the most tangled strategic knots, irrespective of specific historical and political contexts. On the one hand, Bull's proposals read like science fiction set pieces, scenarios that defied reason: shooting satellites into orbit from the surface of the Earth using gigantic cannons (see Figure 8.6). At one point, Bull even suggested using his proposed supergun to target the moon – the fantastical situation that Fisk had elaborated in her work and that Verne had included in books that Bull had idolized in his youth.⁷⁴ On the other hand, his proposals aligned

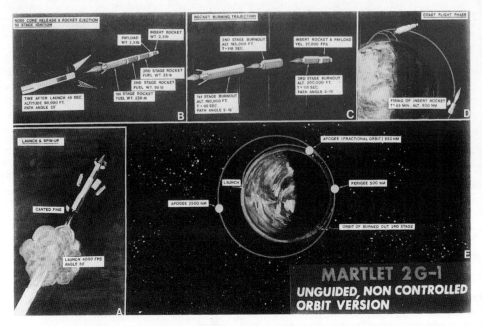

Figure 8.6 Bull's vision of launching satellites into orbit from the surface of the Earth. | © 1988 *Paris Kanonen – the Paris Guns (Wilhelmgeschütze) and Project HARP* – Gerald V. Bull.

philosophically with the same kinds of engineering-based, hyperrational, Cold War schemes that proposed altering global weather patterns, or using atomic weapons for excavation, independent of political or environmental considerations.[75]

Bull wielded this narrow, technocratic rationality as a bludgeon against his critics. When HARP's detractors asserted the impossibility of firing delicate scientific instruments to exo-atmospheric heights, mainly because of the torturous accelerations involved, Bull (like Wohlstetter) riposted with data, diagrams, and equations.[76] When an unnamed fellow professor at McGill University, whom Bull acidly mocked as a "theologian turned historian," read his "fallout" graph as an artillery ranging diagram, Bull rounded on him.[77] Critics of the project were cast as irrational, hostile, and insidious. For Bull, their irrationality was revealed precisely in the gap between narrowly specialized, acontextual, rational expertise and broadly reasonable, contextual, and casual observation. Rather than see their impressions as products of common sense and reasoned interpretation, he suggested that his critics could not be reached by data-based arguments because their motivations were already dishonest; they could not or would not read published articles and submit to scientific data.[78] For Bull, the world of the diagram spoke more clearly and unambiguously than the world in which it operated, a trope of modern scientific representation.[79] Beginning with HARP and relying on diagrams and data, Bull repeatedly argued that his largest guns, the plausible endgame of his ballistics work and weapons consulting, were completely irrational as offensive weapons. The cancellation of HARP in 1967 put his supergun project on hold and deeply embittered Bull, but he would revive the rationale for the technology's ostensibly scientific orientation two decades later in a very different context.

Between March 1966 and July 1967, in what Bull saw as a deep betrayal, both the Canadian and American governments withdrew funding from HARP. Canadian government agents quickly moved to confiscate and destroy his equipment in Barbados and Highwater (as they had done in a different context with the Avro Arrow), but Bull was able to save the assets.[80] Lacking government support, he put aside his designs for a supergun capable of firing projectiles into orbit. In short succession, he resigned from McGill University, obtained US citizenship through a special act of Congress, began contracting for the Pentagon, and started designing, manufacturing, and selling smaller-scale, long-range artillery for some of the world's most controversial conflict zones – Vietnam, Israel, and apartheid South Africa. Bull was eventually jailed for violating 1963 embargo rules regarding South Africa. In an attempt to vindicate himself and to keep his new arms business afloat, he unwittingly became involved in a complex arms-smuggling operation – a clandestine attempt by the Thatcher government to sell arms to Iraq, itself under embargo because of its war with Iran.[81] The plan involved using British intelligence agents to set up or infiltrate existing weapons companies. From their new corporate positions, the agents could simultaneously pursue and obfuscate covert arms deals with Iraq, later forcing out or "eliminating" anyone with sensitive or politically damaging information.[82]

Bull became the fall guy in this complex and covert scheme. Starting in 1987, he had been courted by Saddam Hussein's regime, itself in the midst of a decade-old attempt to modernize the country by fostering connections to the region's ancient, pre-Islamic past, particularly the glory of ancient Babylon.[83] Saddam Hussein, who styled himself after the Babylonian king Nebuchadnezzar, aimed to foster a pan-Arab renaissance based upon technological modernization and cultural cohesion.[84] Using massive investments in history, archaeology, and construction to distract attention from existing

Figure 8.7 A 1964 depiction of Bull's supergun. Note the crude support towers, which Bull would argue made the gun impractical as a weapon. The jeep (mid-left) provides an idea of scale. | © 1988 *Paris Kanonen – the Paris Guns (Wilhelmgeschütze) and Project HARP* – Gerald V. Bull.

ethnic divisions, he (literally) electrified the entire country, modernized its military, and moved to make Iraq the first Arab space power.[85] Bull's company was commissioned to provide regular artillery for the Iraqi army, but in a 1988 meeting with Saddam Hussein's son-in-law, Hussein Kamil, Bull convinced the Iraqis to realize his own ambitions for a new 1,000 millimetre supergun under the title "Project Babylon." The project took the idea of an exciting and unique space program for Canada (in HARP), designed to end the domination of space by the great powers, and repackaged it for a modernizing Iraq.[86]

Although Project Babylon comprised two subprojects designed to produce long-range artillery, its supergun, according to Bull, was always intended as a space launcher. He had already imagined a similar gun about 1964, shown in an artist's rendition (see Figure 8.7). Highlighting its specialized scientific applications and limited cost, he explained how the "complex traverse and elevating mechanisms of gunnery, tube rifling etc. are not

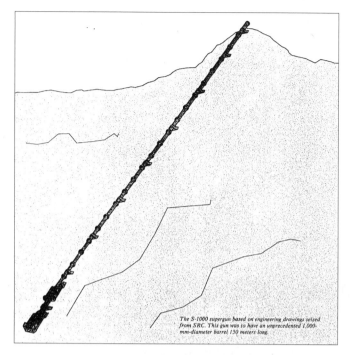

The S-1000 supergun based on engineering drawings seized from SRC. This gun was to have an unprecedented 1,000-mm-diameter barrel 150 meters long.

Figure 8.8 A CIA reconstruction of Bull's supergun for Project Babylon. Based upon engineering files seized from Bull's corporate files, the illustration highlights the gun's immobility. CIA analysts would nevertheless portray it, and the HARP guns, as potential weapons. | © 1988 *Paris Kanonen – the Paris Guns (Wilhelmgeschütze) and Project HARP* – Gerald V. Bull.

required."[87] He cited the brute facts of the object: a barrel 650 feet long; a flame 300 feet long, produced every time it fired and easily visible to surveillance satellites; a launch signature that mimicked ICBM firings and would attract a similar armed response; a loading time of up to eight hours; and sheer immobility, a gun so large that it would have to be built into a mountainside for support (see Figure 8.8). It would be incapable of elevation or training after it was built, which effectively gave it only one line of fire. All of these aspects made the gun irrational as a weapon, Bull claimed. The Iraqis, meanwhile, were intensely interested in the multistage rockets that the gun would fire, which would give them ballistic missile capabilities to strike anywhere in the region.[88] Like the Paris Guns, the psychological threat of the world's largest artillery piece provided unique advantages in a regional arms race.[89]

Two secret CIA reports produced in 1991 and based upon documents captured from Bull's company, Space Research Corporation, underlined the persistent ambiguity of the Iraqi supergun. They illustrated the device built into an imagined mountainside, possibly envisioned for the terrain surrounding the secret Saad 16 weapons complex near Mosul. As one of the documents acknowledged, the device could not be aimed and "would only be able to fire on targets along its fixed firing line."[90] Its immobility would make it vulnerable to retaliatory air strikes and place serious restrictions on its use as a weapon.[91] But CIA analysts nevertheless maintained its possible dual purpose as both a launch platform for satellites and a weapons system for delivering "warheads to much greater distances." Implicitly, the CIA reports, produced in a context very different from Bull's conception of HARP at the height of the Cold War, worked against the hyperrational, decontextualized arguments that characterized his defence of the project in the 1960s. Placing the guns firmly in their thick political and historical contexts, the reports featured photographs and explicit discussions of the technical achievements of HARP and combined them with covert intelligence on Bull's previous arms deals, his designs of two additional (but unbuilt) long-range artillery pieces for the Iraqis, and his 1988 visit to Saad 16. They detailed how "Project Babylon can be traced back to the 1960s joint US-Canadian High Altitude Research Project (HARP), which used high-caliber guns to conduct upper atmospheric research experiments."[92] And they explained how HARP, Bull's avowedly peaceful scientific instrument, had already possessed the ambiguous capacity for either launching objects into low-Earth orbit or sending them "to targets thousands of kilometers downrange."[93]

The CIA documents constructed their own deadly genealogy, one that ran against Bull's postincarceration narratives. Although Bull would continue to espouse his contextually thin visions for the supergun, almost to the brink of the Gulf War, the Cold War rationality that made them plausible had long since unravelled. Where he had attempted to present his archetypal HARP gun as the successor of the Paris Guns, a ballistics instrument stripped of its bellicose purpose, the CIA reports took that technically thick object and layered back precisely its political and military connections to arrive at the supergun of Project Babylon. And the CIA agents were not alone. Tracked for months by Mossad and eventually targeted by British SAS operatives (whose government handlers feared he would expose Britain's covert arms deals), Bull approached his sixth-floor Brussels apartment for the last time one quiet night in late March 1990. The two-man hit squad hired to eliminate him stood in the shadows, waiting to bring his initial foray into ballistics tragically full circle.[94]

Conclusion

Gerald Bull had seen his forays into ballistics as ingenious solutions to technical engineering problems. But in a move that mirrored episode after episode in the history of technology, he also realized that those solutions were at once technical and political.[95] His vision of advanced space research that skirted the state-driven, massively funded, large-scale infrastructures of Cold War engineering was at first an answer to local constraints: inadequate wind tunnels and nonexistent launch facilities. But precisely because those limitations were multiplied around the Cold War globe, in nations anxious about their status and prestige, Bull quickly came to see his inventions as an answer to possibly the most widespread technopolitical problem of the age: how to keep up with the superpowers without being one.

The ambiguity of his guns was crucial to that modernist vision. Throughout his career, Bull traded on the duality of his large cannons, claiming them as scientific instruments while understanding (and benefiting from) their potential military appeal. He was able to make those claims about HARP in particular because of its intersections with the larger histories of projectiles, islands, and reason during the Cold War. As a research program in the time-honoured field of ballistics, situated on a small island nation, and beyond any rational suspicion about its motives, HARP furnished a technical mythology that Bull would repeat to the end of his life. The darker sides of those histories similarly facilitated his weapons sales to Israel and South Africa in the 1970s as he parlayed his ballistics expertise into long-range artillery systems, used his island connections to circumvent trade embargoes, and objectively detached himself from the questionable deployment of his weapons. In the end, it was the commitment to his own mythology for the supergun that led him into a pact with Saddam Hussein on the eve of the Gulf War.

In its broad strokes and in its fine structure, Bull's story seems to provide precisely the kind of counternarrative that Richard Jarrell called on us to look for in the history of Canadian science and technology: an anti–case study, an exemplar of nontypicality that reminds us of the limits of our historical categories.[96] But Bull's fantastical vision of cannon-launched satellites and space probes was precisely of its age. It fit awkwardly into the conservative, stolid universe of postwar government science, but its vision of a space program for the rest of the world spoke directly to both the geopolitics of postwar technology and the central anxieties of the Cold War itself. Although the ambiguity of his superguns was made possible by much larger histories, with origins far beyond Paris, Montreal, or Babylon, Bull brought these seemingly nonparallel categories – projectiles, islands, reason – into alignment and pointed to their explosive conjunction in the late twentieth century. In doing so, he helped to connect his own story not only to the unexpected histories of Canadian science and technology but also to our histories writ large: the genealogies and erasures, the contradictory anxieties and ambivalent objects, of the modern age. And to one of its paradigmatic questions: which historical objects will be connected, remembered, privileged, and preserved, and which will be left dangling in history?

NOTES

1 Bull's lengthy monograph, "A Review and Study of the Case History of Dr. G.V. Bull and the Space Research Corporation and Its Related Companies in North America with Supporting Relevant Documents," was never officially published. His technical genealogy was published as Gerald V. Bull and Charles H. Murphy, *Paris Kanonen – the Paris Guns (Wilhelmgeschütze) and Project HARP: The Application of Major Calibre Guns to Atmospheric and Space Research* (Herford, Germany: E.S. Mittler and Sohn, 1988). Bull strongly opposed the moniker "arms dealer" and instead preferred to see his role as scientific and technical consultant. His indictment in 1980, however, was based upon his company's sale of thirty thousand artillery shells, gun barrels, and technical plans to the South African government, in violation of UN Security Council Resolution 418 (1977), which specifically prohibited both sales and weapons licensing arrangements with South Africa. Part of the aim of this chapter is to question the distinction between weapons as ideas and weapons as material objects.
2 On Paris as the capital of modernity, see David Harvey, *Paris, Capital of Modernity* (New York: Routledge, 2006).
3 For an overview of the Paris Gun, see Henry Willard Miller, *The Paris Gun: The Bombardment of Paris by the German Long Range Guns and the Great German Offensives of 1918* (New York: J. Cape and H. Smith, 1930). For contemporary accounts of the shelling, see *Source Records of the Great War*, ed. Charles F. Horne, vol. 6 (New York: National Alumni, 1928).
4 For an overview of HARP, see Charles H. Murphy and Gerald V. Bull, "A Review of Project HARP," *Annals of the New York Academy of Sciences* 140, 4 (1966): 337–57; and Bull and Murphy, *Paris Kanonen*.
5 Bull and Murphy, *Paris Kanonen*, 146.
6 Ibid.

7   On Bull's meteoric rise and fall, see James Adams, *Bull's Eye: The Assassination and Life of Supergun Inventor Gerald Bull* (New York: Times Books, 1992).
8   On modernity's characteristic sense of loss, see Sumathi Ramaswamy, *The Lost Land of Lemuria: Fabulous Geographies, Catastrophic Histories* (Berkeley: University of California Press, 2004); Eugen Weber, *France, Fin de Siècle* (Cambridge, MA: Belknap Press, 1986); Matt K. Matsuda, *The Memory of the Modern* (New York: Oxford University Press, 1996); and Marshall Berman, *All that Is Solid Melts into Air: The Experience of Modernity* (New York: Viking Penguin, 1988).
9   On the theme of recovery, see, for example, Simon Schaffer, "Newton on the Ganges: Asiatic Enlightenment of British Astronomy," Harry Camp Memorial Lecture, Stanford Humanities Center, Stanford University, January 16, 2008, http://shc.stanford.edu/multimedia/newton-ganges-asiatic-enlightenment-british-astronomy; and Dick Teresi, *Lost Discoveries: The Ancient Roots of Modern Science – from the Babylonians to the Maya* (New York: Simon and Schuster, 2002).
10 Throughout this chapter, I use the term "modernist" to describe self-conscious expressions of what it means to be modern.
11 For examples of Canadian niche responses, see Edward Jones-Imhotep, *The Unreliable Nation: Hostile Nature and Technological Failure in the Cold War* (Cambridge, MA: MIT Press, 2017); Andrew B. Godefroy, *Defence and Discovery: Canada's Military Space Program, 1945–74* (Vancouver: UBC Press, 2012); Richard A. Jarrell, *The Cold Light of Dawn: A History of Canadian Astronomy* (Toronto: University of Toronto Press, 1988); Robert Bothwell, *Nucleus: The History of Atomic Energy of Canada Limited* (Toronto: University of Toronto Press, 1988); Yves Gingras, *Physics and the Rise of Scientific Research in Canada* (Montreal/Kingston: McGill-Queen's University Press, 1991); and John N. Vardalas, *The Computer Revolution in Canada: Building National Technological Competence* (Cambridge, MA: MIT Press, 2001).
12 For a general discussion of high modernism, see James C. Scott, *Seeing like a State: How Certain Schemes to Improve the Human Condition Have Failed* (New Haven, CT: Yale University Press, 1998). For critical responses, see Daniel Macfarlane, *Negotiating a River: Canada, the US, and the Creation of the St. Lawrence Seaway* (Vancouver: UBC Press, 2014), and his chapter in this volume; and Tina Loo, "People in the Way: Modernity, Environment, and Society on the Arrow Lakes," *BC Studies* 142–43 (2004): 161–96.
13 For a discussion of this broader anxiety, see Pankaj Mishra's introduction to *The Time Regulation Institute*, by Ahmet Hamdi Tanpinar, trans. Maureen Freely and Alexander Dawe (1954; reprinted, New York: Penguin, 2014), iv–xx.
14 As James Hull, Dorotea Gucciardo, Jan Hadlaw, Beth Robertson, and David Theodore illustrate in this volume, technologies figured centrally in creating, shaping, and expressing what it meant to be modern in Canada. For Canada's middle-power politics, see Adam Chapnick, *The Middle Power Project: Canada and the Founding of the United Nations* (Vancouver: UBC Press, 2005); and Andrew Fenton Cooper, Richard A. Higgott, and Kim Richard Nossal, *Relocating Middle Powers: Australia and Canada in a Changing World Order* (Vancouver: UBC Press, 1993).

15 There has been much speculation about which foreign agents might have been responsible for the assassination. For speculation on Mossad (Israeli Intelligence) and the CIA, see Adams, *Bull's Eye;* and William Lowther, *Arms and the Man: Dr. Gerald Bull, Iraq, and the Supergun* (Novato, CA: Presidio Press, 1991). For seemingly definitive evidence pointing to British Special Air Service (SAS) operatives, see "Project Babylon: The Iraqi Supergun," CIA Intelligence Summary SW 91–50076X (Washington, DC: Directorate of Intelligence, CIA, 1991), 25.
16 Adams, *Bull's Eye*, 26.
17 Ibid.
18 Lowther, *Arms and the Man*, 49.
19 On the battlefield as a key resource for the sciences of the early modern period, see Kelly DeVries, "Sites of Military Science and Technology," in *The Cambridge History of Science*, vol. 3, *Early Modern Science*, ed. Katharine Park and Lorraine Daston (Cambridge, UK: Cambridge University Press, 2006), 306–19.
20 See Jim Bennet, "The Mechanical Arts," in *The Cambridge History of Science*, 3: 673–95.
21 In 1674, members of the Royal Society of London, including Robert Hooke, carried out gun experiments at Blackheath to test Galileo's findings. See A. Rupert Hall, "Gunnery, Science, and the Royal Society," in *The Uses of Science in the Age of Newton*, ed. John G. Burke (Berkeley: University of California Press, 1983), 111–42. See also A. Rupert Hall, *Ballistics in the Seventeenth Century: A Study in the Relations of Science and War with Reference Principally to England* (Cambridge, UK: Cambridge University Press, 1952).
22 See Sir Isaac Newton, *A Treatise of the System of the World* (London: F. Fayram, 1728), 6. This work, in which Newton presents his example together with an illustration, was originally intended as Book II of the *Principia Mathematica* and was written in late 1684 or early 1685. For the mysteries surrounding the work, its removal from the actual *Principia Mathematica*, and its publication history, see I. Bernard Cohen's introduction in Isaac Newton, *A Treatise of the System of the World* (Mineola, NY: Dover, 2004).
23 On Mach's ballistic experiments, see John T. Blackmore, *Ernst Mach: His Work, Life, and Influence* (Berkeley: University of California Press, 1972), 110; Christoph Hoffmann, "The Pocket Schedule, Note-Taking as a Research Technique: Ernst Mach's Ballistic-Photographic Experiments," in *Reworking the Bench: Research Notebooks in the History of Science*, ed. Frederic Lawrence Holmes, Jürgen Renn, and Hans-Jörg Rheinberger (Dordrecht: Kluwer Academic Publishers, 2003), 183–202.
24 Dorothy Fisk, *Exploring the Upper Atmosphere*, illus. Leonard Starbuck (London: Faber and Faber, 1934), 113–15.
25 See Robert Merton, "Science and the Social Order," and "The Normative Structure of Science," in *The Sociology of Science: Theoretical and Empirical Investigations* (1938; reprinted, Chicago: University of Chicago Press, 1973), 254–56, 265–78. On purification as a modern impulse, see Bruno Latour, *We Have Never Been Modern*, trans. Catherine Porter (Cambridge, MA: Harvard University Press, 1993).
26 On the First World War and modernity, see Modris Eksteins, *Rites of Spring: The Great War and the Birth of the Modern Age* (Boston: Houghton Mifflin, 2000).

27 Lowther, *Arms and the Man*, 50.
28 For histories of DRB research, see D.J. Goodspeed, *A History of the Defence Research Board of Canada* (Ottawa: Queen's Printer, 1958); and Jonathan Turner, "The Defence Research Board of Canada, 1947 to 1977" (PhD diss., University of Toronto, 2012).
29 Adams, *Bull's Eye*, 42–43; Lowther, *Arms and the Man*, 70–71.
30 Lowther, *Arms and the Man*, 58.
31 Ibid.
32 "Space Science and Space Technology – Summary of Points Affecting Canada's Future Position," paper prepared for the Committee of the Privy Council on Scientific and Industrial Research, DRB, December 17, 1958, 9; Library and Archives Canada (hereafter LAC), RG 25, Department of External Affairs (DEA), DRBS 170–80/A16 (CDRB), box 112, vol. 1, file 4145–09–1.
33 Gerald V. Bull and Charles Murphy, "Aerospace Applications of Gun Launched Projectiles and Rockets," paper presented at the American Astronautical Society symposium Future Space Programs and Impact on Range Network Development, New Mexico State University, March 1967, 4.
34 Charles H. Murphy, Gerald V. Bull, and Eugene D. Boyer, "Gun-Launched Sounding Rockets and Projectiles," in *Annals of the New York Academy of Sciences* 187 (1972): 304–23.
35 Bull and Murphy, "Aerospace Applications of Gun Launched Projectiles and Rockets," 2; Bull and Murphy, *Paris Kanonen*, 221.
36 The idea about the inherent neutrality of technologies also underpins modern technological determinism. See Langdon Winner, "Do Artifacts Have Politics?," *Daedalus* 109, 1 (1980): 121–36.
37 See, for example, Marshall Sahlins, *Islands of History* (Chicago: University of Chicago Press, 1985); Pamela H. Smith and Paula Findlen, *Merchants and Marvels: Commerce, Science, and Art in Early Modern Europe* (New York: Routledge, 2002); and Richard Grove, *Green Imperialism: Colonial Expansion, Tropical Island Edens, and the Origins of Environmentalism, 1600–1860* (Cambridge, UK: Cambridge University Press, 1995).
38 Paul M. Kennedy, "Imperial Cable Communications and Strategy, 1870–1914," *English Historical Review* 86, 341 (1971): 728–52; Daniel R. Headrick, *The Invisible Weapon: Telecommunications and International Politics, 1851–1945* (New York: Oxford University Press, 1991), 24, 98; Ruth Oldenziel, "Islands," in *Entangled Geographies: Empire and Technopolitics in the Global Cold War*, ed. Gabrielle Hecht (Cambridge, MA: MIT Press, 2011), 15.
39 This included the Marshall Islands, the Northern Marianas, and Palau. See Oldenziel, "Islands."
40 Thomas Bender, *A Nation among Nations: America's Place in World History* (New York: Hill and Wang, 2006), Chapter 1. For the early modern origins of politicized ocean space, see Elizabeth Mancke, "Early Modern Expansion and the Politicization of Oceanic Space," *Geographical Review* 89, 2 (1999): 225–36.
41 David Vine, *Island of Shame: The Secret History of the U.S. Military Base on Diego Garcia* (Princeton, NJ: Princeton University Press, 2011), 41–43.

42 For the detailed argument, see Oldenziel, "Islands."
43 Ibid.; Bartholomew H. Sparrow, *The Insular Cases and the Emergence of American Empire* (Lawrence: University Press of Kansas, 2006). This ruling helped set the legal framework for CIA black sites such as Diego Garcia and off-site detention centres such as Guantanamo Bay.
44 Oldenziel, "Islands," 20.
45 Lowther, *Arms and the Man*, 72. Northern Quebec was its own site of state-sponsored modernist projects. For the case of hydroelectric power, see Caroline Desbiens, *Power from the North: Territory, Identity, and the Culture of Hydroelectricity in Quebec* (Vancouver: UBC Press, 2014).
46 Stanley Brice Frost, *McGill University for the Advancement of Learning*, vol. 2 (Montreal/Kingston: McGill-Queen's University Press, 1980), 336.
47 On Hare's involvement in aerial photography of the North, see Stephen Bocking, "A Disciplined Geography: Aviation, Science, and the Cold War in Northern Canada, 1945–1960," *Technology and Culture* 50, 2 (2009): 265–90. On McGill University's activities in Barbados, see Kirsten Greer, "The Geographic Tradition in Caribbean Environmental History: David Watts, McGill University, and the Caribbean Project" (paper presented at the 16th International Conference of Historical Geographers, London, UK, July 2015).
48 On the epistemic status of the Tropics, see, for instance, Alexander von Humboldt, *Cosmos: A Sketch of a Physical Description of the Universe*, vol. 1 (London: Longman, 1868); on tropicality, see David Arnold, *The Problem of Nature: Environment, Culture, and European Expansion* (Oxford: Blackwell, 1996). For the relationship between the North and the temperate regions, see Jones-Imhotep, *The Unreliable Nation*; and Patricia Fara, "Northern Possession: Laying Claim to the Aurora Borealis," *History Workshop Journal* 42 (1996): 37–57.
49 Murphy and Bull, "A Review of Project HARP," 341.
50 Ibid.
51 The Yuma facility was essentially a duplicate of the Barbados site. See Bull and Murphy, *Paris Kanonen*, 156.
52 See C.H. Murphy and G.V. Bull, "HARP 5-Inch and 16-Inch Guns at Yuma Proving Ground, Arizona," Ballistic Research Laboratories Memorandum Report 1825, February 1967; and Yuma Proving Ground, "HARP Firing Result," SRI-Technical Note 767, July 1967.
53 On the "hostile regions" of the Cold War, see Matthew Farish, *The Contours of America's Cold War* (Minneapolis: University of Minnesota Press, 2010), 64; P. Whitney Lackenbauer and Matthew Farish, "The Cold War on Canadian Soil: Militarizing a Northern Environment," *Environmental History* 12, 4 (2007): 921; and Jones-Imhotep, *The Unreliable Nation*, Chapter 4.
54 Oldenziel, "Islands."
55 Murphy and Bull, "A Review of Project HARP."
56 On the case of Diego Garcia, see Vine, *Island of Shame*.
57 Christmas Island, for instance, was chosen for live testing of the Atlas missile and warhead because it shifted risk from American citizens to Pacific islanders. See Donald A. MacKenzie, "Missile Accuracy: A Case Study in the Social Processes of

Technological Change," in *The Social Construction of Technological Systems*, ed. Wiebe E. Bijker, Thomas P. Hughes, and Trevor J. Pinch (Cambridge, MA: MIT Press, 1987), 195–222.
58 Lowther, *Arms and the Man*, 82–83.
59 See Paul Erickson, Judy L. Klein, Lorraine Daston, Rebecca Lemov, Thomas Sturm, and Michael D. Gordin, *How Reason Almost Lost Its Mind: The Strange Career of Cold War Rationality* (Chicago: University of Chicago Press, 2013).
60 Ibid.
61 John Rawls has contrasted the "rational" and the "reasonable," casting the first as having to do with theories of rational choice and the second as having to do with moral considerations in deciding whether goals or purposes are valid. John Rawls, "Kantian Constructivism in Moral Theory," *Journal of Philosophy* 77, 9 (1980): 515–72. See also W.M. Sibley, "The Rational versus the Reasonable," *Philosophical Review* 62 (1953): 554–60; and John Rawls, *Political Liberalism*, rev. ed. (New York: Columbia University Press, 1996), 47–88, especially 50–54. Scott, *Seeing like a State*, 152–67, also discusses rationality in the context of high modernism.
62 Erickson et al., *How Reason Almost Lost Its Mind*.
63 Fred Kaplan, *The Wizards of Armageddon* (Stanford, CA: Stanford University Press, 1991); Sharon Ghamari-Tabrizi, *The Worlds of Herman Kahn: The Intuitive Science of Thermonuclear War* (Cambridge, MA: Harvard University Press, 2005). On the characteristics of Cold War rationality, see Erickson et al., *How Reason Almost Lost Its Mind*, 3–4.
64 Justin Vaïsse, *Neoconservatism: The Biography of a Movement* (Cambridge, MA: Belknap Press of Harvard University Press, 2010), 151.
65 Albert Wohlstetter, Fred Hoffman, Robert Lutz, and Henry Rowan, "Selection and Use of Strategic Air Bases," RAND Corporation, Santa Monica, CA, R-266, 1954.
66 Andrew Richter, *Avoiding Armageddon: Canadian Military Strategy and Nuclear Weapons, 1950–63* (Vancouver: UBC Press, 2002), 73.
67 Unlike in Canada, major contributions to strategic thought in the United States were generally made by the civilian defence community rather than by government officials. See ibid. For DRB work in this area, for instance, see "Active Defence for North America," October 16, 1959, LAC, RG 24, vol. 21,754, file 2184.4.D, part 1. A key figure here was Harold Larnder. See Ronald G. Stansfield, "Harold Larnder: Founder of Operational Research," *Journal of the Operational Research Society* 34, 1 (1983): 2–7.
68 See James Lee and David Bellamy, "Dr R.J. Sutherland: A Retrospective by James Lee and David Bellamy," Directorate of History and Heritage of the Department of National Defence, collection 87/253 III, box 6, file 0, 1.
69 Ibid.
70 Joint Ballistic Missile Defence Staff, "The Effect of the Ballistic Missile on the Prevention of Surprise Attack," February 23, 1960, LAC, General Burns Papers, MG 31, G6, vol. 14; emphasis added.
71 This insight led Sutherland to distinguish between "first-" and "second-strike" nuclear forces. See "A Military View of Nuclear Weapons," February 28, 1961, LAC, RG 24, Accession 1983–84/167, box 7373, DRBS 170–80/J56, vol. 3.

72 Ibid.
73 For the impact of city targeting on suburban development, see Peter Galison, "War against the Center," *Grey Room* 4 (2001): 5–33.
74 Lowther, *Arms and the Man*, 26.
75 On the environmental history of the Cold War, see J.R. McNeill and Corinna R. Unger, eds., *Environmental Histories of the Cold War* (Cambridge, UK: Cambridge University Press, 2013).
76 On high-g criticisms, see Bull and Murphy, *Paris Kanonen*, 207–8.
77 Ibid.
78 Ibid. For another example of high modernism's reliance on "facts" and "figures," see Tina Loo and Meg Stanley, "An Environmental History of Progress: Damming the Peace and Columbia Rivers," *Canadian Historical Review* 92, 3 (2011): 399–427.
79 The literature on scientific representation is vast. A few representative works include Martin J.S. Rudwick, "The Emergence of a Visual Language for Geological Science 1760–1840," *History of Science* 14, 3 (1976): 149–95, and *Scenes from Deep Time: Early Pictorial Representations of the Prehistoric World* (Chicago: University of Chicago Press, 1992); Bruno Latour and Steve Woolgar, *Laboratory Life: The Social Construction of Scientific Facts* (Princeton, NJ: Princeton University Press, 1979); Bruno Latour, "Visualization and Cognition," in *Knowledge and Society: Studies in the Sociology of Culture Past and Present*, vol. 6, ed. Henrika Kuklick and Elizabeth Long (Greenwich, CT: Jai Press, 1986), 1–40, and "How to Be Iconophilic in Art, Science, and Religion," in *Picturing Science, Producing Art*, ed. Peter Galison and Caroline Jones (New York: Routledge, 1998), 418–40; Brian Baigrie, *Picturing Knowledge: Historical and Philosophical Problems Concerning the Use of Art in Science* (Toronto: University of Toronto Press, 1996); Michael Lynch, "Discipline and the Material Form of Images: An Analysis of Scientific Visibility," *Social Studies of Science* 15, 1 (1985): 37–66, and "The Externalized Retina: Selection and Mathematization in the Visual Documentation of Objects in the Life Sciences," *Human Studies* 11, 2 (1988): 201–34; Michael Lynch and Steve Woolgar, *Representation in Scientific Practice* (Cambridge, MA: MIT Press, 1990); Catelijne Coopman, Janet Vertesi, Michael Lynch, and Steve Woolgar, eds., *Representation in Scientific Practice Revisited* (Cambridge, MA: MIT Press, 2014); Peter Galison and Caroline Jones, eds., *Picturing Science, Producing Art* (New York: Routledge, 1998); David Kaiser, *Drawing Theories Apart: The Dispersion of Feynman Diagrams in Postwar Physics* (Chicago: University of Chicago Press, 2005); and Luc Pauwels, *Visual Cultures of Science: Rethinking Representational Practices in Knowledge Building and Science Communication* (Lebanon, NH: University Press of New England, 2006).
80 On the Avro Arrow, see Julius Lukasiewicz, "Canada's Encounter with High-Speed Aeronautics," *Technology and Culture* 27, 2 (1986): 223–61; and Donald C. Story and Russell Isinger, "The Origins of the Cancellation of Canada's Avro CF-105 Arrow Fighter Program: A Failure of Strategy," *Journal of Strategic Studies* 30, 6 (2007): 1025–50.
81 See Mark Phythian, *Arming Iraq: How the U.S. and Britain Secretly Built Saddam's War Machine* (Boston: Northeastern University Press, 1997).

82 The details of the operation are contained in a secret CIA document from the period. See "Project Babylon: The Iraqi Supergun," CIA Research Paper SW 91–50076X, Directorate of Intelligence, CIA, 1991.
83 Samira Haj, *The Making of Iraq, 1900–1963: Capital, Power, and Ideology* (Albany: SUNY Press, 1997); Paul W.T. Kingston, *Britain and the Politics of Modernization in the Middle East, 1945–1958*, vol. 4 (Cambridge, UK: Cambridge University Press, 2002), 94–100.
84 Ben Kiernan, "From Bangladesh to Baghdad," in *Blood and Soil: A World History of Genocide and Extermination from Sparta to Darfur* (New Haven, CT: Yale University Press, 2008), 585–87; Jerry M. Long, *Saddam's War of Words: Politics, Religion, and the Iraqi Invasion of Kuwait* (Austin: University of Texas Press, 2009), 52–53; Milton Viorst, "IV: Freedom Is," in *Sandcastles: The Arabs in Search of the Modern World* (New York: Alfred A. Knopf, 1994), 28–46.
85 Ebubekir Ceylan, *The Ottoman Origins of Modern Iraq: Political Reform, Modernization, and Development in the Nineteenth-Century Middle East* (London: I.B. Tauris, 2011), 201–4.
86 Lowther, *Arms and the Man*, 75.
87 Bull and Murphy, *Paris Kanonen*, 231.
88 At the time of his death, Bull was designing two additional guns with barrels, respectively, of 100 feet and 200 feet and capable of being elevated and trained. They would have been the largest artillery pieces in the world, capable of reaching Turkey, Iran, Saudi Arabia, Israel, and potentially some European countries. See Adams, *Bull's Eye*, 261.
89 Kevin Toolis, "The Man behind Iraq's Supergun," *New York Times*, August 26, 1990.
90 "Project Babylon," 4.
91 Ibid.
92 Ibid.
93 Ibid.
94 Ibid., 24–25.
95 For other examples of technopolitical solutions, see, for example, Gabrielle Hecht, *The Radiance of France: Nuclear Power and National Identity after World War II* (Cambridge, MA: MIT Press, 2009); Ken Alder, *Engineering the Revolution: Arms and Enlightenment in France, 1763–1815* (Princeton, NJ: Princeton University Press, 1997); and Paul N. Edwards, *The Closed World: Computers and the Politics of Discourse in Cold War America* (Cambridge, MA: MIT Press, 1997).
96 Jarrell, *The Cold Light of Dawn*.

Percy Schmeiser, Roundup Ready® Canola, and Canadian Agricultural Modernity

EDA KRANAKIS

The turn of the millennium brought a new "poster child" to Canada and the world: Percy Schmeiser. A sixty-eight-year old farmer and grandfather, he acquired this new media status in a battle over one of late modernity's premier technoscientific breakthroughs: genetically modified crops. Biotechnology giant Monsanto sued Schmeiser in 1998, claiming that he had illegally grown canola containing Monsanto's patented "Roundup Ready® gene," which makes the plants resistant to the herbicide glyphosate. For six years, the case made its way through the Canadian federal court system until the Supreme Court made a final ruling in May 2004. In a contentious, 5-4 decision, the court deemed Schmeiser guilty of patent infringement because canola plants containing the gene had been found on his property, even though Canadian law disallowed patents on plants and even though the genetically modified plants had entered his fields through uncontrollable natural processes, such as pollination.

The development and commercialization of Roundup Ready® (RR) canola, and the legal battle that ensued, offer a unique perspective on the experience of late modernity in and beyond the Canadian agricultural sphere. Of course, modernity is not all of a piece; it has intersected the rural sphere in distinct phases, each with its own tensions, challenges, and exaltations, all shaped by local contexts.[1] In the early twentieth century, many Canadian farmers heralded modernity, particularly when it took the form of innovations such as tractors, electrification, or Marquis wheat, but they also tried to steer it in desired ways, through efforts to establish wheat pools, battles over railway freight rates, and so on.[2] In the decades following the Second World War, under the tenets of "high modernism," Canadian agriculture embraced the "green revolution," including its new range of agro-industrial inputs, notably chemical herbicides and insecticides, and a new "made-in-Canada" cash crop – canola – widely praised as a beneficial example of publicly funded, scientific crop breeding.[3]

Yet something changed around the turn of the millennium: the introduction of RR canola in Canada propelled disaffection for the tenets of high modernist agriculture and widening doubts that a society organized around successive scientific and technological solutions to agricultural problems was either inevitable or desirable.[4] In fact, the case of genetically modified canola shows late modernity in disarray in the agricultural sector, corrupted, its premises foundering, its purposes contested, and its future uncertain. Schmeiser became a symbol of this disarray around the world.

Much has been written about the development and impact of genetically modified crops, and this chapter contributes to that body of knowledge in two ways. First, it pays closer attention to the patent that Monsanto used to support the RR system in Canada.[5] Technical and historical analysis of this patent shows, among other things, that it did not disclose or specifically claim the gene that Schmeiser was sued for illegally using. Second, this chapter analyzes the development of RR canola as an episode in the history of high modernist agriculture. Prior studies have examined how genetically modified crops fit into the political economy of capitalism.[6] Exploring the topic as a story about agricultural modernity complements and supplements the political economy approach. It shows how the RR system intersected with high modernism's foundational ideology of science, technology, and progress and how this ideology connected with capitalist dynamics.

The narrative that follows traces the development and deployment of RR canola in Canada, deconstructs the Canadian patent that Monsanto used to protect this new crop system, and discusses crucial points about the Schmeiser trials, showing also how the boundary between nature and invention was redrawn in the patent office and the courts. Along the way, it highlights several important features of this new system of agricultural modernity. For the purpose of economy, these historical sections are not explicitly framed in relation to modernist ideology, but they are followed by a discussion of how the empirical story relates to "the catechism of high modernist agriculture."[7] I show how the RR system and the Schmeiser court cases challenged or contradicted tenets of high modernism and stimulated a more probing, pervasive, and multiform questioning of modernity and high modernist agriculture than was evident earlier in the twentieth century.

Monsanto and the Development of the Roundup Ready® System

Monsanto moved into biotechnology in the 1980s by way of agricultural chemicals.[8] Among its major agrochemical innovations was the weed killer glyphosate, commercialized in 1974 under the trademark Roundup®. Glyphosate has been called "a once-in-a-century herbicide"[9] because of its unique mode of action, the broad spectrum of weeds it controlled, and its comparative environmental safety and low toxicity for humans and animals (though these latter claims have increasingly come into question). The rise of molecular biology opened the possibility of altering plant DNA to make crops resistant to glyphosate. In this way, farmers could control weeds by spraying their growing crops with glyphosate as needed: everything but the crop would die. By the early 1980s, Monsanto was transforming itself into a biotechnology company with this goal. In 1984, it opened the world's largest biotechnology laboratory complex, where, following years of effort and several dead ends, it created commercially viable strains of Roundup Ready® crop plants.[10]

The new "biofacts" that Monsanto created overcame glyphosate's toxicity.[11] Glyphosate works by inhibiting the functioning of EPSPS, an enzyme that all plants, fungi, and bacteria produce in various forms. EPSPS catalyzes reactions that allow plants to produce proteins, hormones, and other substances needed to survive and grow. By binding to EPSPS, glyphosate halts its functioning, killing the plant. Throughout the 1980s, Monsanto researchers tried to alter EPSPS genetically in the laboratory to make it resistant to glyphosate. Yet, despite creating millions of transformed plant cells, they "failed to produce a commercially resistant EPSPS."[12] The eventual key to success lay outside the laboratory: it was a discovery rather than an invention. Beginning in the spring of 1989, the Monsanto team hunted for bacteria beyond the laboratory that had evolved resistance to glyphosate by themselves. Testing bacteria collected from glyphosate-contaminated sludge around Monsanto's chemical production plant in Luling, Louisiana, they found a strain of Agrobacterium tumefaciens, CP4, whose EPSPS had evolved not only enormously high glyphosate tolerance but also very high catalytic efficiency.[13] It was this combination of traits that finally enabled

Monsanto researchers to produce glyphosate-tolerant canola (and other crop plants) suitable for commercialization.

Monsanto researcher Stephen Padgette recalled this discovery as "a great Eureka moment."[14] However, it took about two years of additional work before a patent application could be filed. Monsanto researchers first had to isolate the gene in the CP4 strain that coded for this EPSPS variant. Then they transferred the gene to a virus, used the virus to infect isolated plant cells, regenerated plants from the modified cells, and finally tested offspring of those regenerated plants to determine their vigour and level of glyphosate resistance. At the end of August 1991, they applied for patents in Canada, Europe, and elsewhere by way of a single, international (PCT) filing.[15] The following year Monsanto started field-testing the new line of plants and launched the process of gaining government approval for the commercial release of glyphosate-tolerant seed and plant varieties in both Canada and the United States.[16] The first commercial products reached the market in 1996: canola in Canada and soybean in the United States.

Patenting Roundup Ready® Technology in Canada

Before the 1980s, living organisms could not be patented in Canada, but the commissioner of patents changed this rule in a 1982 decision (*Re Application of Abitibi Co.*). Inspired by a US Supreme Court ruling in 1980 and legal decisions in Australia, Japan, and several European countries, the commissioner reinterpreted Canadian law to allow the patenting of certain living organisms. Given these foreign court rulings, the commissioner explained, it was no longer certain that "a patent for a microorganism or other life forms would not be held allowable by our own courts."[17] Rather than waiting for Canadian courts to decide the matter, however, the commissioner plunged ahead and administratively redrew the line between nature and invention. The decision amounted to a suggestion to the courts to reinterpret existing patent law following the commissioner's guidelines:

> It is of some importance, we think, to recognize how far our recommendation, if accepted, will carry us ... Certainly this decision will extend to all microorganisms, yeasts, molds, fungi, bacteria, actinomycetes, unicellular algae, cell lines, viruses or protozoa; in fact to all new life forms which are produced en masse as chemical compounds are prepared, and are formed in such large numbers that any measurable quantity will possess uniform properties and characteristics ... Whether it reaches up to higher life forms – plants (in the popular sense) or animals – is debatable.[18]

The *Abitibi* decision thus left a question mark on the patentability of plants, and the Canadian Patent Office continued to refuse patent claims for plants.

So how did Monsanto's RR invention mesh with the new patent rules? In Canada, Monsanto used Patent 1,313,830, issued in 1993, to protect its "Roundup Ready® technology." The company's lawsuit against Percy Schmeiser – indeed all its lawsuits against Canadian farmers – cited only this patent. Titled "Glyphosate-Resistant Plants," Monsanto had attempted to include claims to plants in the patent, but the Canadian Patent Office had steadfastly refused these claims.[19] In the end, the patent's claims to living matter were limited to a "chimeric gene" and a plant cell containing the gene.[20]

Close analysis of Patent 1,313,830 shows, surprisingly, that it did not and could not disclose the "Roundup Ready® gene" that Monsanto sued Schmeiser and other Canadian farmers for illegally using. Three criteria support this conclusion. First, Canadian patent law holds that no new subject matter may be added to a patent application once it is submitted.[21] The application for Patent 1,313,830 was submitted in 1986, but, as explained above, Monsanto's research team did not discover and isolate the CP4 gene until May 1989 and after. The CP4 gene therefore could not be added to Monsanto's 1986 patent application, and it is not mentioned there. Second, Monsanto submitted a separate application to the Canadian Patent Office in 1991 that disclosed and claimed the CP4 gene as a (new) invention. Canadian patent law does not allow "double patenting": that is, patenting the same invention twice.[22] Therefore, the CP4 gene could not also be covered by the 1986 application. The 1991 application eventually became Canadian Patent 2,088,661 in 2001, five years after the commercial release of RR canola in Canada and three years after Monsanto launched its lawsuit against Schmeiser.[23] (For convenience, I will call the invention claimed in Patent 2,088,661 the "CP4 invention.") Third, to verify Schmeiser's infringement of the Roundup Ready® gene, the company used a "quick test" to detect the presence of CP4 in Schmeiser's canola.[24] However, since the CP4 gene was not disclosed in Patent 1,313,830, how could Schmeiser have infringed that patent by possessing canola that contained CP4?

Monsanto representatives themselves affirmed that the invention protected by their 1986 patent application did not cover the CP4 gene and that CP4 was the basis of RR crops. In written submissions to the European Patent Office, Monsanto representatives stated that the 1986 invention addressed "the idea of transforming plants ... to obtain sufficient glyphosate tolerance," but the problem remained "unsolved" until the CP4 invention.[25] They also insisted that the earlier patent did not cover the CP4 gene because it did not disclose "any such [DNA] sequence or molecule."[26] Finally, they asserted that the CP4 invention "was the basis for preparing a vast number of herbicide resistant crop plants that are commercially marketed."[27] To underscore this point, they included a table showing the acreage of RR crop plants grown around the world, including all of the acreage of RR canola grown in Canada.

Two questions immediately arise from the preceding discussion. First, if Patent 1,313,830 did not disclose the CP4 gene, then what did it disclose? Second, how is Monsanto's 1998 lawsuit against Schmeiser to be explained? Why would Monsanto sue him before the Canadian patent claiming the CP4 gene was issued? To answer these questions, we need to recognize that there is a profound difference between a patent, considered in a legal sense, and an invention, considered from the perspective of the history of science and technology. A history of science approach shows that what Monsanto researchers "invented" in 1985 – the material basis for the 1986 Canadian patent application – were four "yellow and quite sickly" petunia plants that could withstand modest doses of glyphosate but not the levels needed to control weeds in crop fields.[28] However, when translated into a set of patent claims, this invention became much broader. The magic occurred by ingenious drafting of the claims, which itemize the methods and products that constitute "the invention" over which a temporary monopoly is asserted. Monsanto patent attorney Patrick D. Kelly, who wrote the application for Patent 1,313,830, explained that "the claims of a patent define the property of the inventor."[29]

Monsanto's lawyers deliberately wrote broad claims to maximize their legal scope, avoiding "any limitations that are not absolutely essential."[30] The main reason to limit claims was to avoid *prior art*, not to avoid claiming things not yet invented. In the case of Patent 1,313,830, Kelly achieved these goals through "functional language" that redescribed the petunia invention in generic terms.[31] The narrowest claim in Patent 1,313,830 was for pMON546 (the modified gene construct that Monsanto researchers inserted into the petunia plants to make them glyphosate resistant). This gene was not suitable for commercial-grade RR crop plants and was never used for that purpose. Yet, in Kelly's hands, the pMON546 biofact became a generic claim for a "chimeric plant gene" comprising several functional elements such as "a promoter sequence."[32] All chimeric genes needed analogous elements (just like every car needs a motor, steering mechanism, and braking system), so the generically worded claim potentially covered just about any glyphosate-tolerant gene that might ever be developed in the future![33]

Further understanding of Patent 1,313,830 can be gleaned through comparison with Patent 2,088,661. The latter patent disclosed the CP4 gene, labelled it a new type of "Class II" gene, and contrasted its characteristics with undesirable characteristics of what it now called "Class I" genes, which included those disclosed in the earlier patent. Patent 2,088,661 elaborated how and why "Class II" genes alone could achieve adequate glyphosate tolerance in crop plants.[34] The abstract of that patent also explained that the CP4 invention pertained to weed control in crop fields, and use of the gene for weed control in farming was included in the list of claims.[35] In contrast, Patent 1,313,830 said nothing about farming or weed control in crop fields, and it defined "glyphosate tolerance" in a way that applied only to low tolerance levels needed for laboratory experimentation.

Monsanto took advantage of the generically worded claims of Patent 1,313,830 to include the CP4 gene under its umbrella retroactively, yet by doing so it ran the risk that the patent would be invalidated for lack of "enablement." Under Canadian law, a patent specification must include full disclosure of all details needed to recreate the claimed invention.[36] "If a person skilled in the art can arrive at the same results only through chance or further long experiments, the disclosure is insufficient and the patent is void."[37]

Patent 1,313,830 arguably did not meet the full disclosure requirement for the CP4 gene. Since CP4 was *discovered* in nature rather than created in the laboratory, it could only be enabled in that patent if the specification included either its DNA sequence or a declaration that a physical sample of CP4 had been provided to a public depository (culture collection). But Patent 1,313,830 did neither of these things, nor did it mention where or how to look for a gene like CP4. Finding such a gene required looking for EPSPS variants that combined high glyphosate tolerance with high catalytic efficiency, but the terms "catalytic efficiency" and "kinetic efficiency" do not appear in that patent.[38]

Monsanto ran other risks as well with Patent 1,313,830 because of uncertainty about where Canadian courts would draw the line between "nature" and "invention." A 1989 Supreme Court judgment suggested that Patent 1,313,830 might be deemed invalid. The 1989 case involved a patent application for a new variety of soybean developed by seed company Pioneer Hi-Bred, using traditional cross-breeding techniques. The commissioner of patents had rejected the application on the ground that plant varieties were not patentable under Canadian law. The Federal Court of Appeals and the Supreme Court refused to overturn the commissioner's decision. In their reasoning, the judges considered the adequacy of the disclosure provided in the specification. Justice J. Pratte at the Federal Court of Appeals stressed that the disclosure had to enable someone to regenerate the invention reliably by following the steps outlined. However, in the case of plant breeding, genetic pairings and random genetic variations occurred that remained beyond the inventor's control. The process involved "a degree of luck," so the specification could not "enable others to obtain the same results unless they, by chance, would benefit from the same good fortune."[39] The Supreme Court decision commented that the disclosure had "an important part to play" in distinguishing between "acts of invention" and "products of nature" that depended merely on natural processes, which were not patentable. The Supreme Court ruling contended that Pioneer Hi-Bred's disclosure mainly involved the latter, because the steps disclosed

> do not in any way appear to alter the soybean reproductive process, which occurs in accordance with the laws of nature. Earlier decisions have never allowed such a method to be the basis for a patent. The courts have regarded creations following the laws of nature as being mere discoveries the existence of which man has simply uncovered without thereby being able to claim he has invented them.[40]

These arguments cast doubt on Patent 1,313,830 because Monsanto's "invention" (including the CP4 invention) depended significantly on both luck and natural reproductive processes. Monsanto researchers were never able to insert a modified gene directly into a canola seed or plant. The only way that a canola *seed* could acquire the glyphosate tolerance gene was by natural growth from an RR canola plant. And a canola plant could not be modified directly either. Rather, the modified gene was introduced into a plant virus that infected injured, isolated canola leaf cells in a laboratory container. Monsanto depended on natural growth processes to regenerate plants from the modified (infected) canola cells. These plant regeneration techniques were well known before Monsanto submitted its patent application, so they could not be claimed as an invention.[41]

The disclosure of Patent 1,313,830 also depended on chance because the processes that it discussed could not be fully controlled. To illustrate, Monsanto's RR canola line (GT73) was selected from hundreds of offspring of plants regenerated from genetically modified, isolated plant cells. These offspring expressed the altered gene to differing degrees and in widely differing ways. Monsanto researchers had to test the plants and select those with the desired properties (adequate glyphosate tolerance and vigour).[42] This was analogous to what traditional plant breeders did. The modified gene, moreover, was expressed differently in different parts of each plant. Thus, there was an irreducible element of chance and randomness in Monsanto's creation of genetically modified plants, and the Pioneer Hi-Bred decision implied that the disclosure of Patent 1,313,830 might therefore be inadequate.[43]

In summary, analysis of Patent 1,313,830, considered within the larger context of Canadian patent law, reveals that Monsanto sought to protect its RR system in Canada using a patent that did not cover plants, did not disclose the crucial CP4 gene, and did not even mention seeds, crops, farmers' fields, or weed control. Moreover, the claimed methods depended fundamentally both on chance and on natural processes of growth, both of which, according to the Supreme Court's 1989 Pioneer Hi-Bred decision, thwarted patentability requirements. Even though the claims of Patent 1,313,830 were written generically enough to appear superficially to encompass the "Roundup Ready® gene," Monsanto was clearly taking risks with this patent in Canada. Accordingly, there must have been other factors that drove the company to sue Schmeiser under this patent. The following section shows that those factors were linked to the "value capture technique" that Monsanto implemented for RR technology in Canada.

"Capturing Value" from Roundup Ready® Technology in Canada

As of 1992, Monsanto still had not decided how to commercialize RR crops or which value capture technique to use to maximize the return on this innovation.[44] Seed sellers face a quandary: organisms self-reproduce, and farmers save seeds. A study conducted around the time that RR crops were introduced indicated that 80 percent of the seed used by farmers in the developing world came from farm-saved seed.[45] The proportion was lower in the more highly industrialized countries, but farmers in those countries still regularly traded and saved seed of open-pollinated crops. Percy Schmeiser, for example, grew all of his canola from saved seed. A study carried out in Alberta in 1980 showed that 60 percent of farmers saved seed, and with certain crops, such as wheat, the proportion was reported to be 90 percent.[46] For some crops, notably corn (maize), seed producers overcame competition from saved seed by selling hybrid seed. The latter must be purchased anew each year because the second-generation seed loses the performance qualities of the parent generation. However, the hybrid solution was not implemented for some of the most important open-pollinated crops, including canola, soybean, and cotton, all of which were important RR crops.[47] By saving and replanting some of the RR seed harvested from their own fields, farmers could continue growing RR crops for years without buying any new seed. (A single canola plant, for example, can produce about 2,000 seeds.[48]) As a Monsanto representative explained, with open-pollinated crops, "where a farmer can use the same seed year after year, you have to be careful because if you try to capture too much value he will say, I think you have enough value and I am going to use my own seed."[49]

Monsanto's solution in North America was a multiple-licence approach backed by patents. In Canada, Monsanto licensed seed companies to produce RR versions of those companies' own canola strains, and with the cooperation of both seed producers and retailers it implemented a system that required farmers to sign contracts before buying RR seed. Each farmer who wished to purchase RR canola had to sign two contracts: a "Grower's Agreement" and a "Technology Use Agreement."[50] The former set out a host of rules that transformed Monsanto's relations with farmers. It prohibited saving or sharing RR seed or harvesting any stray RR plants that took root beyond a farmer's field (e.g., in ditches or roadside allowances, where farmers had traditional use rights). Monsanto representatives, moreover, had to be given unfettered access to a farmer's fields and storage bins to sample the crops (even without the farmer being present); this permission extended for three years following the signing of the Grower's Agreement or the planting of any RR crop.[51]

Farmers who signed a Grower's Agreement got a numbered wallet card that told seed retailers that they were certified to purchase RR canola seed, provided that they also signed a Technology Use Agreement (TUA). The latter required the farmer to pay a fee to Monsanto (beyond the price of the seed) for "a license to use the Roundup Ready® gene."[52] The fee was fifteen dollars Canadian per acre. On the TUA, the farmer had to specify the legal land locations of all fields where the farmer would grow RR crops, the specific RR varieties that would be planted, and the total acreage of each variety to be planted. To round out this new system, Monsanto contracted with detective agencies to carry out investigations of seed savers, monitor farmers' compliance with the Grower's Agreement, and follow rumours of RR seed saving or sharing. Monsanto also set up a toll-free line for anonymous tips concerning what it called "seed pirates."[53]

Monsanto's contract system begins to reveal why this phase of high modernist agriculture has bred a sense of disjunction with modernity's "social contract." Modernity was supposed to bring technoscientific advances that made life easier. Science was supposed to create progress for all. But Monsanto's contract system created an onslaught of new red tape. The company required both farmers and seed retailers to be numbered and certified, and separate sections of the TUA had to be completed and signed by the farmer, the seed retailer, and the seed producer, with each receiving a colour-coded copy of the signed form: white for the grower, yellow for the seed retailer, pink for the seed producer, and "goldenrod" for Monsanto.[54] The form also had an "acre reconciliation" section: after planting but prior to July 15, the acres planted with RR seed had to be reconciled with the acres that the farmer had proposed planting when the seed was purchased, with a final determination of the fee owed to Monsanto.

The Grower's Agreement evinced a discourse of threat, control, and penny-ante calculation that also conflicted with the ideology of agricultural modernity. The penny-ante spirit was evident in the prohibition against harvesting RR plants growing in public road allowances, particularly since farmers' use of these spaces was an established agricultural practice. The aspect of control was evident in a warning to seed producers and retailers that only Monsanto had the right to decide who could purchase RR seed. And farmers were warned that violators would face penalties and forever "forfeit any right" to use RR seed. Monsanto reserved the right to require destruction of crops deemed to be illicit, while still requiring the farmer to pay a licensing fee to Monsanto for the destroyed crops and to cover its investigation, testing, and legal costs.[55]

Yet Monsanto's RR contract system fundamentally depended on the strength of Patent 1,313,830. A patent gives its owner a temporary right to prevent others from making, using, diffusing, or selling the claimed invention. Patent holders also have the right to license the sale or use of their inventions, and Monsanto presented the Grower's Agreement and the TUA as a licensing system that allowed farmers to use its patented invention in ways spelled out by Monsanto. Without the patent, Monsanto had no basis to deny farmers the right to save seed and no basis to demand the right to access a farmer's land to investigate or test the crops.

The new patent and contract regime allowed Monsanto to insert itself between farmers and their lands in three new ways: first, by dictating or banning seed-use practices that had previously been the farmer's prerogative; second, by gaining unsupervised access to farmers' fields and storage facilities; and third, by effectively instituting a new form of land rent. This last aspect stems from the way in which Monsanto structured the TUA as a temporary licence to use its modified gene. Since the gene was contained in the seed, one would expect the fees to be assessed on the amounts of seed purchased. Yet Monsanto structured the fees as annual charges per acre, the way that land rents are assessed. The Grower's Agreement further strengthened the land rent character of the licensing fee by making not just the farmer but also the farm itself subject to Monsanto's rules. Finally, the company stipulated that the Grower's Agreement would continue to "be binding and have full force and effect" even if the farm were sold or transferred to a new owner who had not signed a Monsanto contract.[56]

Despite widespread dislike of these contracts, farmers rapidly adopted RR canola. Studies have shown that they did so overwhelmingly with the belief that the RR system would bring easier, cheaper, and more flexible weed control.[57] Clinton Evans helps us to understand farmers' acceptance of RR technology. The commercial farming regime that developed in the Prairies over the twentieth century combined grain monoculture, large farms, and "acute labor shortage."[58] Evans shows that weeds were "cultural artifacts" of the system, simultaneously encouraged and demonized by it.[59] In the post–Second World War period, rather than abandoning this ultimately unsustainable system, governments, industries, and farmers put their faith in science to solve the weed problem. Monsanto's RR technology was designed as a new step in this system. In the context of "the war on weeds," RR technology seemed to promise a "final solution" – a weed control method that would kill everything green except the crop.[60]

Yet Monsanto's biotechnology patent reorganized how profit was generated in this system, with profound consequences. It enabled Monsanto to charge what amounted to an annual rent on farmland that it never owned – nearly 5 million acres in Canada by 2000.[61] It also dictated how that land – and adjoining public land – could and could not be used. This was no longer just a "war on weeds" but a new kind of agricultural empire rooted in intellectual property rather than land ownership.[62]

Monsanto's RR System and "Genetic Drift"

There was an anomaly in Monsanto's RR system, however, particularly as it applied to canola, which helps to explain why the company sued Schmeiser under Patent 1,313,830. In canola, an open pollination crop plant, gene transfer occurs easily through pollen transfer. Genes also move via seed mobility. Canola has tiny seeds in pods prone to shatter before harvest. Canola seeds can be blown by wind, dispersed by animals, or scattered during truck transport.[63] Furthermore, the RR genetic trait is dominant. It is expressed in 100 percent of the first generation of offspring produced by cross-pollination with conventional canola, and in subsequent generations it is expressed in 75 percent of the offspring. RR canola, in short, spreads promiscuously and becomes dominant once established, and its spread cannot be prevented. Empirically, this result has been shown in studies assessing the percentage of RR canola growing among volunteer (stray) canola patches along highways. A major study in North Dakota showed that 80 percent of the volunteer canola growing along 5,600 kilometres of highway was RR canola.[64] Equally important, canola carrying Monsanto's gene cannot be distinguished visually from non-GM canola. The only way that a farmer could determine if a conventional canola crop included RR plants was either by means of expensive, time-consuming gene tests or by spraying some of the growing crop with glyphosate (which would then kill the conventional canola). Not surprisingly, canola has become "the poster child of transgene escape."[65]

Early on, Monsanto experts understood these characteristics. Patent 1,313,830 mentions pollen as a vector for gene transfer.[66] Moreover, Monsanto's field tests of GM crops required special measures to prevent gene escape.[67] In 1990, a Monsanto representative attended a workshop on canola's susceptibility to gene transfer by pollination and seed mobility, with presenters concluding that such transfers were inevitable.[68] In its submission for permission to release RR canola into the environment, Monsanto discussed genetic flow by pollen and seed mobility, concluding that RR canola would behave in the same way as non-GM canola.[69] Since non-GM canola was known to be promiscuous in this regard, Monsanto's documentation attested that RR canola would be too. A Monsanto employee also testified during the Schmeiser lawsuit that the company expected cross-pollination to occur.[70] And, because the RR trait was dominant, genetic drift would lead to eventual dominance of RR canola over conventional canola.

Monsanto's "value capture" system for RR canola thus *had to* consider genetic drift. Its contract system only covered farmers who signed up to grow RR canola, but genetic drift and RR trait dominance meant that RR canola would be increasingly present in the fields of conventional canola growers who did not sign Monsanto contracts. With seed saving, a conventional canola field could become a predominantly RR canola field within a couple of years. The change could happen deliberately, but also inadvertently, since the two types of canola looked identical. Growers who had not signed Monsanto contracts might then trade, give, or sell their saved seed to other farmers.

Conventional canola growers who saved their seed threatened Monsanto's RR canola profit system, so the company wanted to stop seed saving by *all* canola growers, not just RR canola growers. This is where the patent was crucial: only a patent could control farmers who had *not* signed a Grower's Agreement. And, if Monsanto delayed lawsuits by even a few years, the situation could become uncontrollable and weaken the company's legal position. The non-licensed, seed-saving farmers would produce larger and larger quantities of RR seed each year, which could be further replanted and diffused to other farmers. This situation could breed resentment, cheating, or organized opposition from licensed RR farmers who paid stiff annual fees to Monsanto.

These scenarios explain why Monsanto introduced a surveillance system and why it sued Schmeiser without waiting for Patent 2,088,661 to issue.[71] The company could not reasonably sue any conventional canola seed saver in Canada in 1996 or 1997. In 1996, the first year of commercialization of RR canola, no conventional canola farmer could have planted it from saved seed. The first RR canola crops in 1996 were also "identity preserved." Each farmer's RR canola seed harvest in 1996 was kept separate and tracked, with coloured "grainfetti" added to the harvested seed, identifying the farmer who had grown it. Furthermore, farmers growing RR canola had to agree to have their 1996 crops securely transported by designated commercial truckers to specified crushing plants.[72] With these controls in place, seed harvested in 1996 could not be illegally transferred in significant quantities to unlicensed farmers. Monsanto could not mount a convincing lawsuit against Schmeiser in 1997 either because there was inadequate opportunity to acquire enough RR canola seed from the 1996 harvest to plant his fields in 1997. The identity controls were removed for the 1997 growing season, however, and by that time RR canola was spreading by natural means. Monsanto began investigating Schmeiser in the summer of 1997 but sued him only the following summer because the 1998 crop was the first that he could have planted using a significant amount of *saved* RR canola seed.

Monsanto's Legal Victory

Monsanto's calculated but risky lawsuit against Schmeiser succeeded: the courts sided with Monsanto on virtually every issue. The judges who tried the initial case and subsequent appeals ruled that patent rights trump other property rights. They decided that it did not matter if the RR canola genes had entered Schmeiser's fields through uncontrollable natural processes; the result still constituted illegal patent infringement. They also ruled that it did not matter whether Schmeiser used the gene "as intended" – that is, to spray canola crops with glyphosate. (He did not do so.) The Supreme Court case established a further legal principle to Monsanto's benefit: a patent for a genetically modified, isolated plant leaf cell in a laboratory container could be used to control plants in a farmer's field and all their progeny for years to come, even though plant patents were not allowed under Canadian law. That is, if a plant contained a patented gene, the entire plant and its offspring for many generations fell under the full control of the patent holder, despite the time, expense, and labour that the farmer devoted to growing those subsequent generations of plants and seeds. Finally, the Supreme Court declared Patent 1,313,830 to be valid.[73]

Monsanto's trial strategy helps us to understand how these judgments came about. Patrick Kelly, the lawyer who drafted Patent 1,313,830, observed that patents could be used to steer conflict in advantageous ways: "Patents aren't really scientific documents; they're legal and business documents, and the essence of both law and business is conflict ... The goal is not to avoid conflict; the goal is learning to work with conflict and competition, and learning to turn them to your advantage."[74] Monsanto's lawsuit against Schmeiser applied this principle, using Patent 1,313,830 to cement the legal status of the RR system in Canada. Monsanto knew the patent's weaknesses but also knew that it had protection from another principle of Canadian patent law: patents are accorded a legal presumption of validity, so the onus was on the defence to prove invalidity, which, for this abstruse molecular biology patent, would require extensive research and scientific expertise. Monsanto realized that the circumstances of this lawsuit would be unlikely to produce that kind of defence.

The company used its privileged command of molecular biology to steer the lawsuit in desired ways. Molecular biology is a complicated, esoteric science, particularly when applied to the intricacies of genetic engineering, and Monsanto knew that Schmeiser, his lawyer, and the judges deciding the case would be unlikely to have any meaningful command of the subject. Kelly commented that patent judges "usually have no training in science or technology except a couple of freshman science courses in college many years ago."[75] Ian Binnie, one of the Supreme Court judges who upheld Monsanto's claim against Schmeiser, discussed the problem of scientific illiteracy in the courts, observing that Canadian courts "have serious difficulties in digesting and evaluating scientific evidence, even rather crude scientific evidence."[76] Canadian intellectual property expert David Vaver observed that scientific illiteracy affected judges' interpretation of patents: "A patent's meaning is ultimately a question of law, decided by a judge who usually is not skilled in any art or science, let alone the relevant one ... In practice, courts rely heavily on expert evidence to help them understand how those skilled in the art would have understood the language of the patent at its claim date."[77] In most infringement lawsuits, the opposing sides are scientifically knowledgeable companies fighting one another, with judges balancing contradictory scientific evidence that they present. But the Schmeiser case was different: it pitted a deep-pocketed corporation with scientific expertise and a large team of top-flight American and Canadian lawyers against one young Saskatoon lawyer, Terry Zakreski, with no background in science or molecular biology.[78] Zakreski did not attempt a scientific-technical challenge of the patent, and Schmeiser almost certainly could not have borne the cost of doing so.[79] Consequently, the only "expert testimony" on the scientific-technical content of Patent 1,313,830 came from Monsanto employees. The Canadian judges relied on this testimony to decide that the patent was technically valid and infringed.

Monsanto employee and "expert witness" Doris Dixon was careful to present her testimony on Patent 1,313,830 as opinion rather than substantiated fact: "Based upon ... *my reading* of the Patent ... it is *my opinion* that the Defendant's samples contain DNA sequences claimed in Claims 1, 2, 5, and 6 of the Patent."[80] Dixon neglected to mention, however, that those claims do not in fact refer to any particular DNA sequence, nor did she explain that the CP4 gene was not mentioned in Patent 1,313,830 but was specifically disclosed and claimed in Canada as a *new invention* in a subsequent patent application, already in the public domain.

Robert Horsch, one of the inventors named on Patent 1,313,830, also testified. Horsch, with a PhD in genetics, had worked at Monsanto since 1981 and had spent years working on the project to create genetically modified plants resistant to glyphosate. But when questioned by Zakreski, Horsch claimed to have forgotten the identity of the modified gene in Roundup Ready® canola, even though it was one of the company's most important breakthroughs. Zakreski asked him about Monsanto's RR canola line, officially known as RT73: "Do you know what the gene RT73 is?" Horsch replied, "I don't."[81] Then Zakreski asked, "do you know if [RT73] was the name of the gene ... that you introduced into the canola plant to give it glyphosate resistance?" Horsch responded, "I don't remember."[82] When

Zakreski subsequently asked him if he knew the name of the gene that he added to canola, Horsch responded, "I don't remember."[83]

Much more could be said about the Schmeiser cases, but I must limit my focus to the observation that the judicial decisions did not adequately address either the technoscientific or the legal complexities of the case.[84] The judgments bore out the opinion of Binnie in his article on "Science in the Courtroom" that the Canadian legal system "suffers from amateurishness when it deals with scientific matters."[85] The judgments evinced a poor understanding of genetic engineering and a glossing over of the problem of genetic drift.[86] They also failed to address the issues raised in the Pioneer Hi-Bred case, of distinguishing between nature and invention, a problem crucial to the Schmeiser case.[87] On the legal side, the judgments engaged in "cherry picking": they drew from rules and precedents that supported a judgment of infringement but neglected to discuss other equally important rules and precedents that supported the opposite conclusion.

Among the principles that the Supreme Court ignored was "patent exhaustion." It can be explained with a hypothetical example. Suppose you buy a second-hand car and use it to make money as an Uber driver. The car contains dozens, probably hundreds, of patented parts and materials that you inevitably use when driving. Yet, because of the principle of patent exhaustion, the owners of these patents do not accuse you of infringement: once an item with patented parts has been legally transferred to become the property of someone else, the patent holders' claims on that item must cease. Were it not for the principle of patent exhaustion, all of us, as users of a multitude of patented things, would be constantly enmeshed in lawsuits, demands for royalties, and so on. Schmeiser was in the same situation. Monsanto licensed its patented technology to seed companies, which produced RR canola seed then sold to farmers. The RR canola genes that ended up on Schmeiser's land by pollination or wind-blown seed came from crops grown by farmers who had legitimately purchased RR canola seed, so, following the principle of patent exhaustion, Schmeiser – a subsequent user – had the right to use the plants that ended up on his land as he saw fit. Legal scholars Jeremy de Beer and Robert Tomkowicz wrote an article exploring this issue in depth, remarking that it was "puzzling" and "unfortunate" that the Supreme Court ignored the doctrine of exhaustion in its ruling. "The key purpose of the doctrine," they explained, "is to prevent abuses of intellectual property rights from unduly encroaching on rights of classic property owners to freely use their legally acquired property."[88] They concluded that, "if the doctrine of exhaustion would have been applied, Schmeiser's use of Monsanto's patented genes ... would not infringe the right to use the patented invention."[89]

Other analysts found additional omissions and deficiencies in the judgment. It was suggested that the Supreme Court had channelled the law to reach a desired judgment.[90] De Beer and Tomkowicz argued that the court needed to reach a finding of infringement for economic reasons and sacrificed sound legal reasoning to do so: "It appears that Schmeiser was simply an unintended consequence of a specific set of facts and circumstances forcing the Supreme Court to make a decision that may have been economically justified in outcome, even if legally questionable."[91] The Supreme Court judges signalled this economic justification, even though they noted that it was "not directly at issue" in the Schmeiser case: they were worried that seed-saving farmers would gain "future revenue opportunities" by selling RR seed that they harvested to "other farmers unwilling to pay the license fee, thus depriving Monsanto of the full enjoyment of their monopoly."[92] Other legal scholars observed that the court "misunderstood and misapplied" historical case law, gave Monsanto new rights extending beyond those mandated by Canadian patent law, led to absurd legal conclusions, and ignored long-standing, fundamental patent law principles directly relevant to the case.[93] In short, the Schmeiser case unleashed a firestorm of criticism that is still ongoing. Crop scientist E. Ann Clark called the Federal Court's 2001 decision in the Schmeiser case "incomprehensible," while legal scholar Nathan Busch, who reviewed the case from the 2001 decision to the final Supreme Court judgment in 2004, concluded that "the Schmeiser Court has thrown the law of patents as it relates to patents for transgenes into complete disarray."[94]

High Modernist Agriculture: Questions and Doubts

The Schmeiser case did more than throw biotechnology patent law into "disarray": it propelled a deeper, wider questioning of agricultural modernity.[95] Following James Scott's definition, high modernist agriculture was built around a "muscle-bound" faith in scientific and technical progress: "At its core was a supreme self-confidence about continued linear progress, the development of scientific and technical knowledge, the expansion of production, the rational design of social order, the growing satisfaction of human needs, and, not least, an increasing control over nature."[96] The Schmeiser case, because of what it revealed about RR canola and science-based patents, called these ideas into question in new ways, which were then publicized globally.[97]

First, the trial showed science – via a patent – being used not as a beneficial agent of progress and service to society but as a weapon directed at individual farmers. Monsanto's patent represented the cutting edge of science applied to farming, but the company used it – together with surveillance and bullying tactics – to control and intimidate farmers, making practices that they had long engaged in suddenly illegal and transforming their property into Monsanto's property "overnight" via natural processes that farmers could not control or even directly observe.[98] Journalists publicized terms such as "seed police" and "new reign of terror" that Schmeiser and other farmers used to describe Monsanto's use of private investigators and encouragement of rumour mongering to ferret out information.[99] Clearly, this new regime was not just an extension of agricultural modernity's use of science to wage a "war on weeds" for farmers' benefit; it had morphed into a direct assault on farmers and their property rights, using science and invention as justifications. Schmeiser thought that the worst aspect of Monsanto's tactics was the negative effect on social cohesion in farming communities:

> What happens after Monsanto gets a tip or a rumor? They immediately send out two, either their own representatives or two of the ex-RCMP ... They'll say to the farmer, "We have this tip or rumor that you're growing Monsanto's GMO canola or soybeans without a license." ... What do you think goes through a farmer's mind when these gene police as we call them leave a farmer's home? The farmer will think, "Was it this neighbor or the neighbor over here? Or the neighbor down here that has caused me this trouble?" Right away, you get a suspicion, a breakdown of our world social fabric ... My grandparents and my parents had to work together with their neighbors to build our country, our infrastructure, our schools, roads, hospitals, and so on. Now we have the breakdown of our rural social fabric by that contract. I think that is one of the worst things, that suspicion, of not trusting one another, farmers not talking to one another about what they are growing.[100]

Second, the trial revealed a Frankenstein quality in Monsanto's use of science and invention that contradicted high modernist ideology. Rather than bringing nature firmly and advantageously under human control, the Schmeiser trial showed Monsanto heading in the opposite direction. It used science to create a new biofact – RR canola – whose dispersal into the environment could not be controlled, and Monsanto openly used this *lack* of control to extract greater profit from its innovation. Not only were RR canola plants uncontrollable, but they also broke down the distinction between crops and weeds. RR canola plants that took root in a conventional farmer's field became weeds themselves. And the only way to get rid of them, since they were indistinguishable from the conventional plants, was to destroy the entire crop. RR canola even became weeds for farmers who signed the Monsanto contracts because, when they engaged in crop rotation, RR canola plants continued to grow as strays immune to one of the strongest herbicides on the market. And, on top of that, Monsanto's uncontrollable genes – and the expanded use of glyphosate that they fostered – began breeding a new generation of "superweeds" resistant to glyphosate. There are now an estimated million acres or more of these superweeds growing on Canadian cropland, and the population is growing exponentially.[101] Farmers are now going back to older, more toxic herbicides such as 2,4-D ("Agent Orange") to control RR strays and superweeds. This was not science controlling nature; it was science producing disorder, creating new problems, costs, and inefficiencies for farmers and society at large.

Third, the trial showed scientific advances breeding confusion and contradiction rather than promoting clarity and understanding, evidenced, for example, by the line-drawing exercise that undergirded the trials and Monsanto's patent. The project of modernity rested on the assumptions that nature and society are distinct realms and that scientific evidence and reasoning provide the necessary tools to distinguish clearly between these two realms.[102] Yet the line-drawing exercises that occurred in the Schmeiser trials – which decided where innate "nature" ended and human "invention" began – either ignored science or displayed superficial scientific truth, which suggested a declining commitment to decisions based upon careful scientific reasoning. The Canadian judiciary's line-drawing exercise relied on a concept *invalidated* by science – the "ladder of evolution" – that moreover was applied in an unconvincing way. The judges held that a plant is a "higher organism" and a "product of nature" (thus not an invention) but that a plant *cell* can be a human invention. Without offering a credible justification for this conclusion, they invoked claim 22 of Monsanto's patent for a "glyphosate-resistant plant cell comprising a chimeric plant gene."[103]

Yet claim 22 implicitly treated self-reproduction as an inventor-controlled process, even though the Supreme Court had earlier ruled against this approach. In claim 22, Monsanto once again used imprecise, generic wording to increase its reach. The company failed to specify whether the claimed cell was the isolated, injured plant *leaf* cell that its researchers had infected, in a laboratory container, with a virus carrying a modified gene or whether the claim referred to, say, a fourth-generation canola *seed* cell in Schmeiser's field, which could only have come into existence by means of the plant's own self-replication machinery, needing no further input from Monsanto. (Recall that Monsanto researchers themselves could not transfer a modified gene into a canola seed cell.) The judicial minority in the Supreme Court questioned the scope of claim 22, but the majority brushed their concerns away. The Supreme Court's public line-drawing spectacle thus produced further confusion and disarray. Validation of the patent seemed to confirm scientific progress, but the court's reasoning ignored or avoided consideration of relevant scientific evidence.

The Supreme Court judgment and its failure to analyze what Monsanto researchers had "actually in good faith invented" convinced many that the Schmeiser case was more about legitimizing a new form of expropriation in agriculture than about upholding the Patent Act and its principal objective: to balance the interests of inventors and society.[104] Not only did the court's reasoning suggest this conclusion, but the wording of the decision did too. The Supreme Court majority declared that "we emphasize from the outset that we are not concerned here with ... the scope of the respondents' patent or the wisdom and social utility of the genetic modification of genes and cells."[105] The terms "society," "the public," "balance(d)," and "fairness" were absent from the judgment (though the latter three terms appeared in the dissenting minority opinion). What the majority opinion *was* concerned with was whether Schmeiser had "deprived" Monsanto of the "full enjoyment of the monopoly" that its patent conferred. The phrase "full enjoyment of the monopoly" appeared fifteen times in the judgment and was specifically linked to ideas of profit making and protection of business interests. A similarly narrow, pecuniary focus guided the judges' view of farmers and their work: "Farming is a commercial enterprise in which farmers sow and cultivate the plants which prove most efficient and profitable."[106] This was modernist discourse on steroids: a way of thinking that had lost touch with the profusion of nonpecuniary reasons for cultivating plants. Almost entirely absent from the judgment, moreover, were any of the broader ideals that encouraged adherence to the tenets of modernity and high modernism, such as improving lives or regulating society to minimize social conflict and enhance general well-being. The Supreme Court majority judgment projected a modernism that had lost its soul, and it failed to produce a broadly accepted sense of justice or closure, despite the millions of dollars and years of effort that went into the Schmeiser case.

Conclusion

Although framed as an analysis of the experience of modernity in the Canadian agricultural sphere, the story of Roundup Ready® canola reveals numerous links with other spheres and places. As the RR canola system unfolded in Canada, Monsanto was implanting it in other countries and crops – soybean, cotton, and corn in the United States, soybean in Argentina, cotton in India, and so on. Yet the Schmeiser trials marked the first time anywhere that Monsanto's RR system became the focus of a legal challenge, and the trials were discussed in an astounding variety of contexts. Issues raised during the trials diffused not only through the farming world but also into law, journalism, crop science, food retailing, environmental science, cooking, philosophy, theatre, the organic movement, corporate strategies, NGO activities, the health sector, and more.[107] Here are a few of the linkages: other agribusiness corporations emulated Monsanto's successful strategy and began using their own TUAs that prohibited seed saving;[108] the Schmeiser case helped to fuel greater opposition to patents per se;[109] and Monsanto's behaviour and efforts to impose genetically modified crops strengthened the organic food sector, which has become "the most dynamic and rapidly growing sector of the global food industry."[110] Food, agriculture, and GMOs became "hot topics" generating broad new interest among journalists and researchers in the humanities and social sciences; moreover, the peculiarities of patenting self-replicating organisms, which Monsanto's RR system and the Schmeiser case highlighted, helped to fuel wider discussion of the limits and dangers of reductionist science in the biological realm.[111]

There was also an enormous geographical spread of knowledge and information from the Canadian experience, which affected groups around the world: farmers, of course, but also lawyers, scientists, politicians, food importers, and others. For example, a plan by BASF to test genetically modified potatoes in Ireland was cancelled in 2008 following pressure from over one hundred "farm and food industry groups," and a press release issued on this occasion by GM-Free Ireland referred to Percy Schmeiser and the Canadian Supreme Court decision, conveyed a direct message of congratulations from Schmeiser, and noted that he would soon be a keynote speaker at a conference "to discuss Ireland's GMO policy."[112] It is impossible to know exactly how much the Canadian story has influenced political strategies and actions regarding GMOs in other countries, but Schmeiser's case has been studied and invoked all around the world. In North America, it helped to mobilize resistance that succeeded in halting the planned introduction of RR wheat.[113] And, throughout most of Europe, in Japan, and in many other areas of the world, the growing of GM crops has not occurred, in part because strict rules have been imposed to prevent genetic contamination and to make producers or certified users of GMOs financially and legally responsible for their unwanted spread.

Finally, it is worth reflecting on the significance of Schmeiser as a "poster child." He acquired this role in two contradictory respects. Monsanto publicized him as a sinner, a cheat who tried to get the company's technology for free. Even before the first trial began, Monsanto deliberately released negative publicity about Schmeiser to the press.[114] To this day, Monsanto's main US website devotes a page to him, asserting that "Percy Schmeiser is not a hero. He's simply a patent infringer who knows how to tell a good story."[115] The company's telling of his story implies that Schmeiser illegally obtained and planted (bags of) RR canola seed – basically that he was a thief. In contrast, the Irish example discussed above presents Schmeiser as a saintly symbol of resistance to the progression of high modernist agriculture. And this new movement of resistance does not seek merely to tweak modernist agriculture in desired ways (as in struggles over freight rates decades earlier). Rather, it seeks to overthrow the premises underlying high modernist agriculture, to defy and overturn the entire system, and to do so both locally and globally. This incommensurable duality of what Schmeiser represented, combined with the depth of opposition to high modernist farming that he came to symbolize, together explain why we can see him as an exemplar of agricultural modernity in disarray.

Investigating the individual behind the image reinforces this perspective: we see a farmer who spent nearly all of his life working within the high modernist tradition, reinforcing it. Schmeiser practised commercial monoculture, used pesticides, and chem-fallowed his fields. He owned a vehicle repair garage and gas station. And, from the 1940s on, he was involved in the farm equipment business, also helping to develop a new kind of rotary combine. By the 1980s, he had established an independent farm machinery dealership as well. Schmeiser also served as the mayor of his town for many years and as a member of the Saskatchewan legislature.[116] He was, in short, a successful, establishment, green-revolution farmer. Yet Monsanto's allegations deeply affronted him, and he refused to be cowed. Schmeiser was thereby thrust into a public position of deepening philosophical opposition to the culture of farming that he had followed his whole life. In the end, high modernism devoured Schmeiser the green-revolution farmer. And, when modernity begins to devour its own children in such a direct, forceful, and public way, what is that but a corrupted modernity, a modernity in disarray?

NOTES

1 Devlin Kuyek argues that Canadian agriculture has passed through three distinct "seed regimes" since the late nineteenth century, which we can also interpret as phases of agricultural modernity. Devlin Kuyek, "Sowing the Seeds of Corporate Agriculture: The Rise of Canada's Third Seed Regime," *Studies in Political Economy* 80 (2007): 31–54. See also Deborah Fitzgerald, *Every Farm a Factory: The Industrial Ideal in American Agriculture* (New Haven, CT: Yale University Press, 2003).

2 Rod Bantjes, "Modernity and the Machine Farmer," *Journal of Historical Sociology* 13, 2 (2000): 121–41; Rod Bantjes, *Improved Earth: Prairie Space as Modern Artefact, 1869–1944* (Toronto: University of Toronto Press, 2005); J.W. Morrison, "Marquis Wheat – a Triumph of Scientific Endeavor," *Agricultural History* 34, 4 (1960): 182–88; H.S. Ferns, "Landholding Systems and Political Structures: Causes and Consequences," *Business History Review* 67, 2 (1993): 297–98.

3 James C. Scott, *Seeing like a State: How Certain Schemes to Improve the Human Condition Have Failed* (New Haven, CT: Yale University Press, 1999), 270; Devlin Kuyek, *Good Crop/Bad Crop: Seed Politics and the Future of Food in Canada* (Toronto: Between the Lines, 2007); Brewster Kneen, *The Rape of Canola* (Toronto: NC Press, 1992); and Clinton Lorne Evans, *The War on Weeds in the Prairie West: An Environmental History* (Calgary: University of Calgary Press, 2002). Canola is a type of rapeseed with low erucic acid developed by crop scientists at Agriculture Canada and the University of Manitoba, making it edible for humans and livestock.

4 Statistically, Canadian canola growers rapidly and widely adopted RR canola. However, the adopters represented only a portion of the total number of farmers in Canada, and farmers as an occupational group make up no more than 2 percent of the total population. Moreover, crop farmers were not the only people who had an interest in the RR system. Finally, farmers' reasons for adopting RR crops were more complex than simple admiration for the product. They involved pressing economic issues, labour demands, and expectations shaped by Monsanto's advertised promises. On Canadian farmers' motives for adopting RR canola, see Gabriela Pechlaner, *Corporate Crops: Biotechnology, Agriculture, and the Struggle for Control* (Austin: University of Texas Press, 2012).

5 No previous study has analyzed Monsanto's patents in their historical technoscientific context. One study that does look carefully at Monsanto patents, but from the perspective of molecular biology rather than from the perspective of the history of science and technology, is Nathan A. Busch, "Genetically Modified Plants Are Not Inventions and Are, Therefore, Not Patentable," *Drake Journal of Agricultural Law* 10, 3 (2005): 387–482.

6 See, for example, William Peekhaus, *Resistance Is Fertile: Canadian Struggles on the BioCommons* (Vancouver: UBC Press, 2013); Kuyek, *Good Crop/Bad Crop*; Pechlaner, *Corporate Crops*; Jack Ralph Kloppenburg Jr., *First the Seed: The Political*

Economy of Plant Biotechnology, 2nd ed. (Madison: University of Wisconsin Press, 2004); and Richard C. Lewontin, "The Maturing of Capitalist Agriculture: Farmer as Proletarian," *Monthly Labor Review* 50, 3 (1998): 72–84.
7   Scott, *Seeing like a State,* 270.
8   A broad review and critique of Monsanto is Marie-Monique Robin, *The World According to Monsanto: Pollution, Corruption, and the Control of the World's Food Supply,* trans. George Holoch (New York: New Press, 2010).
9   Stephen O. Duke and Stephen B. Powles, "Glyphosate: A Once-in-a-Century Herbicide," *Pest Management Science* 64, 4 (2008): 319–25.
10 Rick Weiss, "Seeds of Discord," *Washington Post,* February 3, 1999, https://www.washingtonpost.com/archive/politics/1999/02/03/seeds-of-discord/c0f613a0-02a1-476f-b54d-af25413844f5/.
11 See Nicole C. Karafyllis, "Growth of Biofacts: The Real Thing or Metaphor?," in *Tensions and Convergences: Technological and Aesthetic Transformations of Society,* ed. Reinhard Heil, Andreas Kaminski, Marcus Stippak, Alexander Unger, and Marc Ziegler (Bielefeld: Verlag, 2007), 141–52. Karafyllis introduced the neologism *biofact* (combining *bios* and *artifact*) to refer to "a being that is both natural and artificial," one that is "brought into existence by purposive human action, but exists by processes of growth" (145). The term is helpful because it fosters inquiry into "the differences between 'nature' and 'technology'" in the domain of "biotechnical design activities" (142) and how or where this line is drawn to distinguish what counts as the "natural" versus the "invented" aspects of these new organisms.
12 Stephen R. Padgette, Diane B. Re, Gerard F. Barry, David E. Eichholtz, et al., "New Weed Control Opportunities: Development of Soybeans with a Roundup Ready™ Gene," in *Herbicide-Resistant Crops: Agricultural, Economic, Environmental, Regulatory, and Technological Aspects,* ed. Stephen O. Duke (Boca Raton, FL: CRC Press, 1996), 66. The quotation is from Gerald M. Dill, "Glyphosate-Resistant Crops: History, Status, and Future," *Pest Management Science* 61, 3 (2005): 220. See also Daniel Charles, *Lords of the Harvest: Biotech, Big Money, and the Future of Food* (Cambridge, MA: Perseus, 2001), 60–69.
13 Stephen R. Padgette, Gerard F. Barry, and Ganesh M. Kishore, "Glyphosate Tolerant 5-Enolpyruvylshikimate-3-Phosphate Synthases," Canadian Patent 2,088,661, filed August 28, 1991, and issued December 18, 2001, 13; Laurence E. Hallas, William J. Adams, and Michael A. Heitkamp, "Glyphosate Degradation by Immobilized Bacteria: Field Studies with Industrial Wastewater Effluent," *Applied and Environmental Microbiology* 58, 4 (1992): 1215–19.
14 Padgette et al., "New Weed Control Opportunities," 67.
15 PCT refers to the Patent Cooperation Treaty. It permitted a single patent application (usually written in English) to become a series of national applications, processed separately by each national patent office. See Eda Kranakis, "Patents and Power: European Patent-System Integration in the Context of Globalization," *Technology and Culture* 48, 4 (2007): 689–728. Monsanto's PCT patent application for the CP4 invention, with international publication number WO 92/04449, is available at https://worldwide.espacenet.com/publicationDetails/originalDocument?CC=WO&NR=9204449A1&KC=A1&FT=D&ND=3&date=19920319&DB=&locale=en_EP.

16 See Government of Canada, Canadian Food Inspection Agency, "DD1995-02: Determination of Environmental Safety of Monsanto Canada Inc.'s Roundup® Herbicide-Tolerant Brassica Napus Canola Line GT73," March 1995, http://www.inspection.gc.ca/plants/plants-with-novel-traits/approved-under-review/decision-documents/dd1995-02/eng/1303706149157/1303751841490#a3. See also Monsanto Company, "Petition for Determination of Nonregulated Status: Soybeans with a Roundup Ready™ Gene," September 14, 1993, https://www.aphis.usda.gov/brs/aphis docs/93_25801p.pdf. The approval by the US Department of Agriculture, Animal and Plant Health Inspection Service, is available at https://www.aphis.usda.gov/brs/aphisdocs2/93_25801p_com.pdf.
17 *Re Application of Abitibi Co.* (1982), 62 CPR (2d).
18 Ibid.
19 This statement is based upon my examination of the application file for this patent held at the Canadian Patent Office.
20 Dilip M. Shah, Stephen G. Rogers, Robert B. Horsch, and Robert T. Fraley, "Glyphosate-Resistant Plants," Canadian Patent 1,313,830, filed August 6, 1986, and issued February 23, 1993.
21 This rule applies not only in Canada but is also a basic principle of patent law globally.
22 The prohibition against double patenting is not only a Canadian patent law provision but also a standard provision of patent law around the world.
23 Padgette, Barry, and Kishore, "Glyphosate Tolerant 5-Enolpyruvylshikimate-3-Phosphate Synthases," Canadian Patent 2,088,661.
24 Monsanto employee Aaron Mitchell demonstrated the quick test during the initial Schmeiser trial and confirmed that it specifically tested for the presence of the CP4 gene. See *Monsanto Canada Inc. v Schmeiser*, FCC A-367–01, document 21, trial transcript, testimony of Aaron Mitchell, 511–22.
25 Monsanto Company, written submission in preparation to/during oral proceedings, February 11, 2000, European Patent Office, European Patent Register, https://register.epo.org/application?number=EP91917090&lng=en&tab=doclist. Monsanto's submissions to the European Patent Office were made in response to a legal proceeding initiated by companies that opposed the European patent for Monsanto's CP4 invention. (Recall that Monsanto's European and Canadian patent applications for the CP4 invention came from a single PCT filing.)
26 Monsanto Company, reply of the patent proprietor to the notice(s) of opposition, April 30, 1998, 14, European Patent Office, European Patent Register, https://register.epo.org/application?number=EP91917090&lng=en&tab=doclist.
27 A. Menges, letter to the European Patent Office re Monsanto Patent 0546 090, March 6, 2005, 11, 91917090–2005–03–06-APPEAL-ORAL-Letter dealing with oral proceedings during the appeal procedure, European Patent Office, European Patent Register, https://register.epo.org/application?number=EP91917090&lng=en&tab=doclist.
28 These were the words used by one of the inventors for Patent 1,313,830, Robert Fraley, quoted in Josh Flint, "Roundup Ready Retrospective: 15 Years Post Release," *Dakota Farmer*, July 2011, 27, http://magissues.farmprogress.com/DFM/DK07Jul11/dfm027.pdf. See also Charles, *Lords of the Harvest*, 65–69. Monsanto researchers

published a scientific account of this invention in 1986: Dilip M. Shah, Robert B. Horsch, Harry J. Klee, Ganesh M. Kishore et al., "Engineering Herbicide Tolerance in Transgenic Plants," *Science* 233, 4762 (1986): 478–81.
29 Patrick D. Kelly, "Drafting a Patent Application," *Bent of Tau Beta Pi* 93, 4 (2002): 23, https://www.tbp.org/pubs/Features/F02Kelly.pdf.
30 Ibid. According to Canadian patent authority David Vaver, "the game for patentees, especially in highly competitive industries, is to reveal as little and to claim as much as possible." David Vaver, *Intellectual Property Law: Copyright, Patents, Trade-Marks* (Toronto: Irwin Law, 1997), 139. On the trend toward "speculative patenting" in biotechnology, see I. Mgbeoji and B. Allen, "Patent First, Litigate Later! The Scramble for Speculative and Overly Broad Genetic Patents: Implications for Access to Health Care and Biomedical Research," *Canadian Journal for Law and Technology* 2, 2 (2003): 83–98.
31 Kelly, "Drafting a Patent Application," 22.
32 Shah et al., "Glyphosate-Resistant Plants," Canadian Patent 1,313,830, 68.
33 Patent 1,313,830 is analogous to the infamous Selden automobile patent. In each case, the wording of the claim was deliberately and carefully structured to cast as wide a net as possible over existing and future technical practice in its domain. The Selden patent was eventually invalidated for lack of "enablement." See William Greenleaf, *Monopoly on Wheels: Henry Ford and the Selden Automobile Patent*, rev. ed. (Detroit: Wayne State University Press, 2011).
34 Padgette et al., "Glyphosate Tolerant 5-Enolpyruvylshikimate-3-Phosphate Synthases," Canadian Patent 2,088,661.
35 Ibid., 110.
36 Canadian Intellectual Property Office, *Manual of Patent Office Practice*, Chapter 9, Section 9.02.05.
37 Vaver, *Intellectual Property Law*, 138–39.
38 High "catalytic efficiency" was essential for crop plants to maintain vigorous growth along with glyphosate tolerance. The petunias of Monsanto's earlier patent application were yellow and sickly because their (modest) level of glyphosate tolerance came together with their lowered catalytic efficiency.
39 *Pioneer Hi-Bred Ltd. v Canada (Commissioner of Patents)*, [1987] 3 FC 8, 77 NR 137.
40 *Pioneer Hi-Bred Ltd. v Canada (Commissioner of Patents)*, [1989] 1 SCR 1623.
41 Busch, "Genetically Modified Plants Are Not Inventions."
42 Padgette et al., "Glyphosate Tolerant 5-Enolpyruvylshikimate-3-Phosphate Synthases," Canadian Patent 2,088,661, 55–63.
43 Evidence from a series of Monsanto patents (not all discussed in this chapter) and from research articles published by Monsanto scientists shows that all of the phenomena described in this paragraph were at play in the development of RR canola. The reliance of genetic engineering on chance and luck is analyzed further in Kathleen McAfee, "Neoliberalism on the Molecular Scale: Economic and Genetic Reductionism in Biotechnology Battles," *Geoforum* 34, 2 (2003): 203–19.
44 Kneen, *Rape of Canola*, 64–65; Michael E. Gorman, Jeanette Simmonds, Caetie Ofiesh, Rob Smith, and Patricia H. Werhane, "Monsanto and Intellectual Property," *Teaching Ethics* 2, 1 (2001): 91–100.

45 Jeroen van Wijk, "Plant Breeders' Rights Create Winners and Losers," *Biotechnology and Development Monitor* 23 (1995), http://biotech-monitor.nl/2306.HTM. See also Weiss, "Seeds of Discord."
46 Gorman et al., "Monsanto and Intellectual Property," 94.
47 It is not impossible to develop hybrid varieties of these crops, but for various reasons, such as cost and complexity, this was not done, at least not until recently.
48 *Monsanto Canada Inc. v Schmeiser*, FCC A-367–01, document 21, trial transcript, testimony of Aaron Mitchell, 565.
49 Quoted in Kneen, *Rape of Canola*, 65.
50 *Monsanto Canada Inc. and Monsanto Company v Percy Schmeiser*, FCC (Trial Division), Case 1593–98, document 8, Monsanto motion record, affidavit of Carolyn Chambers, August 6, 1998.
51 *Monsanto Canada Inc. and Monsanto Company v. Percy Schmeiser*, FCC (Trial Division), Case 1593–98, document 57, exhibit A, affidavit of Aaron Mitchell, August 10, 1999.
52 Ibid., exhibit B.
53 Monsanto published a *Seed Piracy* newsletter, which included its hotline number. See a sample at http://www.monsanto.ca/newsviews/Documents/seed_piracy_newsletter_oct_2009.pdf.
54 *Percy Schmeiser et al. v Monsanto Canada Inc. et al.*, FCA A-367–01, document 20, appeal book, vol. 2, exhibit 4.
55 *Schmeiser v Monsanto*, FCC, Case 1593–98, affidavit of Aaron Mitchell, exhibit A. (This exhibit is a copy of the Grower's Agreement.)
56 Ibid.
57 Pechlaner, *Corporate Crops*; Michael Stumo, "Down on the Farm: Farmers Get the Biotech Blues," *Multinational Monitor* 21, 1–2 (2000): 17–22.
58 Evans, *The War on Weeds*, 108.
59 Ibid., 16.
60 Among Canadian crop plants, canola was known as "one of the more difficult crops to grow with respect to weed control" because it had many weedy relatives. Their genetic similarities to canola meant that weed killers that controlled them would also harm the canola. See Pechlaner, *Corporate Crops*, 80.
61 *Monsanto Canada Inc. v Schmeiser*, 2001 FCT 256, para 17.
62 It is not my intention to suggest that Monsanto alone transformed the political economy of agriculture through biotechnology, though the company did push these systems in new ways, with significant consequences. Nevertheless, Monsanto operates within the broader contexts of corporate capitalism, industrial agriculture, intellectual property systems, and the rise of biotechnology, and ultimately these are the systems that have been transforming agricultural regimes around the world. Several studies help to place Monsanto's activities within these larger contexts. In addition to the works cited in note 6, see also (in chronological order) Cary Fowler, Eva Lachkovics, Pat Mooney, and Hope Shand, "The Laws of Life: Another Development and the New Biotechnologies," *Development Dialogue*, 1988, 1–2, http://www.daghammarskjold.se/publication/laws-life-another-development-new-biotechnologies/; Cary Fowler and Pat Mooney, *Shattering: Food, Politics, and the Loss*

of Genetic Diversity (Tucson: University of Arizona Press, 1990); Robert Bud, *The Uses of Life: A History of Biotechnology* (Cambridge, UK: Cambridge University Press, 1994); Pat Roy Mooney, "The Parts of Life: Agricultural Biodiversity, Indigenous Knowledge, and the Role of the Third System," *Development Dialogue*, 1996, 1–2, http://www.daghammarskjold.se/publication/parts-life-agricultural-bio diversity-indigenous-knowledge-role-third-system/; Brewster Kneen, *Farmageddon: Food and the Culture of Biotechnology* (Gabriola Island, BC: New Society, 1999); Fitzgerald, *Every Farm a Factory*; and Abby J. Kinchy, *Seeds, Science, and Struggle: The Global Politics of Transgenic Crops* (Cambridge, MA: MIT Press, 2012).

63 Norwich BioScience Institutes, "Shatter-Resistant Pods Improve Brassica Crops," *Science Daily*, May 27, 2009, https://www.sciencedaily.com/releases/2009/05/0905 27151134.htm.

64 Meredith G. Schafer, Andrew A. Ross, Jason P. Londo, Connie A. Burdick et al., "The Establishment of Genetically Engineered Canola Populations in the US," *PLoS One* 6, 10 (2011): e25736.

65 See Lindsey Konkel, "The Great Escape: Gene-Altered Crops Grow Wild," *Environmental Health News*, January 27, 2012, http://ehn.org/ehs/news/2012/gm-crops-escape-and-grow-wild.

66 Shah et al., "Glyphosate-Resistant Plants," Canadian Patent 1,313,830, 10.

67 The Canadian Food Inspection Agency rules for "confined research field trials" are posted on its website at http://www.inspection.gc.ca/plants/plants-with-novel-traits/approved-under-review/field-trials/eng/1313872595333/1313873672306.

68 Sally L. McCammon and Sue G. Dwyer, eds., *Workshop on Safeguards for Planned Introduction of Transgenic Oilseed Crucifers: Proceedings* (Ithaca, NY: Cornell University, 1990), https://archive.org/stream/CAT10750314?ui=embed#page/n1/mode/2up.

69 Government of Canada, Canadian Food Inspection Agency, "DD199502: Determination of Environmental Safety of Monsanto Canada Inc.'s Roundup® Herbicide-Tolerant Brassica Napus Canola Line GT73," March 1995, http://www.inspection.gc.ca/plants/plants-with-novel-traits/approved-under-review/decision-documents/dd1995-02/eng/1303706149157/1303751841490#a3.

70 *Monsanto Canada Inc. v Schmeiser*, FCC A-367–01, document 21, trial transcript, testimony of Aaron Mitchell, 564.

71 Patent 2,088,661, issued in 2001, fell under a new Canadian rule: it was valid for twenty years from the date of application. Monsanto could not sue anyone under this patent until it was issued. If the company had waited until 2001, it would have lost six years of income from the patent – nearly a third of the patent's lifespan. In principle, Monsanto could have sued retroactively after the patent was issued, but by then RR canola would have spread extensively enough to affect the company's ability to control the legal situation (e.g., by class action lawsuits or less lenient judicial decisions), because many conventional farmers' canola fields would have been contaminated.

72 *Monsanto Canada Inc. and Monsanto Company v Percy Schmeiser*, FCC, Case 1593–98, document 57, affidavit of Aaron Mitchell, August 10, 1999, 5; *Monsanto Canada Inc. v Schmeiser*, FCC A-367–01, document 21, trial transcript, testimony of Aaron Mitchell, 472–73, 560.

73 *Monsanto Canada Inc. v Schmeiser*, 2001 FCT 256; *Monsanto Canada Inc. v Schmeiser*, 2002 FCA 309; *Monsanto Canada Inc. v Schmeiser*, [2004] 1 SCR 902, 2004 SCC 34.
74 Kelly, "Drafting a Patent Application," 23.
75 Ibid., 18.
76 Ian Binnie, "Science in the Courtroom: The Mouse that Roared," *University of New Brunswick Law Journal* 56 (2007): 320.
77 Vaver, *Intellectual Property Law*, 140.
78 Schmeiser says that on one occasion Monsanto had nineteen lawyers in court. See the talk that Schmeiser gave at Western Washington University, October 21, 2011, https://www.youtube.com/watch?v=RUwuNR42qP0.
79 A technical challenge of Patent 1,313,830 would have required a legal team expert in biotechnology patent law. Such a case could have been mounted because crucial evidence was available in 1998. Most importantly, the Canadian application for the CP4 invention (which later issued as Canadian Patent 2,088,661) was published and publicly available years before Monsanto sued Schmeiser.
80 *Monsanto Canada Inc. v. Schmeiser*, FCC A-367-01, document 21, trial transcript, testimony of Doris Dixon, 239; emphasis added.
81 *Monsanto Canada Inc. v Schmeiser*, FCC A-367-01, document 21, trial transcript, testimony of Robert Horsch, 70.
82 Ibid., 70–71.
83 Ibid., 71.
84 One important issue that this chapter neglects because of space is the relationship between the Supreme Court's 2004 decision in the Schmeiser case and its 2002 decision in the Harvard University OncoMouse case. The OncoMouse decision, which ruled that Harvard's patent claim for a "higher life form" was unpatentable subject matter, opened the door to Schmeiser's Supreme Court appeal. *Harvard College v Canada (Commissioner of Patents)* [2002] 4 SCR 45, 2002 SCC 76.
85 Binnie, "Science in the Courtroom," 311.
86 Several legal scholars critiqued the court's handling of genetic drift in the Schmeiser case. See, for example, Bruce Ziff, "Travels with My Plant: *Monsanto v. Schmeiser* Revisited," *University of Ottawa Law and Technology Journal* 2, 2 (2005): 493–509; and Katie Black and James Wishart, "Containing the GMO Genie: Cattle Trespass and the Rights and Responsibilities of Biotechnology Owners," *Osgoode Hall Law Journal* 46, 2 (2008): 397–425.
87 Busch, "Genetically Modified Plants Are Not Inventions," discusses aspects of the courts' failure to grapple adequately with where to draw the line between nature and invention in the Schmeiser case.
88 Jeremy de Beer and Robert Tomkowicz, "Exhaustion of Intellectual Property Rights in Canada," *Canadian Intellectual Property Review* 25, 3 (2009): 15, 23.
89 Ibid., 16.
90 Busch, "Genetically Modified Plants Are Not Inventions," 445; Jeremy de Beer, "The Rights and Responsibilities of Biotech Patent Owners," *UBC Law Review* 40, 1 (2007): 343–73; Sean Robertson, "Re-Imagining Economic Alterity: A Feminist Critique of the Juridical Expansion of Bioproperty in the Monsanto Decision at the Supreme Court," *University of Ottawa Law and Technology Journal* 2, 2 (2005): 227–53.

91 De Beer and Tomkowicz, "Exhaustion of Intellectual Property Rights in Canada," 23.
92 *Monsanto Canada Inc. v Schmeiser*, [2004] 1 SCR 902, 2004 SCC 34, para 85.
93 The quotation is from Robert Stack, "How Do I Use This Thing? What's It Good for Anyway? A Study of the Meaning of Use and the Test for Patent Infringement in the *Monsanto Canada Inc. v. Schmeiser* Decisions," *Intellectual Property Journal* 18 (2004): 277. The dissenting opinion in the Supreme Court judgment cited crucial patent law principles that the majority decision ignored. Examples of critiques by scholars (additional to those already cited) are Martin Phillipson, "Giving Away the Farm? The Rights and Obligations of Biotechnology Multinationals: Canadian Developments," *King's Law Journal* 16, 2 (2005): 362–72; de Beer, "The Rights and Responsibilities of Biotech Patent Owners"; Wendy A. Adams, "Confronting the Patentability Line in Biotechnological Innovation: *Monsanto Canada Inc. v. Schmeiser*," *Canadian Business Law Journal* 41 (2005): 393–412; Wendy Adams, "Determinate/Indeterminate Duality: The Necessity of a Temporal Dimension in Legal Classification," *Alberta Law Review* 44, 2 (2006): 403–28; and Wilhelm Peekhaus, "Primitive Accumulation and Enclosure of the Commons: Genetically Engineered Seeds and Canadian Jurisprudence," *Science and Society* 75, 4 (2011): 529–54.
94 E. Ann Clark, "The Implications of the Percy Schmeiser Decision," *Synthesis/Regeneration* 26 (2001): 1, http://www.greens.org/s-r/26/26-08.html; Busch, "Genetically Modified Plants Are Not Inventions," 389.
95 Bruno Latour, *We Have Never Been Modern*, trans. Catherine Porter (Cambridge, MA: Harvard University Press, 1993), argues, in effect, that modernity has always been in disarray because it is based upon the false ideology that nature and society are entirely separate realms. Yet we remain "modern," he says, as long as we ignore the discrepancy between this ideology and lived reality. Viewed from this perspective, the story told in this chapter shows how the Schmeiser case and RR canola made the discrepancy increasingly visible and therefore more problematic.
96 Scott, *Seeing like a State*, 89–90.
97 Following the launch of the Schmeiser lawsuit, Monsanto undertook a series of analogous lawsuits in the United States, and they similarly called into question the tenets of agricultural modernity.
98 The quotation is from a speech that Schmeiser gave at the Canadian Health Food Association (CHFA) Expo West, n.d., http://old.globalpublicmedia.com/transcripts/237. In a talk that he gave at Western Washington University (see note 78), he emphasized that he had no idea prior to the lawsuit that patents on plant parts even existed.
99 Donald L. Barlett and James B. Steele, "Monsanto's Harvest of Fear," *Vanity Fair*, May 2008, https://www.vanityfair.com/news/2008/05/monsanto200805; Weiss, "Seeds of Discord."
100 Schmeiser, speech at CHFA Expo West, http://old.globalpublicmedia.com/transcripts/237.
101 Stratus Ag-Research, "One Million Acres of Glyphosate Resistant Weeds in Canada: Stratus Survey," blog, May 29, 2013, http://www.stratusresearch.com/newsroom/one-million-acres-of-glyphosate-resistant-weeds-in-canada-stratus-survey.

102 Latour explains in *We Have Never Been Modern* how scientists make this division by defining what constitutes "nature" and bracketing it off conceptually from everything else. Think of scientists' role, for example, in removing magic from the domain of nature.
103 The Supreme Court decision offered a superficial parallel between Lego blocks and plant cells to support the conclusion that patenting a canola plant cell gave control over the entire plant, but the analogy ignored the crucial point that the plants have various types of cells that function differently and interact in complex yet crucial ways and that Monsanto only put the modified gene into one type of plant cell.
104 The quotation is from *Monsanto Canada Inc. v Schmeiser*, [2004] 1 SCR 902, 2004 SCC 34, para 35, quoting an earlier Supreme Court decision. The Supreme Court has repeatedly affirmed that the fundamental objective of the Patent Act is to promote the development of *society* by giving inventors a temporary monopoly on their creations in return for their sharing all of the new knowledge and know-how that they embody. See, for example, *Pioneer Hi-Bred Ltd. v Canada (Commissioner of Patents)*, [1989] 1 SCR 1623.
105 *Monsanto Canada Inc. v Schmeiser*, [2004] 1 SCR 902, 2004 SCC 34, para 2.
106 Ibid., para 90.
107 This assertion is based upon a diversity of sources consulted over the past decade; it would be a disproportionate and tedious exercise to list them here.
108 For example, Syngenta, Dow, and Bayer now use grower agreements like Monsanto's; some can be found on the Internet. Bayer's agreement prohibits seed saving and allows Bayer to examine and copy relevant financial records of the grower. A copy of Bayer's "Grower Trait Licensing Agreement" is available at https://legendseeds.net/wp-content/uploads/2014/07/2010_Bayer_GTL.pdf.
109 See, for example, Scott Prudham, "The Fictions of Autonomous Invention: Accumulation by Dispossession, Commodification, and Life Patents in Canada," *Antipode* 39, 3 (2007): 406–29; Peekhaus, "Primitive Accumulation and Enclosure of the Commons"; Robertson, "Re-Imagining Economic Alterity"; and Wenwei Guan, "The Poverty of Intellectual Property Philosophy," *Hong Kong Law Journal* 38, 2 (2008): 359–97.
110 This statement can be found on the Agriculture and Agri-Food Canada website, http://www.agr.gc.ca/eng/industry-markets-and-trade/statistics-and-market-information/by-product-sector/organic-products/organic-production-canadian-industry/?id=1183748510661.
111 I am arguing here not that Monsanto's GMOs or the Schmeiser case *caused* these developments but that they contributed to their diffusion and impact. For example, there has been an upsurge of studies critiquing the reductionist science promoted by the biotechnology industry, such as McAfee, "Neoliberalism on the Molecular Scale"; and Prudham, "The Fictions of Autonomous Invention."
112 GM-Free Ireland Network, "News Release, Dublin, 24 May 2006 – BASF Admits Defeat of GMO Potato Experiment," http://www.global-vision.org/press/GMFI/GMFI26.pdf.
113 Ted Nace, "Breadbasket of Democracy," in *Money and Faith: The Search for Enough*, ed. Michael Schut (Denver: Church, 2008), 117–27; Kelly Bronson, "What We Talk

about When We Talk about Biotechnology," *Politics and Culture*, 2009, 2, 80–94; Sean Pratt, "Schmeiser Decision May Help Organic Case," *Western Producer*, May 27, 2004, https://www.producer.com/2004/05/schmeiser-decision-may-help-organic-case/. An in-depth analysis of the Canadian movement to stop RR wheat is presented in Emily Eaton, "Getting behind the Grain: The Politics of Genetic Modification on the Canadian Prairies," *Antipode* 41, 2 (2009): 256–81.

114 *Monsanto Canada Inc. v Schmeiser*, FCC A-367-01, document 21, trial transcript, testimony of Percy Schmeiser, 877–78.
115 "Percy Schmeiser," Monsanto, https://www.monsanto.com/newsviews/pages/percy-schmeiser.aspx.
116 *Monsanto Canada Inc. v Schmeiser*, FCC A-367-01, document 21, trial transcript, testimony of Percy Schmeiser, 813–15.

PART 3

ENVIRONMENTS

10 Landscapes of Science in Canada
Modernity and Disruption

STEPHEN BOCKING

In one of his first publications – a book of primary sources coedited with Trevor Levere – Richard Jarrell demonstrated the diverse ways in which Canadians have known the world through science.[1] This collection also illustrated how this diversity extended to the places of science: the territories of inventory science; agricultural, forestry, and fisheries research stations; sites of experimentation in field and lab. In these places, science became, as my colleagues in this volume have argued, central to the history of Canada: contributing to the formation of the state and to the development of an industrial economy, while shaping Canadians' relations with their environment.[2]

In this chapter, I complement the analyses of my colleagues by evaluating the historical roles of science in these relations. My agenda is informed, in part, by the recent work of historians of science. Widening their view beyond the formation of knowledge itself, and informed by postcolonial and constructivist perspectives, historians have examined scientists' practices in the field and elsewhere; the influence of diverse social, institutional, and geographical contexts; the roles of a wider range of participants; and the relations and boundaries among science and other forms of knowledge.[3] Science has also captured the attention of environmental historians: they have explored its environmental consequences, its roles in guiding or justifying manipulation of the environment, and the material and intellectual relations between science and landscape, as expressed through the practices of science, contention concerning what counts as "natural," and the reciprocal interactions between scientific and environmental change.⁴ To some extent, these perspectives have converged. Historians of science and environmental historians now share an interest in "putting science in its place": understanding it – as knowledge, practice, or instrument – in relation to where it is practised.

Canadian environmental historians, like their counterparts elsewhere, have noted these historical roles of science. Recent historiographical analyses have considered, for example, the role of scientific cultures and institutions in forming national and regional identities, and in defining the Canadian landscape as wild or domesticated. They have also identified a persistent confidence, aligned with the formation of state agencies, in science as a basis for agricultural development, natural resource exploitation, and administration of urban environments. Historians have also drawn attention to the movement of knowledge – often in association with mobile nature, people, capital, or commodities – and to the relations among mobile knowledge and the multiplicity of spaces and places in the Canadian landscape. Finally, these analyses have identified instances of contrast, contention, and resistance in the relations between scientific expertise and locally situated ways of knowing (particularly Indigenous knowledge), or between knowledge applied by the state and by nonstate agencies, and in the diverse forms of knowledge linked to differences in gender, class, and work.⁵

This chapter contends that these diverse observations of the weaving of science into people's relations with nature can be interpreted in terms of Canada's experience of the condition of modernity. Among the essential features of modernity, in Canada and elsewhere, have been the primacy of the nation-state as the basis of social organization and the accompanying transformation of people into citizens. Within this framework, science, as an instrument of both the state and economic actors, has provided the foundation for rational action through the systematic gathering of facts and the formation of expert advice and the application of this advice to state and social priorities. Science provided universal descriptions of nature, implying a universal model of society – as was evident in theories of economic development through technological modernization, with action, ambition, and agency defined in terms of the individual and the public good defined in terms of efficiency. Science therefore became the driving force of modernization, justifying its unique status, ample funding, and institutional security. In contrast, subjective ethical considerations became matters of the private sphere.⁶

These ideas regarding science and modernity have been founded upon assumptions regarding the physical world: that it could be understood through reason, with events being the products of causes that have predictable consequences consistent with the rules of probability. Objective science and individual agency, acting in concert with a supportive nation-state, could therefore provide the basis for confidence in continual progress through the prediction and control of nature. However, modernity has also always contained within it the potential for resistance, particularly from those social groups that have dissented from these assumptions and so became defined as outside modernity. These groups have included Indigenous people, communities resisting capitalistic transformation of the landscape, and others wary of the implications of modernity.[7] Of particular relevance to the history of science and environmental history is that this resistance has been expressed, in part, through the assertion of other ways of knowing as well as by nature itself, through its own unpredictability, accelerated in recent decades by climate change and the advent of the Anthropocene.

These general features of modernity have been expressed in different ways in particular places, and any analysis must acknowledge this specificity of the experience of the modern. But how can we do this in the Canadian context? I will address this question through a historiographic synthesis of the environmental history of science in Canada, teasing out essential findings and concerns from the collective work of Canadian historians and historical geographers. This synthesis will be framed in terms of a central dynamic of the modernist enterprise: the relations among scientists seeking (in concert with other agents) to understand and manipulate an often unruly nature, and an array of other actors pursuing their own – and often contrasting – aims. Nonhuman actors have also been parts of these relations, forming a variable and complex Canadian environment that has often frustrated human ambitions to know and to control.

Recent work by historians of science and environmental historians can provide much guidance as we explore the relations between scientists and modernity in Canada. I will examine these relations in terms of four themes, each central to the history of modernity. The first is territory: the extension and contestation of the authority of the state, economic interests, or science itself over landscapes. The second is transformation, as landscapes became raw material for the modern economy. The third is administration: how the environment itself became a focus of controversy and regulation. And the fourth is how ideas and assumptions about science and human relations with nature have been disrupted. Although I will discuss each of these themes in sequence, they are only loosely chronological – indeed, as we will see, they do not succeed each other so much as represent the accretion of new values, priorities, and forms of knowledge onto pre-existing relations between Canadians and the landscape. All remain significant today.

Territory

In the 1840s, in Mississauga territory that is now part of central Ontario, David Fife bred from a few European grains a wheat variety that became known as Red Fife. This hardy variety enabled grain farmers to extend their cultivation into the north and west of Canada – expressing in biological form one relation between knowledge and territory. Such relations are a staple of postcolonial scholarship: discovering, naming, and mapping have served as a basis for colonial authority, defining conquered territory and how it must be understood. In Canada, science reinforced views of territory as "empty" space onto which modernist relations between society and nature, like those embodied in Red Fife wheat, could be projected, with apparent limits – of ignorance, nature, or pre-existing inhabitants – merely obstacles to be overcome. Thus, surveyors mapped settlement grids that accommodated property rights but disregarded local environmental conditions. Hardy grain varieties exemplified the role of experimental farms in extending agricultural territory. Beginning in 1842, the Geological Survey of Canada advanced inventory science by mapping the nation's rocks and resources. In 1880, John Macoun reported optimistically on the potential of the arid southern prairie region, reinforcing the view of Canada as a transcontinental nation open to European settlement. Two decades later a new station for fisheries and marine research extended into salt water the territory of the state and of science.[8] Together these institutions epitomized the relation between science and state territorial authority essential to modernity.

In northern Canada, these relations between knowledge and territory took distinctive forms. After 1900, scientific activities served as instruments for asserting Canada's northern sovereignty. These activities included expeditions – most famously the Canadian Arctic Expedition of 1913–18, discussed by Andrew Stuhl in this volume – as well as field studies and attempts to domesticate reindeer and other wildlife. As Tina Adcock explains in her chapter, the 1925 ordinance requiring scientific fieldworkers to obtain permission to study in the Northwest Territories was also a means of asserting sovereignty. After the Second World War, new national agencies enlisted scientists in extending Canada's epistemic sovereignty by asserting its capacity not only to regulate scientists of other countries but also to produce its own knowledge of its landscapes. Instead of relying on foreign experts to interpret the results of the Canadian Arctic Expedition (as Stuhl explains in Chapter 11), Canadian scientists of the Polar Continental Shelf Project were now producing knowledge of atmospheric or geophysical phenomena, and the Canadian Wildlife Service and the Fisheries Research Board, not the Hudson's Bay Company and Oxford's Bureau of Animal Population, were now the leading agencies for northern wildlife and fish research.[9]

The territorial implications of science became particularly evident during the Cold War. The Defence Research Board supported northern science because of its value in guiding military activity and because having Canadian scientists in the North was a cheap way of asserting sovereignty. Strategic imperatives also provoked a novel relation between science and territory by demanding total surveillance through aerial surveys of land, sea, ice, and atmosphere, geophysical surveys, and satellite reconnaissance. The consequences included new research objects, such as sea ice, weather, and radio waves in the upper atmosphere.[10] Research in geology, oceanography, and atmospheric chemistry also encompassed the "vertical territory" beneath rock, ice, and water. Most recently, geological surveys of the continental shelf have supported Canada's case for jurisdiction over part of the Arctic Ocean.[11]

Yet this modernist view of state and science extending in concert authority over territory – once a historiographic staple – is also incomplete.[12] The Canadian landscape has been a multiplicity of territories, shifting with evolving views of places and species. At first, these were Indigenous spaces, known through experience, skilled practice, and social ties.[13] Although at first newcomers relied on the knowledge of Indigenous peoples, they soon redefined these landscapes as wilderness frontiers awaiting settlement.[14] Fish and wildlife regulations imposed state power over rural and Indigenous folkways and practices: particularly in northern Canada, the extension of territorial authority implied transformation of Indigenous peoples' relations with the land and with other species.

Furthermore, in Canada, the state is not a unitary entity. Scientists have been implicated in issues of provincial and federal territory. As provinces assumed jurisdiction over land and water, they also developed scientific capacities. The Ontario Fisheries Research Laboratory was created in 1920 to align university scientists with provincial fisheries managers; other provinces made similar arrangements. Provincial authority over forests was accompanied by science-based regimes; in British Columbia, forestry expertise justified the division, beginning in the 1940s, of the province into management units that ignored local and Indigenous territories. In the 1960s, scientific surveys of hydropower and other resources extended provincial authority into northern Quebec. Scientists themselves enlisted nature in constitutional interpretation: early in the twentieth century, the definition of agricultural pests as a federal or provincial concern hinged on how entomologists described outbreaks.[15] In the 1960s, studies in British Columbia of logging near salmon streams exposed another jurisdictional tangle: fish were a federal responsibility, but the streams and surrounding forest were provincial responsibilities. In practice, however, even scientific evidence of the impacts on salmon could not persuade the federal government to regulate the forest industry.[16]

Economic authority over territory has also meant epistemic authority, exercised by forest companies, power utilities, and mining companies. In recent decades, science has served to extend corporate territorial authority. Geologists (working for both the state and private interests) have used airborne observation, photography, and detection of magnetic anomalies to push mineral exploration and exploitation into new territories. Resource surveys identifying oil, gas, and other minerals, including diamonds, have helped to integrate the North into the national economy, with novel engineering technologies extending oil and gas exploitation into ever-more-difficult northern environments.

Landscapes themselves can assert authority, demanding responses from scientists and reconsideration of the modernist relation between science and the state's territorial authority. In the 1950s, McGill University geographers developed aerial strategies for vegetation surveys, and biologists at the University of British Columbia used tags to track migrating salmon because they believed that these strategies were necessary to understand phenomena on those scales.[17] The history of field science presents many instances of local ecological conditions shaping opportunities for research.[18] Difficult working conditions along stormy coastlines or in the North have also imposed requirements on scientists' practices, training, and exchanges with local people. But in return scientists have asserted their own intellectual dominion over landscapes. University science faculties, agencies' requirements that staff have scientific training, and access to modern technology, including airplanes, have all helped to extend this dominion. Canadian scientists train in foreign universities elsewhere and join international scientific communities; as a result, assertions of national epistemic authority have often drawn on scientific frameworks developed elsewhere – as when

Canadian Wildlife Service biologists applied what they had learned in American schools of wildlife science. This indifference to political boundaries has also been evident in scientists' spatial perspectives on nature. One such perspective is the life zone concept that Stuhl discusses in his chapter. More recently, descriptions of the Arctic as particularly sensitive to global warming have reinforced perceptions of the area as international territory. These spatial perspectives have been the basis for "disciplinary spaces" within which scientists can assert their status as authoritative experts.[19] Scientists have also argued that their knowledge should determine the geography of governance so that it corresponds to the "natural" organization of nature. Soon after its creation in 1909, the International Joint Commission recommended that a single organization be responsible for shared waters. In the more than a century since then, scientists have argued that many issues – such as migrating species, contaminants, and climate change – require cooperation across borders.[20] Yet this extension of scientific authority has been an ambiguous process. Scientists in the field have often relied on people with local knowledge (even if they have not always been acknowledged).[21] Local knowledge has remained essential to networks of power and mobility: "triumphs" of engineering science, such as big dams and the St. Lawrence Seaway, were built upon a foundation of knowledge of local environments.[22]

Transformation

In 1913, Bernhard Fernow, dean of the University of Toronto's Faculty of Forestry, considered the Trent Watershed in central Ontario. And he was appalled:

> At the present time, the pine timber, at least, is practically gone from this watershed. A forest cover still exists, but, with the present commercial value almost entirely extracted interest in its condition is gone; fires have swept through it repeatedly, each time causing further deterioration of the forest cover, until, finally, the bare rock condition or man-made desert is the result.

The irrationality of it all was equally evident:

> If the present policy of indifference and neglect continues, what might have been a continuous source of wealth will become not only a useless waste, but, through the changes which the water conditions will undergo, may also

prove a menace to industries which have been developed to utilize the water-powers of this watershed.[23]

The Trent Watershed survey was published as a report of the Commission of Conservation. Until its demise in 1921, the commission sought to apply science to the relations between people and nature. The survey also addressed the commission's immediate priorities: to survey water power, forests, and other resources and to find ways to avoid waste and pursue efficiency, as defined in terms of the requirements of a modern industrial economy.[24]

The survey's conclusions were also rooted in this place: degraded forests prone to fires, thin soils atop unforgiving granite, a declining timber industry and rural economy. Nearby Algonquin Provincial Park (then only twenty years old) was demonstrating one approach to regulating access to land: allowing timber harvesting and wilderness recreation, excluding Indigenous hunting and agricultural settlement. It and other forest reserves (e.g., Laurentides Park in Quebec) were considered exemplars of conservation ideals: they owed their existence to timber interests seeking to exclude settlement, sporting interests opposed to subsistence hunting, and city dwellers' desires to get away from it all.[25] Conservation was as much about access and control as it was about scientific management. It also reflected larger changes under way. Canada was becoming modern: urbanized and industrialized, with expanding cities and resource industries, the reach of both being extended by roads, railways, and electricity networks. Private property was primary, state and capital collaborated in resource exploitation, and the dispossession of Indigenous peoples was considered inevitable.

In this context, the modernist concept of transformation provided discursive and material links between nature, society, and science. Nature was transformed into natural resources: rivers were merely water to drive turbines, and forests were merely timber. As James Hull explains in his chapter in this volume, a central feature of the Second Industrial Revolution during these decades was the incorporation of science into technology and industry. Science was similarly invoked to guide and justify transformations of nature. In 1881, Samuel Wilmot, pioneer fish culturalist, lauded the "science of artificial fish-breeding."[26] Confidence in science and technology became a dominant theme in fisheries management: on both coasts, the Biological Board of Canada (after 1938 the Fisheries Research Board) applied science to counting, catching, and processing fish, seeking sustained yields and stable coastal economies.[27] Technology, it was hoped, could enact modernity, moderating the irregularities of nature, even resolving social conflicts over resources.

The hope that experts could ensure efficient and stable resource use and thereby a modern economy became evident in landscapes across the country. Experiments in scientific management were tried on the interior rangelands of British Columbia. In the Prairies, drought in the 1930s encouraged scientists of the new Prairie Farm Rehabilitation Administration to work with farmers on methods and tools to conserve soil. In Algonquin Provincial Park, multiple-use management – land-use zoning, in effect – was applied to enable the timber industry and recreational interests to coexist. Even the fur trade was to be made "scientific": fur farms, guided by animal husbandry and veterinary science, would become modern business enterprises, eschewing the unpredictability of fur-bearer populations in the wild.[28]

Similar hopes were held for cities facing demands for clean water and waste disposal. Expertise provided blueprints for organizing governments to provide these services. Public health officials were among a wave of professionals who attempted, beginning in the 1880s, to assert control over social and physical infrastructure, promising efficiency and reliability – thereby creating the modern city. One way of doing so was by redefining water quality to be a matter not of the senses but of the laboratory. As Hull notes in his chapter in the context of industry, this strategy provided a veneer of objectivity while shifting control from the water-drinking public to scientists. Experts became evidence of municipal competence and a source of civic pride, especially when their advice was consistent with political priorities. And especially if it was cheap: by 1915, sanitary engineers' advice to add chlorine to drinking water was being applied universally since it cost so much less than treating sewage (unfortunately, it also provided an excuse to allow local water to deteriorate).[29] After the Second World War, Canada's biggest cities transformed themselves in order to apply expertise more effectively: regional governments were created in Vancouver and Montreal, and Toronto created a new level of metropolitan government with the capacity to act on plans for pipes and highways. Engineers and other technical experts advised on reshaping urban spaces: reconfiguring waterways as urban waste systems and building networks for water, people, or vehicles.[30] The management of motion became their central task.

Human-wildlife relations have raised many issues about science and its relations to access, class, and ethics, subsistence versus sporting values, and the relations between urban and rural knowledge and interests, thereby illustrating some of the tensions accompanying the imposition of modernity. Historians have presented diverse perspectives on these issues. One view emphasizes the civil servants and scientists who established a national park system, the Convention for the Protection of Migratory Birds (1916), and other early conservation initiatives. Another focuses on local naturalist organizations and fish and game clubs. Some have described the formation of wildlife science, its application to forming a modern wilderness at the expense of local values and Indigenous ways of life, and the role of local knowledge in challenges to state initiatives such as game reserves and closed seasons.[31]

Among those who gained fame writing about nature – Ernest Thompson Seton, Jack Miner, Grey Owl, Roderick Haig Brown – some embraced science, while others drew more on their own experiences and antimodern sentiments. Government and university scientists also spoke for nature, sometimes expressing skepticism about modernity. Percy Taverner, James Munro, and other federal scientists; Arthur Coventry and his colleagues, who began in 1931 to push for protected areas through the Federation of Ontario Naturalists; and Ian McTaggart-Cowan, by the 1940s one of British Columbia's leading ecologists, were guided not just by scientific ideas, such as those of Aldo Leopold and Charles Elton, but also by their ethic of care for local places.[32]

Although early Canadian ecologists found their colleagues and teachers elsewhere (e.g., Elton in Great Britain and Leopold and Joseph Grinnell in the United States), in most regions of Canada their discipline became established by demonstrating its relevance to local issues: ranchlands and insect pests in the BC Interior, inland and marine fisheries management, reindeer and range productivity in the North, wildlife management in the Rocky Mountain national parks.[33] These ecologists shared a concern with variability: insect plagues, fluctuating fish stocks, cycles in populations of mammals and other species were all obstacles to stable resource economies. Accordingly, the factors influencing variability became central to ecological research agendas. Ecology also had local roots that reflected, as Adcock discusses in her chapter, the blurred distinctions between field science (including ecology) and other outdoor activities, such as sport and resource harvesting. McTaggart-Cowan, for example, took care to learn from trappers and wardens as he pursued his studies of ecology and biogeography.[34]

In the postwar era, enhancing the status of wildlife management meant making it modern: that is, redefining it as a science. When provincial fish and game departments hired biologists, degrees now outweighed hunting and trapping skills, and the Canadian Wildlife Service imposed stricter requirements for scientific training.[35] This widening distinction between local and scientific knowledge also framed policy debates as divergent views – for example of predator control – became matters of professional identity.[36]

Forests – the country's most extensive habitat and source of anxieties regarding waste and lost opportunities – prompted, as Hull notes in his chapter, the application of science to industrial innovation, through the Commission of Conservation, university forestry programs, and the Canadian Forest Product Laboratories. They also provoked efforts to use science to transform nature into natural resources. Silviculturalists bred high-yielding trees to reduce the variability inherent in natural forests. In the 1920s, Ontario forest companies concerned about wood supplies hired foresters to manage their lands, taking action even as the province remained reluctant to do so. In 1947, British Columbia adopted a "sustained yield" policy based upon annual allowable cuts, imparting a transformation imperative to corporate territory. In the same decade, entomologists in New Brunswick began a study of how to protect the forest industry from the spruce budworm. For several decades, scientists described the massive spraying of chemicals, including DDT, as necessary to restore "balance" to the forests; when the practice came under debate, both sides invoked science to support their views.[37] In each case, close relations formed between universities and industry, exemplifying how modernist assumptions regarding science had become central to Canadian resource development.

As Hull explains in this volume, Canada and the United States shared a continental pool of technological science. Scientists also considered themselves part of a transnational community of expertise focused on relations with nature. By the 1880s, there was substantial exchange between the United States and Canada on conservation practice, with Europe providing some models. Individuals such as Fernow served as cross-border conduits. Geological Survey and Department of Agriculture scientists formed ties with American colleagues and modelled the survey and experimental farms on American examples. Canadian fisheries scientists gained early inspiration from their Scottish counterparts before coming in 1914 under the sway of the eminent Danish biologist Johan Hjort. Wildlife scientists learned from their British counterparts, especially Elton, before shifting in mid-century to American models.[38]

Scientific perspectives crossed borders easily, pushed by scientists through their networks, and pulled by demands for rational, quantitative, and simplified descriptions of nature that avoided local subjective judgment – attributes valued throughout the formation of modern state administration in Europe.[39] These perspectives were particularly attractive in Canada, where natural systems tend toward extremes: unpredictable movements of fish and wildlife, countless local variations in forest productivity and seasonal water flow. Only by quantifying and simplifying this complexity, it was thought, could it be managed. Accordingly, fisheries biologists studied fish populations in terms of their MSY (maximum sustained yield), a concept requiring only demographic information. Foresters reduced forests to the growth curves of a few tree species. Engineers viewed rivers as merely physical matter describable with a few variables.[40]

Upon this scientific foundation, nature was transformed into natural resources: units of production that could be evaluated and standardized for maximum efficiency. With everything uncountable or uncontrollable excluded – including the histories of and values attached to places – this transformation became a technical problem to be solved using techniques that provided the necessary quantitative or visual information concerning allowable cuts, catch and bag limits, or flows of water. Nature became a modern technological product: trees bred for maximum productivity and arranged in even-aged, single-species tree farms; salmon produced in hatcheries; water "harnessed" by dams and turbines; "pure" water produced through engineering; waterbodies serving as organic machines by assimilating humanity's wastes.[41] By defining nature in these terms, science promised the predictability that resource industries demanded, aiding the expansion of capitalism into local environments across the country. Similarly, planners reduced the complexity of cities to a few basic principles – separation of activities, private property, and mobility – producing communities that converged on a single model, especially in the suburbs. Resource and urban landscapes alike existed in abstract spaces, disengaged from local environments, tied instead to networks that imposed new geographies of industrialization and consumption.[42] The consequences included the near erasure of local knowledge, drowned under reservoirs, displaced by forest management regimes, or denied by biologists attempting to transform Indigenous ways of life.

Science (as Daniel Macfarlane and other authors in this volume have noted) was therefore essential to the expression of modernity in Canada. It imparted an optimistic confidence in the capacity to transform nature, bringing it under rational control to serve the economy and society. By underpinning the competence of centralized management at the expense of local knowledge and experience, science also contributed to the formation of the state. Scientists and technical experts also served as mediators between Canadians and their environment, taking on the task of knowing nature on behalf of society, by providing those commodities – water, wood, energy, food – that Canadians themselves had once obtained through daily engagement with the natural world. Sensory experience became less important, as did experiential knowledge, such as that of farmers. And, finally, by linking modernist imperatives to local circumstances, science shaped their environmental consequences.[43] In return, science gained a secure status in federal and provincial agencies and in universities.

Yet the significance of science to modernity can also be exaggerated. Landscape transformations were driven not by science but by industrial expansion and its flows of material, energy, and investment. Elements of nature left untransformed – rivers flowing freely, fish uncaught, or forests existing past their prime – were "wasted." Scientists redefined nature accordingly, taking their cue from the interests that employed them, legitimizing the production-oriented approaches that industrial expansion required.[44] And, when they warned of unintended consequences, they were often ignored, sometimes on the ground that their advice was uncertain. Fish in particular felt the effects. During the nineteenth century, accounts of sawdust pollution in eastern rivers were overruled by the forest industry. This pattern continued in the twentieth century, as observations of the effects of forest cutting and log running on salmon and other species and an accompanying history of failure to act on the advice of Pacific fisheries scientists illustrate.[45] (That the Fraser, the biggest salmon river in British Columbia, remained free of dams was a rare exception.[46]) Entomologists studying BC rangelands attributed grasshopper infestations to overgrazing, yet poisons (arsenic, then DDT) were deemed to be the more expedient solution.[47]

Efforts to transform nature also had to respond to local circumstances. These efforts often involved improvisation, as became evident in contexts ranging from northern wildlife management, to dam construction in Banff National Park, to (as Macfarlane discusses in his chapter) the engineering of the St. Lawrence Seaway.[48] Big dams epitomized these tensions by combining modernist ideals of transformed rivers, accommodations to local environments, and provincial ambitions – British Columbia's desire to industrialize and assert western self-determination, Ontario's electrification campaigns, and Quebec's amalgam of rural traditions and nationalist aspirations.[49] These combinations illustrated modernity's capacity to adapt, taking on different forms in response to local circumstances. This capacity would also become evident as environmental concerns became more prevalent in Canadian society.

Administration

It is one of the most famous images in environmental science: an aerial view of Lake 226, with one half dark green, the other half light green. In 1973, limnologists had separated the two sides of this lake in the Experimental Lakes Area and fertilized both sides with carbon and nitrogen but only one side with phosphorus. Rich algae blooms promptly formed on the latter side, indicating that phosphorus was the limiting nutrient. It was also, therefore, the likely cause of algae blooms afflicting the Great Lakes, especially Lake Erie. The demonstration helped to resolve an ongoing controversy about the Great Lakes environment. It did so through a novel scientific strategy: experimentation using an entire ecosystem. This research, part of a larger federal investment in environmental science, also made the case that science could be a public matter. Clearly, the meanings of science in Canadian society were changing.

By the late 1960s, the state of the Great Lakes was one environmental issue that had captured public attention. There were many others. Resource industries became a focus of criticism, evident in conflicts between the forest industry and other values: coastal scenery and salmon streams in British Columbia, recreational opportunities in Algonquin Park.[50] A century of industrial expansion had left depleted fish populations on both west and east coasts; mines and their tailings were scattered across the Canadian Shield and elsewhere; forests had been transformed into tree farms and rivers into reservoirs; and the oil and gas industry was extending its reach into the North. Highways had reshaped regional geographies, providing access to once remote places. The unexpected consequences of dams and the St. Lawrence Seaway were demonstrating how rivers could resist modernist ambitions: as engineers around the world were finding out, rivers were more than plumbing systems to be rearranged at will.[51]

By describing its consequences, science became central to critiques of modernism. A few scientists adopted this role long before the supposed emergence of environmental values in the 1960s – illustrating the need to reconsider chronologies of the "origins" of environmentalism. By the 1950s, some scientists were working with the recreational and conservation community, including organizations such as Ducks Unlimited Canada and the Audubon Society, to present an ecological perspective on policy, with positions on issues such as habitat protection, pollution, and pesticides. Their views were well represented at the Resources for Tomorrow conference in 1961.[52] In other instances, scientists worked with interest groups on specific local concerns. Sportfishers encouraged biologists from the Fisheries Research Board station at St. Andrews and the University of New Brunswick to study the impacts of DDT on salmon, thus challenging the spruce budworm spraying program. In Toronto, real estate and other interests enabled scientists to study local air pollution. And in 1965 study of the impacts of a log drive on salmon habitat challenged the BC forest industry.[53] These initiatives illustrate the need to interpret the relations between science and the environment not only through the lens of 1960s environmentalism but also in the context of the relations among various interests and their ways of understanding and experiencing nature.

Scientists also identified a host of new issues. Sometimes the novelty came from new geographies of environmental change: the presence of contaminants in distant places, including the Arctic; the realization that pollution affected the entire Great Lakes, not just waters near cities or industries; the observation that land-use activities were also sources of water pollution; and the capacity of cities to modify their own climates through the heat island effect.

New disciplines and areas of interdisciplinary study, such as ecotoxicology and air pollution chemistry, emerged to make sense of these observations. The "environment" itself became an object of study. Research agendas shifted accordingly: Canadian Wildlife Service biologists now studied not only economically important but endangered species, and the effects of both hunting and contaminants, while Fisheries Research Board scientists studied not only fish populations but other marine species and their habitats. New research institutions such as the Canada Centre for Inland Waters and the Experimental Lakes Area were formed. Governments found science to be a useful way of demonstrating their environmental resolve.

For environmental organizations, science was a helpful source of authority. Although Pollution Probe made only modest use of scientific information, its origins in the University of Toronto's Department of Zoology imparted credibility to the young organization.[54] Knowledge and conviction also led some scientists to become activists. Douglas Pimlott and other biologists in Ontario urged expansion of provincial parks and a new perspective on wolves. David Suzuki and CBC Television began their long collaboration in communicating science and environmental awareness. David Schindler called attention to the consequences of Alberta's resource industries, from pulp mills to the oil sands. In the early 1970s, many ecologists urged protection of the "fragile" northern environment – reversing the image of the Arctic as harsh and challenging.[55]

The irony of these scientists' critiques, of course, was that science remained part of the apparatus of modernism now being criticized. Trust in resource science was being eroded by perceptions that experts lacked awareness of the wider consequences of transformation, had become advocates of their industries, or were too focused on narrow economic objectives. Experience with dams became one focus of critique, such as the devastation at Manitoba's South Indian Lake and among its Indigenous community, the disruption of the Peace-Athabasca Delta far downstream of the Bennett Dam, and mercury contamination in reservoirs created by the James Bay Project.[56] Scientists contributed their own accounts of these collisions between modernist ambitions and local ecologies.

Critiques of science and resource transformation concerned not only the substance but also the process by which science was applied. Citizens and civil society organizations questioned foundational assumptions of modernism: deference to technical experts tied to state agencies and economic interests and restrictions on those considered qualified to participate in decisions. In the 1960s, discussions on controlling detergents (a source of nutrient pollution) in Ontario were restricted at first to business and government; this closed-door approach was then challenged.[57] The perceived need to pose such challenges was sharpened by the sense – especially in provinces where one resource industry was dominant (e.g., forestry in British Columbia and New Brunswick, hydroelectricity in Quebec) – that science was being used less to balance conflicting interests than to justify trampling on the rights of others. Modernist efficiency and focus, once lauded, were now seen as hindering the capacity of scientists to reform practices that they had themselves constructed. Instead, it often fell to "outside" scientists, particularly those in universities, to push for substantial changes: questioning the assumptions of resource management, proposing new approaches such as ecosystem management and adaptive management.

The most substantial responses by governments to environmental concerns were the new agencies and regulations through which the environment itself was made an object of administration. From the outset, science was central to these initiatives. Much of Environment Canada's budget was (and remains) devoted to research, monitoring, and other science-based activities.[58] Universities established environmental science programs, and consulting firms proliferated in response to demand from industry and government for help in meeting environmental impact assessment and other regulatory requirements. The mechanisms of environmental administration – risk assessment, impact assessment, negotiation of acceptable releases of contaminants – have large appetites for scientific advice, both to guide decisions and to reinforce their authority. In response, a new kind of practice developed that became known as regulatory science. Seeking neither theoretical knowledge (as in basic research) nor solutions for industry (as in applied science), regulatory science involved collecting the copious data on environmental trends and impacts that legal and administrative procedures demanded. However, the purpose of administration was not only to allay environmental concerns but also to forestall the controversies that, in a new era of activism, threatened to derail development. Administration became a means of imposing a rational, managerial approach, thereby defusing conflict and ensuring that decisions would remain predictable. While accommodating environmental values considered moderate and reasonable, this approach remained consistent with modernist practices. Industry increasingly influenced the practice and application of regulatory science, shaping the questions and insisting on high standards of proof of harm before taking action. This marked the formation of environmental modernism, in which the practices and artifacts of modernism were modified to correspond more closely to nature – as seen, for example, in regulations imposed on industry and in dams modified in response to environmental concerns, such as the Libby Dam in Washington State just south of British Columbia.[59]

Environmental administration has had some benefits. Regulations have restrained egregious polluters, and impact assessments have provided opportunities to modify projects for the better. The Great Lakes are cleaner now than they were in the late 1960s (even the once "dying" Lake Erie), with this effort guided, in part, by evidence such as that produced at Lake 226. But modernist projects and activities have continued: dams, highways, industrial forestry. While acknowledging environmental concerns, the mechanisms of environmental administration have ensured that most projects could proceed and that pollution controls would not prove to be too onerous for most industries. Science has been essential to the view that economic development and environmental protection can coexist. Environmentalism – ostensibly challenging modern industrial society – was readily accommodated, illustrating modernity's capacity to accommodate challenges.[60] Eventually, however, other changes – in society, science, and nature – would disrupt its assumptions.

Disruption

In the late 1980s, Inuit in northern Quebec were shocked to learn of contaminants in their bodies.[61] The discovery disrupted both their relations with their landscape and the assumptions of modernist environmental administration. Presence of contaminants so far north demonstrated the novel relations between humans and nature created by industry and the vulnerability of some social groups. Over time, as Inuit became able to evaluate themselves the risks that they faced, assumptions about knowledge and who is qualified to make decisions were revised. Inuit bodies became one among many sites of disruption in Canadian science and environmental affairs. The consequences have included new ways of doing science and a reconsideration of modernist perspectives on knowledge.

As northern contaminants illustrate, environmental change is now being observed at novel scales. In the 1970s, the study of pollution from smelters in Sudbury found that its impacts extended much farther than had been thought. Understanding this required new research methods and the definition of a novel environmental syndrome – acid rain.[62] In the Great Lakes and elsewhere, invasive species have led scientists to adopt a larger view of the factors influencing ecosystems. And, on the largest scale, climate change has required new research strategies and new ways of managing human-nature relations. For example, protected areas (the epitome of modernist conservation, with the assumption that nature can be kept separate from humans) are no longer sufficient to conserve vulnerable habitats and species. Novel scales also extend to the smallest aspects of nature. Trace contaminants such as persistent organic pollutants, endocrine disrupters, and nanoparticles have consequences even when present at only a few parts per billion. And, because they affect cellular and molecular processes, they have forced a redefinition of the environment to include the spaces within as well as between bodies.[63]

Nature is now seen as having a complexity and unpredictability – even an element of chaos – that renders knowledge unavoidably uncertain. Scientists now talk of the disruption of knowledge and environments, as seen, for example, in an Arctic experiencing melting sea ice and permafrost, invasive species, and other novelties. One scientist's comment regarding Pacific fish stocks could be echoed elsewhere: "If you're asking me what's going on ... the answer is, I don't know, and I don't think anybody does."[64] So much for modernist confidence in a knowable nature.

Since the 1970s, the new politics of Indigenous knowledge has also disrupted modernist definitions of reliable knowledge, forcing attention to oral traditions and daily experiences, responsibilities toward other species, and views of landscapes as homelands, not wilderness areas. The practice of science has also changed, especially in the North, to include the obligation to collaborate with and report results back to communities. These changes have occurred at the insistence of Indigenous communities and organizations, and they have been accompanied by descriptions by anthropologists and other scholars of the empirical and cultural significance of Indigenous knowledge. Thomas Berger's Mackenzie Valley Pipeline Inquiry (1974–77) and subsequent land claims provided opportunities to document this knowledge, with the Inuit Land Use and Occupancy Study in the 1970s as an early model and the eventual basis for the 1993 Nunavut agreement.[65] Indigenous organizations have used their knowledge to challenge the state's authority over the environment, and its use has now become required in northern decision making, particularly for wildlife management and impact assessment. Yet in these processes Indigenous perspectives – sometimes codified as traditional ecological knowledge – have often been reframed to be consistent with scientific and bureaucratic processes: meeting formal requirements for consultation but dismissing the complex social and moral relations between Indigenous peoples and their lands.[66]

The modernist vision of knowledge limited to a coterie of official experts has also been disrupted. Civil society groups have accumulated their own scientific capacities. Bigger organizations such as World Wildlife Fund Canada and the David Suzuki Foundation now pursue their own scientific priorities by funding and sometimes conducting their own research. Citizen science has also flourished since organizations see it as a way both to obtain data and to involve their members. Its consequences have included a blurring of the boundary between expert and lay knowledge. A new mode of scientific activity has emerged, known as advocacy science, that supports alternative environmental strategies, challenges to policies and industrial practices, and claims of environmental injustice.

While advocacy organizations have become more prominent in Canadian environmental science, governments have retreated, reducing support for their own research and monitoring activities and "muzzling" their scientists. This retreat has been most closely associated with the federal Conservative government of Stephen Harper of 2006–15, yet its decisions only intensified a trend under way, with occasional interruptions, since the 1970s, when the Liberal government of Pierre Trudeau began encouraging the private sector to take on a wider range of research and monitoring tasks.

Several factors, therefore, have challenged the primacy of the state in environmental and resource science and policy. The "old" politics of modernity, based upon objectivity and efficiency, is being disrupted by a "new" politics of pluralism.[67] One consequence has been more diverse roles for scientists. Many remain in the lab, eschewing involvement in public affairs. Some orient their work toward activism. Others work within resource and environmental administration, focusing on solving problems and maintaining production. Scientists in each of these categories are sometimes skeptical of those in the others, viewing them as neglecting their social responsibility, being unprofessional in their advocacy, or not being critical enough of industry.

Conclusion

In this chapter, I have argued that the history of science, the environment, and Canadian modernity can be understood in terms of a few key themes. Science has been implicated in the diverse territories asserted by the state, economic interests, nature, and scientists themselves. As territory was transformed into raw materials for an industrial economy – a central feature of the Canadian experience with modernity – science also became essential to this transformation, providing the means of turning a complex and variable environment to human purposes. Assessing the consequences of this transformation eventually became the task of environmental administration, which again called on scientific expertise to provide the basis for rational decisions – extending the precepts of modernity. Finally, though, changes in nature, in knowledge (particularly the reassertion of Indigenous knowledge), in who does and uses science, and even regarding the possibility of certainty in the face of a chaotic nature have disrupted the modernist assumptions underpinning the relations between science and environmental affairs. The outcomes of these developments are not yet clear.

Each of these themes has been and remains essential to Canadian modernity as expressed through the relations between science and the environment. In exploring these themes, we can draw on the history of science and environmental history. As historians of science have explained, science has asserted its own priorities through disciplines and professions and international networks (at first predominantly imperial British and subsequently North American) that have defined research methods and the nature of authoritative knowledge. Perspectives on authoritative knowledge have encompassed its changing locations, such as from the farm to agricultural science facilities or from hunting camps and trapping lines to university departments of wildlife science and state management agencies. And, as environmental historians have suggested, the environment has been present throughout this history of science. Animals, air, water, and pollutants have crossed borders, challenging claims of territorial authority and compelling scientists to respond accordingly. Spatial arrangements that humans have imposed on nature have shaped scientific activities, such as the designation of protected areas. Nature has often failed to behave as scientists expected, and their models have failed to capture its dynamism, complexity, and unpredictability. More recently, scientists have considered how to embrace, rather than ignore, this unpredictability.

When we consider the place of science in the evolving relations between Canadians and their environment, we can begin by noting scientists' consistent positioning of themselves at the centre of these relations: as an instrument for asserting territorial authority, transforming nature, administering the environment, or even disrupting accepted ways of relating to nature. Throughout, science has contributed to (or sometimes disrupted) efforts to manage a variable and unpredictable environment so that it fulfills its role in a modern industrial society. Environmental change has often influenced human affairs only when it has been described by science. Conversely, when scientists have neglected the experiences of actors – human and nonhuman – marginalized by modernity, changes experienced by those actors have been rendered politically invisible. The consequences of science, including material changes in the environment, neglect of other ways of understanding and appreciating nature (including daily sensory experience), and assertions of relations of power, particularly by linking institutions through a common language and set of assumptions and values, can thus be understood as aspects of the history of modernism in Canada. Modernism justified support for science by demonstrating that it was consistent with dominant values and ways of seeing. In return, science supported modernism by asserting not only that it represented the preferences of an interest group but also that it was rooted in nature – and that social issues, such as conflicting claims to nature, were therefore amenable to technical solutions. Science also provided the basis for connecting the general imperatives of modernism – territorial authority, economic growth, transformation of nature – to specific places, adapting them to local conditions.[68]

This history also suggests two more general observations about Canadian modernity. First, it illustrates the distinctive place of the North in the history of science and modernity: this region has been the site both of ambitious expressions of modernity, such as the extension of territorial authority and huge technological projects, and of challenges to it, including alternative forms of knowledge and accelerated climate change and other environmental disruptions. Attention to science and the environment therefore demonstrates how the North, too often considered at the margin of Canadian history, has in fact been central to Canadians' encounter with modernity. Second, this history reminds us that modernity is a variable phenomenon, asserted and contested by diverse actors. For example, attention to the interplay between knowledge and the environment demonstrates how the territorial dimensions of modernity have gone beyond, and even challenged, the central institution of modernity: the nation-state. The implication for historical practice is that, when we seek to locate modernity in environmental history and the history of science, the task is not to evaluate the past in relation to a single template of modernity but to be alert to its diverse forms and consequences.

Much, of course, remains to be understood. We need to consider, for example, the relations among the geographies of science and of territorial control and transformation; the significance of social and environmental injustices to the construction and disruption of science and modernism; and the prospects for constructive roles for the state in this context of disruption. Historical analysis of science and the environment in the context of modernity is also relevant to discussions about the Anthropocene: the impending unpredictability of nature as a result of climate change promises to upset modernist assumptions about cause, effect, and prediction itself. As we turn to these and other issues, a generation of Canadian historians of science, environmental historians, and historical geographers has given us much to consider.

NOTES

1 Trevor H. Levere and Richard A. Jarrell, eds., *A Curious Field-Book: Science and Society in Canadian History* (Toronto: Oxford University Press, 1974).
2 In this context, I refer to "science" as historically constructed forms of knowledge and associated practices, conducted by individuals with formal training, distinguished in both social and intellectual terms from other ways of knowing, such as Indigenous or local knowledge. A wide range of scientific disciplines is relevant to the relations between people and the environment, including field sciences (e.g., ecology and oceanography), disciplines relevant to resource management (e.g., forestry and fisheries science), as well as laboratory sciences (e.g., toxicology). These scientific fields are also tied to the history of technology – consider, for example, harvesting technologies, airplanes for aerial surveys, and pollution control equipment.

3 For an overview, see, for example, Bernard Lightman, ed., *A Companion to the History of Science* (Chichester, UK: Wiley Blackwell, 2016).
4 Matthew Evenden, "A View from the Bush: Space, Environment, and the Historiography of Science," *Scientia Canadensis: Canadian Journal of the History of Science, Technology, and Medicine* 28 (2005): 27–37; Diarmid A. Finnegan, "The Spatial Turn: Geographical Approaches in the History of Science," *Journal of the History of Biology* 41, 2 (2008): 369–88.
5 This brief discussion references a large literature on the historiography of the Canadian environment. An entry into this literature is provided by the eight essays that discuss *The Landscape of Canadian Environmental History*, introduced by Alan MacEachern, "The Text that Nature Renders?," *Canadian Historical Review* 95, 4 (2014): 545–54.
6 There is, of course, a massive literature on modernity. A useful overview is Carol Gluck, "The End of Elsewhere: Writing Modernity Now," *American Historical Review* 116, 3 (2011): 676–87.
7 See, for example, Tina Adcock, "Many Tiny Traces: Antimodernism and Northern Exploration between the Wars," in *Ice Blink: Navigating Northern Environmental History*, ed. Stephen Bocking and Brad Martin (Calgary: University of Calgary Press, 2017), 131–77.
8 Morris Zaslow, *Reading the Rocks: The Story of the Geological Survey of Canada 1842–1972* (Toronto: Macmillan, 1975); Suzanne Zeller, *Inventing Canada: Early Victorian Science and the Idea of a Transcontinental Nation* (Toronto: University of Toronto Press, 1987); W.A. Waiser, *The Field Naturalist: John Macoun, the Geological Survey, and Natural Science* (Toronto: University of Toronto Press, 1989); Jennifer Hubbard, *A Science on the Scales: The Rise of Canadian Atlantic Fisheries Biology, 1898–1939* (Toronto: University of Toronto Press, 2006). This station eventually became the basis for the Biological Board of Canada, established in 1912.
9 Richard C. Powell, "Science, Sovereignty, and Nation: Canada and the Legacy of the International Geophysical Year, 1957–1958," *Journal of Historical Geography* 34 (2008): 618–38; Stephen Bocking, "Science and Spaces in the Northern Environment," *Environmental History* 12 (October 2007): 868–95. Sovereignty over knowledge could not be asserted overnight. For example, until the 1950s, Ducks Unlimited and the US Fish and Wildlife Service channelled American expertise to support waterfowl conservation in the Prairies since the Canadian Wildlife Service was not yet able to fulfill this role; see Shannon Stunden Bower, *Wet Prairie: People, Land, and Water in Agricultural Manitoba* (Vancouver: UBC Press, 2011).
10 Matthew Farish, "Frontier Engineering: From the Globe to the Body in the Cold War Arctic," *Canadian Geographer* 50, 2 (2006): 177–96; Edward Jones-Imhotep, "Communicating the North: Scientific Practice and Canadian Postwar Identity," *Osiris* 24, 1 (2009): 144–64; Powell, "Science, Sovereignty, and Nation"; Simone Turchetti and Peder Roberts, eds., *The Surveillance Imperative: Geosciences during the Cold War and Beyond* (New York: Palgrave Macmillan, 2014).
11 Klaus Dodds and Mark Nuttall, *The Scramble for the Poles* (Cambridge, UK: Polity, 2016); Elizabeth Riddell-Dixon, *Breaking the Ice: Canada, Sovereignty, and the Arctic Extended Continental Shelf* (Toronto: Dundurn, 2017).

12 A prominent example of this perspective is Morris Zaslow, *The Northward Expansion of Canada, 1914–1967* (Toronto: McClelland and Stewart, 1988).
13 Hugh Brody, *Maps and Dreams: Indians and the British Columbia Frontier* (Vancouver: Douglas and McIntyre, 1988).
14 Numerous studies have described this process in specific places. See, for example, Jocelyn Thorpe, *Temagami's Tangled Wild: Race, Gender, and the Making of Canadian Nature* (Vancouver: UBC Press, 2012); and John Thistle, *Resettling the Range: Animals, Ecologies, and Human Communities in British Columbia* (Vancouver: UBC Press, 2015).
15 Richard A. Rajala, *Clearcutting the Pacific Rain Forest: Production, Science, and Regulation* (Vancouver: UBC Press, 1998); Bruce Braun, *The Intemperate Rainforest: Nature, Culture, and Power on Canada's West Coast* (Minneapolis: University of Minnesota Press, 2002); Caroline Desbiens, *Power from the North: Territory, Identity, and the Culture of Hydroelectricity in Quebec* (Vancouver: UBC Press, 2013); Stéphane Castonguay, "Naturalizing Federalism: Insect Outbreaks and the Centralization of Entomological Research in Canada, 1884–1914," *Canadian Historical Review* 85, 1 (2004): 1–34.
16 Richard A. Rajala, "'This Wasteful Use of a River': Log Driving, Conservation, and British Columbia's Stellako River Controversy, 1965–72," *BC Studies* 165 (2010): 31–74. A similar federal reluctance to infringe on provincial jurisdiction has been evident regarding species at risk.
17 Stephen Bocking, "A Disciplined Geography: Aviation, Science, and the Cold War in Northern Canada, 1945–1960," *Technology and Culture* 50, 2 (2009): 320–45; Matthew Evenden, "Locating Science, Locating Salmon: Institutions, Linkages, and Spatial Practices in Early British Columbia Fisheries Science," *Environment and Planning D: Society and Space* 22 (2004): 355–72.
18 Robert E. Kohler and Jeremy Vetter, "The Field," in *A Companion to the History of Science*.
19 Bocking, "Science and Spaces in the Northern Environment."
20 Jennifer Read, "'A Sort of Destiny': The Multi-Jurisdictional Response to Sewage Pollution in the Great Lakes, 1900–1930," *Scientia Canadensis: Canadian Journal of the History of Science, Technology, and Medicine* 22, 51 (1998–99): 103–29; Lynne Heasley and Daniel Macfarlane, eds., *Border Flows: A Century of the Canadian-American Water Relationship* (Calgary: University of Calgary Press, 2016).
21 Robert Kohler noted the continuing historical relevance of local knowledge to the field sciences. Robert E. Kohler, *All Creatures: Naturalists, Collectors, and Biodiversity, 1850–1950* (Princeton, NJ: Princeton University Press, 2006).
22 Daniel Macfarlane, *Negotiating a River: Canada, the US, and the Creation of the St. Lawrence Seaway* (Vancouver: UBC Press, 2014); Tina Loo and Meg Stanley, "An Environmental History of Progress: Damming the Peace and Columbia Rivers," *Canadian Historical Review* 92, 3 (2011) 399–427.
23 B.E. Fernow, "Conditions in the Trent Watershed and Recommendations for Their Improvement," *Trent Watershed Survey: A Reconnaissance*, ed. C.D. Howe and J.H. White (Ottawa: Commission of Conservation, 1913), 4.

24 Michel Girard, *L'ecologisme retrouvé: Essor et declin de la Commission de la Conservation du Canada* (Ottawa: Presses de l'Université d'Ottawa, 1994).
25 Darcy Ingram, *Wildlife, Conservation, and Conflict in Quebec, 1840–1914* (Vancouver: UBC Press, 2013); Tina Loo, "Making a Modern Wilderness: Conserving Wildlife in Twentieth-Century Canada," *Canadian Historical Review* 82, 1 (2001): 1–18.
26 Quoted in Joseph E. Taylor III, "Making Salmon: The Political Economy of Fishery Science and the Road Not Taken," *Journal of the History of Biology* 31 (1998): 38; see also William Knight, "Samuel Wilmot, Fish Culture, and Recreational Fisheries in Late 19th Century Ontario," *Scientia Canadensis: Canadian Journal of the History of Science, Technology, and Medicine* 30, 1 (2007): 75–90. However, as Taylor explains, by the 1920s scientists were becoming increasingly skeptical of fish culture, and by 1937 their influence had led to the shutdown of all federal hatcheries.
27 Hubbard, *Science on the Scales.*
28 Thistle, *Resettling the Range;* Gerald Killan and George Warecki, "J. R. Dymond and Frank A. MacDougall: Science and Government Policy in Algonquin Provincial Park, 1931–1954," *Scientia Canadensis: Canadian Journal of the History of Science, Technology, and Medicine* 22, 51 (1998–99): 131–56; George Colpitts, "Conservation, Science, and Canada's Fur Farming Industry, 1913–1945," *Histoire sociale/Social History* 30, 59 (1997): 77–107.
29 Read, "'A Sort of Destiny.'"
30 Arn Keeling, "Sink or Swim: Water Pollution and Environmental Politics in Vancouver, 1889–1975," *BC Studies* 142–43 (2004): 69–101; Michèle Dagenais, *Montréal et l'eau: Une histoire environnementale* (Montréal: Boréal, 2011); Stephen Bocking, "Constructing Urban Expertise: Professional and Political Authority in Toronto, 1940–1970," *Journal of Urban History* 33, 1 (2006): 51–76.
31 Janet Foster, *Working for Wildlife: The Beginning of Preservation in Canada*, 2nd ed. (Toronto: University of Toronto Press, 1998); Loo, "Making a Modern Wilderness"; Tina Loo, *States of Nature: Conserving Canada's Wildlife in the Twentieth Century* (Vancouver: UBC Press, 2006); Ingram, *Wildlife;* John Sandlos, *Hunters at the Margin: Native People and Wildlife Conservation in the Northwest Territories* (Vancouver: UBC Press, 2007); Peter Kulchyski and Frank James Tester, *Kiumajut (Talking Back): Game Management and Inuit Rights, 1950–70* (Vancouver: UBC Press, 2007).
32 See, for example, Robert McDonald and Arn Keeling, "'The Profligate Province: Roderick Haig-Brown and the Modernizing of British Columbia," *Journal of Canadian Studies* 36, 3 (2001): 7–23; Loo, *States of Nature;* J. Alexander Burnett, *A Passion for Wildlife: The History of the Canadian Wildlife Service* (Vancouver: UBC Press, 2003); and Briony Penn, *The Real Thing: The Natural History of Ian McTaggart Cowan* (Victoria: Rocky Mountain Books, 2015).
33 See, for example, Thistle, *Resettling the Range;* and Wendy Dathan, *The Reindeer Botanist: Alf Erling Porsild, 1901–1977* (Calgary: University of Calgary Press, 2012). Pierre Dansereau, who founded the study of plant ecology in Quebec, was one exception to this relation between ecology and practical issues; see Stephen Bocking,

"Dansereau, Pierre Mackay," in *New Dictionary of Scientific Biography* (New York: Charles Scribner's Sons, 2008), 2: 234–37.

34 Penn, *The Real Thing*. Kohler, *All Creatures*, discusses in detail the significance of local knowledge and environments to ecological research practice in the early twentieth century; his approach awaits application to the early history of Canadian ecology.

35 Burnett, *A Passion for Wildlife*.

36 Alan MacEachern, "The Sentimentalist: Science and Nature in the Writing of H.U. Green, a.k.a. Tony Lascelles," *Journal of Canadian Studies* 47, 3 (2013): 16–41.

37 Mark Kuhlberg, "'We Have "Sold" Forestry to the Management of the Company': Abitibi Power & Paper Company's Forestry Initiatives in Ontario, 1919–1929," *Journal of Canadian Studies* 34, 3 (1999): 187–209; Mark J. McLaughlin, "Green Shoots: Aerial Insecticide Spraying and the Growth of Environmental Consciousness in New Brunswick, 1952–1973," *Acadiensis* 40, 1 (2011): 3–23; Jeremy Wilson, "Forest Conservation in British Columbia, 1935–85: Reflections on a Barren Political Debate," *BC Studies* 76 (1987–88): 3–32.

38 Bocking, "Science and Spaces in the Northern Environment"; Hubbard, *Science on the Scales*; Richard A. Jarrell, "Science and the State in Ontario: The British Connection or North American Patterns?," in *Patterns of the Past: Interpreting Ontario's History*, ed. Roger Hall and Anthony Westell (Toronto: Dundurn Group, 1996), 238–54; John Sandlos, "Nature's Nations: The Shared Conservation History of Canada and the USA," *International Journal of Environmental Studies* 70, 3 (2013): 358–71.

39 Theodore M. Porter, *Trust in Numbers: The Pursuit of Objectivity in Science and Public Life* (Princeton, NJ: Princeton University Press, 1996).

40 Rajala, "'This Wasteful Use of a River'"; Jamie Linton, *What Is Water? The History of a Modern Abstraction* (Vancouver: UBC Press, 2010); Desbiens, *Power from the North*; Hubbard, *Science on the Scales*; Dean Bavington, *Managed Annihilation: An Unnatural History of the Newfoundland Cod Collapse* (Vancouver: UBC Press, 2010).

41 On such waterbodies, see Arn Keeling, "Charting Marine Pollution Science: Oceanography on Canada's Pacific Coast, 1938–1970," *Journal of Historical Geography* 33 (2007): 403–28.

42 Braun, *The Intemperate Rainforest*.

43 Erik Swyngedouw, "Modernity and Hybridity: Nature, Regeneracionismo, and the Production of the Spanish Waterscape 1890–1930," *Annals of the Association of American Geographers* 89, 3 (1999): 443–65; James C. Scott, *Seeing like a State: How Certain Schemes to Improve the Human Condition Have Failed* (New Haven, CT: Yale University Press, 1999); Tina Loo, "High Modernism, Conflict, and the Nature of Change in Canada: A Look at Seeing Like a State," *Canadian Historical Review* 97, 1 (2016): 34–58.

44 Rajala, *Clearcutting the Pacific Rain Forest*.

45 R. Peter Gillis, "Rivers of Sawdust: The Battle over Industrial Pollution in Canada, 1865–1903," *Journal of Canadian Studies* 21, 1 (1986): 84–103; Taylor, "Making Salmon"; Rajala, "'This Wasteful Use of a River.'"

46 This was due more to the availability of other sites for dams than to the arguments of fisheries scientists – indeed, it was hoped that they could come up with a technological solution that would allow both dams and salmon. See Matthew D. Evenden, *Fish versus Power: An Environmental History of the Fraser River* (Cambridge, UK: Cambridge University Press, 2004).
47 Thistle, *Resettling the Range*.
48 Sandlos, *Hunters at the Margin;* Matthew Evenden, *Allied Power: Mobilizing Hydro-Electricity during Canada's Second World War* (Toronto: University of Toronto Press, 2015).
49 Loo and Stanley, "An Environmental History of Progress"; Desbiens, *Power from the North*.
50 Rajala, "'This Wasteful Use of a River'"; George M. Warecki, *Protecting Ontario's Wilderness: A History of Changing Ideas and Preservation Politics, 1927–1973* (Bern: Peter Lang, 2000).
51 Macfarlane, *Negotiating a River*.
52 Darcy Ingram, "Governments, Governance, and the 'Lunatic Fringe': The Resources for Tomorrow Conference and the Evolution of Environmentalism in Canada," *International Journal of Canadian Studies* 51 (2015): 69–96.
53 McLaughlin, "Green Shoots"; Owen Temby, "Trouble in Smogville: The Politics of Toronto's Air Pollution during the 1950s," *Journal of Urban History* 39, 4 (2013): 669–89; Rajala, "'This Wasteful Use of a River.'"
54 Ryan O'Connor, *The First Green Wave: Pollution Probe and the Origins of Environmental Activism in Ontario* (Vancouver: UBC Press, 2015). Although numerous historians have alluded to the role of science in the emergence of environmentalism in Canada, a focused study of this topic has not yet appeared.
55 Bocking, "Science and Spaces in the Northern Environment."
56 James Waldram, *As Long as the Rivers Run: Hydroelectric Development and Native Communities in Western Canada* (Winnipeg: University of Manitoba Press, 1988).
57 Jennifer Read, "'Let Us Heed the Voice of Youth': Laundry Detergents, Phosphates, and the Emergence of the Environmental Movement in Ontario," *Journal of the Canadian Historical Association* 7, 1 (1996): 227–50.
58 G. Bruce Doern, Graeme Auld, and Christopher Stoney, *Green-Lite: Complexity in Fifty Years of Canadian Environmental Policy, Governance, and Democracy* (Montreal/Kingston: McGill-Queen's University Press, 2015).
59 Philip Van Huizen, "Building a Green Dam: Environmental Modernism and the Canadian-American Libby Dam Project," *Pacific Historical Review* 79, 3 (2010): 418–53.
60 S.N. Eisenstadt, "Multiple Modernities," *Daedalus* 129, 1 (2000): 1–29.
61 David Leonard Downie and Terry Fenge, eds., *Northern Lights against POPs: Combatting Toxic Threats in the Arctic* (Montreal/Kingston: McGill-Queen's University Press, 2003).
62 Don Munton, "Using Science, Ignoring Science: Lake Acidification in Ontario," in *Science and Politics in the International Environment*, ed. N.E. Harrison and G.C. Bryner (Lanham, MD: Rowman and Littlefield, 2004), 143–72.

63 Nancy Langston, *Toxic Bodies: Hormone Disruptors and the Legacy of DES* (New Haven, CT: Yale University Press, 2010).
64 Quoted in Terry Glavin, *Dead Reckoning: Confronting the Crisis in Pacific Fisheries* (Vancouver: Douglas and McIntyre, 1996), 18.
65 Milton M.R. Freeman, "Looking Back – and Looking Ahead – 35 Years after the Inuit Land Use and Occupancy Project," *Canadian Geographer* 55, 1 (2011): 20–31.
66 Carly A. Dokis, *Where the Rivers Meet: Pipelines, Participatory Resource Management, and Aboriginal-State Relations in the Northwest Territories* (Vancouver: UBC Press, 2015).
67 Sheila Jasanoff, *Designs on Nature: Science and Democracy in Europe and the United States* (Princeton, NJ: Princeton University Press, 2005), 14.
68 Loo, "High Modernism, Conflict, and the Nature of Change in Canada."

11

"For Canada and for Science"
Transnational Modernity and the *Report of the Canadian Arctic Expedition 1913–1918*

ANDREW STUHL

For nearly a century, the Canadian Arctic Expedition of 1913–18 has held symbolic meaning in understanding modern Canada. In 1926, the federal government placed a tablet at the Dominion Archives Building in memory of the sixteen members who perished on the expedition. According to an editorial in the *Ottawa Journal* commemorating the event, the "glorious pursuit of knowledge" in the North repaid those who jeopardized their lives because scientific discovery promised societal uplift. Had Columbus not "faced the terrors of wide seas," the author wrote, America "might yet be the habitat of the savage." The expedition thus marked the nation's devotion to progress. A pride in country would be stirred in every visitor who paused to read the inscription on the tablet: "For Canada and for Science."[1]

Even as that plaque no longer graces Library and Archives Canada, historians have maintained the meaning of the Canadian Arctic Expedition in their narratives of national development. For historian Trevor Levere, this venture was "truly modern", not only because of the advanced graduate training of hired fieldworkers but also because the prime minister championed their research as political and economic expansion. The act of surveying distant landscapes and their natural resources bespoke a Canadian state eager to solidify the legal status of territories transferred from England in the late 1800s and to reap profits from commodities that they might contain.[2] According to anthropologist Gísli Pálsson, representations that emerged after the expedition in newspaper articles and travelogues of a "howling, exotic wilderness" and a "semi-domestic, 'friendly' space" in the Arctic reflected a nation "modernizing at home" and "creating its own space abroad."[3] The Canadian Arctic Expedition thus positioned northern lands and peoples as "object" and "other," a requisite for the modernization of the state apparatus and the Canadian mind.

There are more ways in which science, nature, and nation inflected one another in this particular expedition, however. Importantly, historians have focused on either the experiences of travellers in the "field" – along the coast of the Beaufort Sea to the start of the Arctic Archipelago – or the discussions among bureaucrats, elected officials, explorers, and scientists that orbited explorer Vilhjalmur Stefansson's two books based upon the expedition – *The Friendly Arctic* (1921) and *The Northward Course of Empire* (1922).[4] In this chapter, I approach the Canadian Arctic Expedition from a series of reports stemming from it that have attracted little scholarly attention. This project involved seventy-five authors from the United States, Canada, and Europe who produced ninety separate reports in fourteen volumes under the title *Report of the Canadian Arctic Expedition 1913–1918* (hereafter, the *Report*; see the appendix). Appearing serially between 1919 and 1926, the *Report* included 2,500 pages on Arctic biology, anthropology, and geology and featured more than 600 coloured plates, photographs, and line drawings.[5] Scientists with the federal government distributed the *Report* to distinguished academics, research laboratories, and libraries from Australia to Siberia. As testament to its reception, the renowned American polar explorer Adolphus Greeley labelled the *Report* the "most valuable scientific contributions relative to Polar Canada ever published."[6] In sum, the *Report* was an international project of considerable magnitude – "for Canada and for science." It was distinct both from the collecting activities of the expedition and from the lecture tours and popular travelogues that some expedition members launched when they returned south. The *Report* thus ought to be understood as a unique event in Canadian and global history.

Transnational Modernity and the Circulation of Science

What makes the *Report* significant for my investigation in this chapter is that its production makes visible a set of transnational exchanges through which Canada came to be understood as modern. This transnational modernity through science and exploration is not without precedent, even in the Canadian Arctic.[7] In a study of print culture about the Arctic in Canada, historian Janice Cavell demonstrates that the Arctic was Canadian space between 1890 and 1930 only to the extent that Canadians marked the traces of nineteenth-century explorers who had failed time and again to secure trade routes through the Northwest Passage. Government surveyors, for instance, staked claims to Arctic islands by placing cairns near landmarks of the Franklin Expedition as popular periodicals and historical monographs cultivated nostalgia for the heroic era of exploration. Materially and symbolically, then, Canadian writers at the turn of the twentieth century pursued modernity in the Far North by nurturing ambivalent legacies of the British Empire.[8]

The *Report*, in contrast, signalled Canadian modernity because the nation produced legitimate research through and for a broad scientific community, one not anchored in England. The practitioners called on to develop the *Report* were stationed across North America and Europe, with museums and government scientific bureaus in the United States acting as centres of gravity for specimen identification and analysis. Moreover, the volumes contained views of nature that contrasted sharply with how politicians, historians, and Canadian citizens thought of northern environments at the time. Based upon their newfound understanding of abundant flora and fauna – and their geographical distribution – authors positioned the areas covered by the expedition as parts of a circumpolar zone or a region that had striking parallels with Antarctica. Such expressions of a global Arctic both challenged and informed nationalist ideas of the Canadian North as a stage for performing sovereignty or a "second frontier" whose landscape thwarted the industries of westward expansion.[9] In short, "for Canada and for science" were two pursuits that involved a wide geography of forces and actors. Teasing apart their relations through the *Report* helps us to comprehend modern Canada as a transnational phenomenon.

My analysis takes its impulse from a generation of historical scholarship on circulation or the tracking of people and objects as a means of analyzing the "sociology of translation" in material ways.[10] As historian Lissa Roberts points out, circulation allows scholars to connect local encounters to global economic flows and international social movements. What were once termed collection and codification in the "core" and dissemination to the "periphery" have been reinterpreted as dynamic, iterative processes of "going forth and coming back" across varied epistemic spaces.[11] Few Canadian case studies have served as bases for understanding circulation. Instead, scholars have concentrated on science in the Atlantic world, which sometimes includes attention to British North America, as does Efram Sera-Shriar's chapter in this volume. When considering the twentieth century, studies of circulation have centred on science and American imperialism in the Caribbean and the Pacific world.[12]

Fortunately, the archive of the *Report* is well suited for this line of inquiry. Rudolph Martin Anderson, who served as editor of the *Report*, maintained meticulous files on its creation and distribution. In what follows, I examine his correspondence with scientists with the Canadian government, with their superiors in federal departments, and with authors in the United States and Europe. In the first three sections, I consider the organization of research, the intellectual values, and the distribution of the *Report*. Each element required practices and perspectives that transcended national borders. In turn, these elements shaped notions of modernity, as evidenced in discussions of Canadian science and Canada's claims in the Arctic. In the fourth section, I examine biological research in the *Report* to draw out tensions between imperial and national interpretations of life in the Arctic. In concluding, I gesture to additional possibilities for the concept of "transnational modernity" to inform historical and contemporary understanding in Canada.

The Organization of Research in the *Report*

In January 1917, just a few months after the southern party of the expedition returned to Ottawa, the deputy ministers of the Department of Naval Service and the Department of Mines formed the Arctic Biological Committee to transform specimens and artifacts gathered in the Arctic into research publications. As historian Robert Kohler notes, such committees played important roles in distinguishing "finders" from "keepers" – whereas the former selected, extracted, recorded, and transported objects from the field to safe storage, the latter ordered and classified them to give them purpose and permanence in the scientific record.[13] The committee's mandate foreshadowed the scope of labour involved in this process. The committee was to select specialists "qualified to work up and report on" the collections, supervise the specialists, make recommendations to the departments on the details of publications, and make recommendations for the distribution of "any surplus material" not needed by the National Museum. The committee quickly identified a plan to publish sixteen volumes on biology, anthropology, geology, and geography, which required several dozen specialists (see the appendix).[14]

In composition alone, the Arctic Biological Committee signalled the importance of the *Report* to Canadian science in the interwar period. Between 1917 and 1930, when the committee held regular meetings, nearly all of its members occupied executive positions in federal scientific institutions. The initial group consisted of A.B. Macallum of the Biological Board of Canada; J.M. Macoun, chief of the Biology Division for the Geological Survey of Canada; C. Gordon Hewitt, the dominion entomologist; Rudolph Anderson of the Geological Survey; and Chairman E.E. Prince, the commissioner of fisheries. The committee secured an order in council in 1918 that covered the costs of publishing at least three thousand copies of each separate report and one thousand copies of each bound volume. Committee members articulated this financial relationship between scientists and the federal government (rather than individual departments) as setting "a new standard" in Canada. In their presentation before the Subcommittee of Council on Government Printing, Prince and Hewitt underlined the "valuable character" of the proposed reports. Many of the specimens had "never been collected anywhere in the western Arctic area" and were "from districts and localities which are practically unrepresented in the existing collections." Given these conditions, and the substantial sums already spent on fieldwork, the government ought to support scientists to properly generate the results and ensure that they were "given to the world."[15]

Yet, even with a commitment of nearly $40,000 from the federal government for printing the *Report*, the Arctic Biological Committee could not carry out its plan by turning only to Canadian scientists. Between January and March 1917, members of the committee wrote to botanists, entomologists, and marine biologists across the country. They found only nineteen colleagues who had the necessary training. "We have so few men in Canada who can really be called specialists in any line," Macoun wrote to Chairman Prince, "that I think we who are in the government service should do everything we can to encourage and assist workers in every field."[16] In this statement, one can read both the promise of the *Report* as a site for developing research capacities and young scholars at home and the necessity of looking beyond the nation's borders to produce rigorous knowledge. By 1920, the committee had recruited more than eighty scientists stationed in institutions across the United States, England, Scotland, Norway, and Denmark (see Table 11.1).[17] Specialists were leading thinkers and established scholars: they were employed by scientific bureaus at the federal and state levels, universities, and national museums (see Table 11.2). "Assistants" were additional staff called on not by the committee but by specialists. They either worked at the same institution as their supervisors, as advisees, or held some role in the specialist's career – such as a colleague from a previous institution.

TABLE 11.1

The geographical distribution of scientists recruited to analyze Canadian Arctic Expedition collections

	Canada	United States	United Kingdom	Norway
Specialists	19	34	5	0
Assistants	4	17	1	1

Note: United Kingdom refers to England and Scotland.
Sources: This information is taken from the affiliations listed by authors of the official reports of the Canadian Arctic Expedition and "Statement, April 1918, Canadian Arctic Expedition, 1913–1916," file 23a, box 53, CMNA. For some reports, authors did not list national affiliations, and they could not be identified in the archival records consulted.

TABLE 11.2

The institutional affiliation of scientists hired to analyze Canadian Arctic Expedition collections

	Canada	United States	United Kingdom	Norway
National museum	1	12	1	0
Federal bureau	13	1	0	0
State or provincial bureau	0	7	0	0
University	6	21	4	1
Private institution	0	1	0	0
Unknown	3	9	1	0

Sources: This information is taken from the affiliations listed by authors of the official reports of the Canadian Arctic Expedition and "Statement, April 1918, Canadian Arctic Expedition, 1913–1916," file 23a, box 53, CMNA. For some reports, authors did not list institutional affiliations, and they could not be identified in the archival records consulted.

Importantly, two-thirds of the researchers brought on to create the *Report* hailed from the United States. This number reflects less the geographic proximity to Ottawa than the potential for knowledge production of using American institutions and experts. Writing in the *Canadian Field-Naturalist* in 1919, Arctic Biological Committee member Rudolph Anderson praised American collections as "unsurpassed by those of any other country in the world." Accordingly, the relative completeness of collections therein would "permit increased attention to be paid to the study of life-histories" of the species represented in Canadian Arctic Expedition specimens.[18] For historians of science, Anderson's testimony signals the rise of the United States as a centre of gravity for research on the natural history of the North American Arctic, a position that England had maintained since the late 1400s.[19] In these ways and for these reasons, the organization of the research for the *Report* involved both national and transnational exchanges particular to the early twentieth century.

The Intellectual Values of the *Report*

Why might so many scholars outside Canada have committed their time and energy to materials gathered during the Canadian Arctic Expedition? Perhaps surprisingly, the Arctic Biological Committee did not pay specialists and assistants in cash. Rather, hired scientists were compensated with 250 free copies of the reports that they authored.[20] Specialists' motivations for participating in the *Report* thus further reveal how the values of an international research community shaped the idea and practice of modern science in Canada during the early 1900s.

As historian Lynn K. Nyhart writes, the multivolume bound report of the expedition was a well-established venue by the early twentieth century. The high-end standard for such reports had been set in the early 1800s in Paris through the *Description de l'Egypte* and *Voyage de Humboldt et Bonpland*. England employed the same organization of research in *The Zoology of the Voyage of the H.M.S. Beagle (1838–1843)* and the *Report on the Scientific Results of the Voyage of H.M.S. Challenger*, which involved sixty-seven authors, fifty tomes, and thirty thousand pages between 1880 and 1895. Over the 1800s, then, the form came to be seen in Europe as a benchmark for young nations and a proving ground for aspiring scientists. After unification in 1871, Germany immediately began sponsoring its own scientific voyages and printing the results in the tradition set by the British and French Empires. Nyhart notes that, at least until 1940, scientists around the world engaged in a "system of social organization" in which specialists "voluntarily chose to undertake [their] investigation[s], perhaps instead of a different one," and committed to publishing research in expedition volumes rather than academic journals. Especially since the western Arctic appeared to be relatively unknown in the scientific record, specialists must have seen the *Report* as an excellent opportunity not only to contribute new knowledge but also to enhance their reputations.[21] Indeed, there is little evidence in the archive of the *Report* to suggest that the Arctic Biological Committee needed to explain to specialists the task at hand.

This shared understanding between the committee and the scientific community of the value of the *Report* contrasts with how Canadian elected officials viewed this publication. In May 1925, as the minister of the Department of Marine and Fisheries stood before the House of Commons to ask for a final appropriation for printing the *Report*, he was challenged by a member of Parliament on its "practical good":

> MR. BAXTER (MP): What practical good are we going to get out of this that will help a soul in Canada?
>
> MR. CARDIN (MINISTER OF MARINE AND FISHERIES): I am informed that scientific people throughout Canada attach a very great importance to this publication.
>
> MR. BAXTER: The scientific people may appreciate it very much, but is there any practical value to science? Is anybody going to take a boat into these regions? Are we locating icebergs? Are we gaining territory of any value?[22]

In one respect, Baxter's position echoes the conclusions of historians of Canadian science. Between the world wars, Canadian scientists struggled to earn federal support and faced constant pressure to pursue "applied science" rather than "pure" or "basic" science.[23] Here oceanography and cartography appeared to the member of Parliament as useful and thus worthy recipients of taxpayers' dollars. The disciplines of systematic biology – like entomology, marine biology, botany, mammalogy, and ornithology – held only intellectual value. Setting Baxter's perceptions of science aside, though, this debate in Parliament verified the significance of the *Report* to the "scientific people throughout Canada": it helped them to wield leverage in halls of power. In 1918, leading scientists in Canada struggled to find enough researchers within the country capable of working up the results of the Canadian Arctic Expedition. After eight years of appropriation (1918–26) and the publication of many volumes, the Arctic Biological Committee had made progress on its goal of growing Canadian science through the *Report*.

The Distribution and Application of the *Report*

Through the *Report*'s distribution and application, scientists continued to shape Canadian and global affairs in material ways, even if certain politicians could not recognize this. Rudolph Anderson's correspondence as editor demonstrates that the *Report* enjoyed a wide distribution among the international scientific community, with practical outcomes in Canada and beyond.

Between 1919, when the first separate report was published, and 1938, when his archival trail ends, Anderson received requests for the report from a range of scientists and scientific institutions. Individual university professors; representatives of government departments in North America, Europe, and Asia; elite academic societies in Europe; and a variety of museums all wanted copies of the *Report*.[24] Sometimes researchers wrote to Anderson to complete the set of volumes in their libraries. In other instances, they had a particular project in mind. Foster H. Benjamin of the US Department of Agriculture requested the volume on *Diptera* for a taxonomic project on the Mediterranean fruit fly, while Francis Pospisil requested the volume on Eskimo string figures to complement his ethnographic studies at the Moravian Museum.[25] In these ways, the *Report* informed the elaboration of disciplines and research questions that transcended Canadian government interests in the North. The *Report* also returned benefits directly to federal agencies, though politicians might not have considered them "practical." Anderson parlayed interest in the *Report* to open and maintain lines of communication with other scientific institutions. In sending copies of volume 14 to the National Library of Saxony in 1928, he hoped to keep the National Museum "on good terms with the Dresden people." Later he would be able to request from them "some anthropological literature." Circulation of the *Report* thus sustained flows of knowledge between Ottawa and many scientific institutions around the globe.[26]

Envisioning a wide circulation for the *Report*, the Arctic Biological Committee viewed the set of documents as an instrument for elevating the international profile of Canadian science. As the first separate report to be published, William H. Dall's manuscript received careful scrutiny from Anderson as editor. Because Dall had not followed the International Rules of Zoological Nomenclature in 1913, Anderson reset the proofs and ensured that the remainder of reports followed this protocol.[27] Although members of the committee recognized that this would cause a delay in printing Dall's piece, they could see little point in publishing outmoded materials.[28] Meeting the standards of the scientific community also required the committee to lobby for funds to pay for maps, coloured plates, helios, and line drawings. Especially because few other researchers had been to the western Arctic, these supplements were necessary to convince readers of the validity of the evidence presented in the text. In the early 1920s, Anderson noted that Diamond Jenness, an anthropologist of the Canadian Arctic Expedition, had faced criticism in *American Anthropologist* for publishing "summaries and synopses without adequate figures to back his statements." Anderson helped to procure funding to include more than 450 line drawings in the four volumes on anthropology.[29] In these two cases, then, the medium was the message: formatting of the *Report* conveyed Canada as a nation aware of standards for research and capable of meeting them. In a word, Canada was modern because its science was.

Along the same lines, distribution and application of the *Report* likely solidified Canada's intellectual claims in the Arctic. The Arctic Biological Committee decided against providing separate titles for volumes that came from the northern and southern parties (see Figure 11.1).[30] The ultimate goal, committee members concluded, was to connect the total body of knowledge of the *Report* with Canada and the Canadian Arctic Expedition, regardless of the field scientist who collected the specimens and even though Canadian scientists did not author the bulk of the content. This aspiration, in effect, was complementary to the memorial placed at the Dominion Archives Building – the *Report* was a circulating monument that displayed Canada's service to humanity. It communicated to scientists the tight associations among nation, Arctic nature, and modernity. Two different titles might interfere with these associations: researchers might incorrectly cite the *Report* in their publications, and others might have difficulty tracking down references to it. Indeed, searches in the Web of Science for "Canadian Arctic Expedition" and "Report of the Canadian Arctic Expedition" show citations of the *Report* across the 1920s, with irregular but occasional references extending to 2015.[31] The *Report* enjoyed reception beyond scholarly audiences as well, and Anderson fielded requests from public libraries, concerned citizens, and grade school teachers.[32] For both scientific and lay audiences, the visual aesthetic of the *Report* must have underscored its symbolic meaning. The cover of the bound volumes featured the title along with a simple design of the Arctic region from the Bering Strait to Greenland. This image positioned the majority of the northern territories of the western hemisphere as Canadian, accentuated by sprawling the word CANADA across the map.

That a member of Parliament dismissed "scientific people throughout Canada" in an appeal to commercial applications of science in the North in 1925 conceals the work performed by distributing publications of "basic" research. The *Report* constituted a venue for the elaboration of Canadian science, in the manner of a laboratory, a field site, or a museum. It also sustained relations between scientific institutions in Canada and an international research community and helped readers around the world to recognize Canada as a modern, Arctic nation. Circulation of the *Report* thus provides an example from the interwar Arctic of the relations between

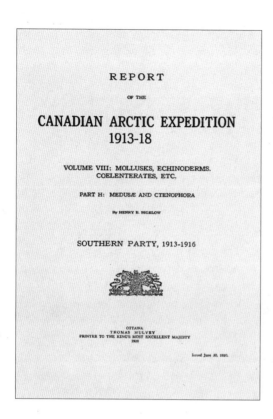

Figure 11.1 The title page of the *Report of the Canadian Arctic Expedition 1913–1918*, volume 8, part H. The committee favoured a single title for the *Report* to prevent confusion among the scientific community and to ensure that all future citations would reference one body of work. Printed on "superior paper" and bound in green cloth, the bound volumes were delivered in protective cardboard boxes. These choices were meant to enhance the durability of the *Report* both in transit to its destination and through years of use. | *Wikimedia Commons*.

science and environmental authority that historian Adrian Howkins has described for British claims on Antarctica and that historian Stephen Bocking describes in this volume.[33] Both at home and abroad, a cross-section of society came to know modern Canada through reference to the *Report*.

Biology and the Politics of Arctic Imaginaries in the *Report*

Still other political and economic implications of the *Report* become visible when one examines the disciplinary conversations within it and flowing from it. As historians Peter Bowler, Janet Browne, and David Livingstone have argued, understanding the geographical characteristics of life on Earth – a project known as "biogeography" – fused the modern scientific and colonial enterprises.[34] Bowler has noted that European and American biogeographers in the early 1900s were "fascinated by the attempt to carve the world up into biological provinces which matched their own political and economic zones of influence."[35] Although historians of the Canadian Arctic

Expedition have located the venture as an exercise of national sovereignty in the North, the *Report* places Canada more firmly in line with this interpretation of science and global empire.

For instance, depending on how biologists treated Canadian Arctic Expedition specimens, a particular nation's status could be highlighted or diminished. Entomologists noted that abundance in the Arctic stemmed from adaptations to its seasonal and topographical conditions. Canada certainly was more "northern" in these ways than the other countries in North America, but these physical conditions were not restricted to Canada. The landscapes from which collected objects originated could be imagined as a circumpolar area, which spotlighted the connections between the eastern and western hemispheres, or as bipolar, with links to Antarctica. Moreover, many botanists were convinced that life at high latitudes was best understood alongside life at the tops of mountains, which suggested some amount of continuity across the varied landscapes of all continents.[36]

The political and economic stakes of this situation are most evident when contrasting the Arctic imaginaries of biologists working with the American and Canadian governments in the early 1900s. Consider the work of Clinton Hart Merriam, one-time chief of the Division of Economic Ornithology for the US Department of Agriculture. At the end of the nineteenth century, Merriam created the "life zone" concept, a categorization of biomes of North America according to latitudinal and altitudinal sequences from the Equator to the North Pole. For each category, he produced lists of plant and animal species to allow Department of Agriculture officers to tailor the development of territories based upon their known regional attributes. Given the abundance of new knowledge produced by explorers and scientists heading to Alaska, northern Canada, and Greenland, Merriam identified a "boreal zone" that comprised much of northern North America. He partitioned this life zone into three subdivisions. The Canadian and Hudsonian areas, characterized by spruce and fir stands, defined much of the Canadian North and Alaskan interior. Above these biological provinces, forming a ribbon along the Arctic Ocean no more than 100 miles wide, was a third biological province, dubbed Arctic-Alpine. Here, at the edge of the continent (and at the tops of its highest peaks), lichens and grasses dominated a mostly treeless terrain.[37]

This strategy was itself a paradigm shift that mirrored a shift in the history of imperialism. At the time, the existing reference works on this region, *Flora Boreali-Americana* (1840) and *Fauna Boreali-Americana* (1829–31), came from British scientists. Descriptions of natural history in these publications often considered northern plants and animals in tandem with counterparts in Antarctica. This approach owed as much to the British Empire's expeditions in northern and southern oceans as it did to questions about distribution that defined natural history in the middle of the nineteenth century.[38] In a similar vein, Merriam's theories built upon American expeditions in Latin America, the Pacific islands, and Alaska – and the infrastructure of American military and economic networks in those places. The organization of plant and animal life from the Equator to the North Pole along the American continents, as Merriam did through his life zone concept, reflected the rise of an imperial power in the Western hemisphere.

Unfortunately, for American settlers, Merriam wrote, the Arctic-Alpine zone was "far too cold for agriculture," the principal means of frontier settlement. His disappointment found its opposite in Rudolph Anderson, editor of the *Report*, who argued for development of the western Canadian Arctic *because* of its life zones. In so doing, he articulated a Canadian claim to knowledge and nature by disputing accepted scientific theory. Anderson challenged Merriam, based upon more than seven years of fieldwork in the western Arctic and the ten volumes on Arctic biology in the *Report*. In an article in the *Canadian Field-Naturalist*, Anderson claimed that "tongues" of the Hudsonian zone stretched far above where Merriam had limited it. This was not a trivial point: the Hudsonian zone was typified by a variety of fur-bearing animals and at least three different species of trees that did not exist on the tundra. Timber and fur could be useful resources for northern development, Anderson implied. The same resources were considered "staples" in the contemporary Canadian economy.[39]

As the examples of Merriam and Anderson indicate, biological research in the *Report* helped to forge national identities and conceptions of modernity. Historians of the Canadian Arctic Expedition, like environmental historian Adam Sowards, have also come to this conclusion, but they have done so by emphasizing federal efforts to demarcate northern resources as precursors to economic expansion.[40] Classification strategies also forwarded Canadian interests in the North and development, but these strategies were embedded in the transnational exchanges of imperial relations. Anderson considered his vision of the Arctic as a pathway for modern Canada precisely because he juxtaposed his conclusions with those of a respected American biogeographer. Although he never said as much in his formal publications, Anderson did drop hints in private letters about his desires for the impact of the *Report* on international recognition of Canadian science and Canadian northern development. "When the reports [of the Canadian Arctic Expedition] are all in," he wrote in 1916, "we shall have made quite a showing."[41]

Taking this position on life zones and the potential of the northern fur economy also forced Anderson to articulate his differences with Vilhjalmur Stefansson on how to modernize Canada. A self-proclaimed "prophet" of the North, Stefansson saw the tundra as "desirable country" that, like the West, could be transformed from desert to garden by pioneering citizens and interested companies that understood it was not perpetually dark and cold. He advocated colonizing Wrangell Island in the Chukchi Sea, testing airplanes and submarines in the Arctic Ocean, and introducing reindeer across the North.[42] Anderson thought Stefansson crazy for encouraging notions of a welcoming Arctic, especially given the loss of life during the Canadian Arctic Expedition. Yet, regardless of his anti-Stefansson sentiments, Anderson never allowed his statements as a Canadian zoologist to be conflated with antidevelopment. Importantly, he did not disagree with Stefansson's plan to introduce reindeer. He admitted to relatives that such a scheme would prove to his government superiors the "tangible useful results of several years of Arctic travel."[43] Anderson was more concerned with the model of modernization that Stefansson promoted – that mix of settler colonialism and corporate capitalism.

Anderson wanted government scientists to steer the northern economy rather than leave it to foreign corporations, explorers, or independent fur trappers. He preferred the approach taken by Denmark in Greenland, where colonial officials controlled exploration through a scientific licensing program and only dispatched state agents who held "the utmost rigid moral and physical qualifications."[44] He also highlighted the benefits of scientific managers by pointing out the advances made through the *Report* in knowledge of fur-bearing creatures. Anderson spoke and wrote about how little Canadian researchers and game wardens knew about these northern species and suggested that this was an artifact of leaving the duties of knowledge production "to the trapper."[45] He emphasized how the *Report* had begun to correct this gap in knowledge by documenting new species and creating identification keys. Because scientists on the expedition stayed longer than a trapping season and were not interested in turning in pelts for cash, they extended what was known about the behaviours, habitats, and life cycles of northern animals. In Anderson's opinion, this was "a striking example of how little fragments of knowledge gleaned in any by-path of science may furnish data" for any field.[46] Here "for Canada and for science" had a different connotation, one that pinned the future development of the nation to particular agents, policies, and research practices.

In sum, like other specialists working on the *Report*, Anderson assembled his view of Arctic life through a transnational network. Although his plans for northern economic development accorded with other nationalist agendas in the House of Commons at the time, they were formulated through a wider set of historical forces: an understanding of scientific theory, a grasp of the biogeography of the circumpolar region, and assessments of natural resource management regimes in the United States and Greenland.[47] After completing his work as editor, Anderson was promoted to chief of biology in the Geological Survey of Canada. Through the 1940s, he was a leading voice in the conservation of caribou and muskox through hunting regulations, the creation of parks and sanctuaries in the Arctic islands, and the introduction of reindeer to the western Arctic. Whether or not these activities conserved or developed the North is up for debate. Without question, though, they were conscious efforts to modernize a region and a nation that owed as much to Anderson's role within the federal government as to his experience navigating many political, scientific, and ideological spaces.[48]

Toward Transnational Modernities

Attending to the circulation of the *Report on the Canadian Arctic Expedition 1913–1918* improves our understanding of science and modernity in several ways. First, it provides an example from twentieth-century Canada of the relationship between the transnational movement of scientific objects and the formation of national knowledge, identities, and representations of nature. As an epistemic site, the *Report* was a constellation rather than a core, linking researchers and institutions across North America and Europe. At the same time, it served as a global ambassador for a modern, scientific Canada. Production of the *Report* highlights particularly Canadian phenomena – such as the timing of the development of its university system, the influence of high-ranking scientists in federal agencies on the national research scene, and a surge of political and economic interest in the western Arctic in the early 1900s. It also brings to the foreground changes in global science and empire between the mid-1800s and the Second World War. Despite considerable museum holdings and resident scientific expertise on polar matters, England was not a major player in the production of the *Report*. The United States claimed the majority of specialists and assistants working on Canadian Arctic Expedition materials. This reality reflected the histories of scientific exploration in the Arctic as well as on the frontiers of American imperialism. The pursuit of circulation across the history of Canadian science will likewise continue to inform interpretations of Canadian and world history.

Second, a transnational analysis of a document such as the *Report* contributes to our growing understanding of northern history. Scholars have recently linked the evolution of the modern Canadian state to events in the interwar Arctic, and the Canadian Arctic Expedition has been identified as a key event in this process. Janice Cavell and Jeff Noakes have shown how the Department of External Affairs elaborated a "sector theory" in the 1920s to assert Canadian sovereignty against the claims of Denmark, the United States, and the Soviet Union. Similarly, historian John Sandlos has demonstrated that the federal government constructed an ideology of a northern nation and its wildlife management apparatus after the First World War by displacing Inuit, Dene, and Metis through natural resource conservation programs.[49] These scholars confirm that Canadians have had to exert considerable political effort to make the Arctic a region of Canada. The *Report* was an agent of this process as well. Even as they came into conflict, Merriam, Anderson, and Stefansson drew from expedition collections and the *Report* to project intellectual authority over the Arctic and to provide a scientific rationale for modernization schemes.

Third, notions of the Arctic emerging through the *Report* had long careers in northern and Canadian history. The circumpolar Arctic, or the Arctic that shared attributes with Antarctica and the Tropics, would come to define geopolitical and scientific imaginaries of the Canadian North during the Second World War and the Cold War era. These later periods can be profitably examined and situated in narratives of continuity and change by taking seriously the life of the *Report* and similar publications. Although often buried in technical language and bibliographic references, encyclopedias, manuals, accelerated base-line studies, and databases produced about the Arctic in the second half of the twentieth century contain links to histories of exploration and knowledge production. These sources bring useful nuances to persistent conceptions of the North as a frontier, whether intellectual, political, or economic.[50]

Fourth, and related to the last point, the *Report* reminds historians of science that our scholarship ought to inform contemporary public understanding. Attention to the Canadian Arctic Expedition has been renewed in the lead-up to its centennial anniversary, through museum exhibits in Ottawa, accompanying websites, articles in *Canadian Geographic*, and a new book-length history of the expedition.[51] As literary scholar Adriana Craciun has shown for the vessels of the third Franklin expedition, commemorations of historic scientific travel are central to geopolitical manoeuvres and articulations of nationalism.[52] Craciun's observation applies to recent coverage of the Canadian Arctic Expedition, even as scholars and authors disrupt simple notions of science and progress by acknowledging experiences of northerners and Inuit. Biologist David Gray, for instance, has profiled the lasting impacts of the expedition on Canadian geography and ideas of the North as well as on Inuit communities of the Yukon and Northwest Territories. Although these treatments go some length toward recognizing the colonial aspects of the expedition, they represent them as evidence of Canada's claims in the North. As I have in this chapter, historians of science add friction to these representations as we detail the transformation of collected specimens into scientific publications and the global processes behind interpretations of nature and nation. In so doing, we prepare civil society to think more critically about expressions of "for Canada and for science," in the past, present, and future.

APPENDIX

The Plan for Reports of the Canadian Arctic Expedition, as of December 1920

Notes: Actual publication dates are listed in parentheses. Authors who were also expedition members are listed in italics. If known, the country in which the author lived while working on the report is listed in brackets.

Volume 1: **Narrative of the Expedition (never published)**
  Part A: Northern Party, 1913–1918 by *Vilhjalmur Stefansson* [USA]
  Part B: Southern Party, 1913–1916 by *Rudolph Martin Anderson* [Canada]

Volume 2: **Mammals and Birds (never published)**
  Part A: Mammals of Western Arctic America by *Rudolph Martin Anderson* [Canada]
  Part B: Birds of Western Arctic America by *R.M. Anderson* [Canada] and P.A. Taverner [Canada]

Volume 3: Insects (1919–1922)
             Introduction by C. Gordon Hewitt [Canada]
             Part A: Collembola by Justus W. Folsom [USA]
             Part B: Neuropteroid Insects by Nathan Banks [USA]
             Part C: Diptera by Charles P. Alexander [USA], Harrison G. Dyar [USA], and J.R. Malloch [USA]
             Part D: Mallophaga and Anoplura by A.W. Baker [Canada], G.F. Ferris [USA], and G.H.F. Nuttall [England]
             Part E: Coleoptera by J.M. Swaine [Canada], H.C. Fall [USA], C.W. Leng [USA], and J.D. Sherman Jr.
             Part F: Hemiptera by Edward P. Van Duzee [USA]
             Part G: Hymenoptera and Plant Galls by Alex D. MacGillivary [USA], Charles T. Brues [USA], F.W.L. Sladen [Canada], and E. Porter Felt [USA]
             Part H: Spiders, Mites, and Myriapods by J.H. Emerton [USA], Nathan Banks [USA], and Ralph V. Chamberlin [USA]
             Part I: Lepidoptera by Arthur Gibson [Canada]
             Part J: Ortoptera by E.M. Walker [Canada]
             Part K: Insect Life on the Western Arctic Coast of America by *Frits Johansen* [Canada]

Volume 4: Botany (1921–1924)
             Part A: Freshwater Algae and Freshwater Diatoms by Charles W. Lowe [Canada]
             Part B: Marine Algae by F.S. Collins [USA]
             Part C: Fungi by John Dearness [Canada]
             Part D: Lichens by G.K. Merrill [USA]
             Part E: Mosses by R.S. Williams [USA]

Volume 5: Botany (1921–1924)
             Part A: Vascular Plants by James M. Macoun [Canada] and Theo. Holm [USA]
             Part B: Contributions to the Morphology, Synonymy, and General Distribution of Arctic Plants by Theo. Holm [USA]
             Part C: General Notes on Arctic Vegetation by *Frits Johansen* [Canada]

Volume 6: Fishes, Tunicates, Etc. (1922)
             Part A: Fishes by *Frits Johansen* [Canada] (never published)
             Part B: Ascidians, Etc. by A.G. Huntsman [Canada]

Volume 7: Crustacea (1919–1922)
  Part A: Decap Crustaceans by Mary J. Rathburn [USA]
  Part B: Schizopod Crustaceans by Waldo L. Schmitt [USA]
  Part C: Cumacea by W.T. Calman [England]
  Part D: Isopoda by P.L. Boone [USA]
  Part E: Amphipoda by Clarence R. Shoemaker [USA]
  Part F: Fycnogonida by Leon J. Cole [USA]
  Part G: Euphyllopoda by *Frits Johansen* [Canada]
  Part H: Cladocera by Chauncey Juday [USA]
  Part I: Ostracoda by R.W. Sharpe [USA]
  Part J: Freshwater Copepoda by C. Dwight Marsh [USA]
  Part K: Marine Copepoda by A. Willey [Canada]
  Part L: Parasitic Copepoda by Charles B. Wilson [USA]
  Part M: Cirripedia by H.A. Pillsbury [USA]

Volume 8: Mollusks, Echinoderms, Coelenterates, Etc. (1919–1924)
  Part A: Mollusks, Recent and Pleistocene by William H. Dall [USA]
  Part B: Cephalopoda and Pteropoda by S.S. Berry [USA] and W.F. Clapp [USA]
  Part C: Echinoderms by Austin H. Clark [USA]
  Part D: Bryozoa by R.C. Osburn [USA] and H.K. Harring [USA]
  Part E: Rotatoria by A.G. Huntsman [Canada]
  Part F: Chaetognatha by A.E. Verrill [USA]
  Part G: Alcyonaria and Actinaria by H.B. Bigelow [USA]
  Part H: Medusae and Ctenophora by C. McLean Fraser [Canada]
  Part I: Hydroids by C. McLean Fraser [Canada]
  Part J: Porifera (no author listed)[53]

Volume 9: Annelids, Parasitic Worms, Protozoans, Etc. (1919–1924)
  Part A: Oligochaeta by Frank Smith [USA] and Paul S. Welch [USA]
  Part B: Polychaeta by Ralph V. Chamberlin [USA]
  Part C: Hirudinea by J.P. Moore [USA]
  Part D: Gephyrea by Ralph V. Chamberlin [USA]
  Part E: Acanthocephala by H.J. Van Cleave [USA]
  Part F: Nematoda by N.A. Cobb [USA]
  Parts G–H: Trematoda and Cestoda by A.R. Cooper [USA]
  Part I: Turbellaria by A. Hassell [USA]
  Part J: Gordiacea (no author listed)[54]
  Part K: Nemertini by Ralph V. Chamberlin [USA]
  Part L: Sporozoa by J.V. Mavor [USA]
  Part M: Foraminifera by J.A. Cushman [USA]

Volume 10: Plankton, Hydrography, Tides, Etc. (1920)
    Part A: Plankton by Albert Mann [USA] (never published)
    Part B: Marine Diatoms by L.W. Bailey [Canada] (never published)
    Part C: Tidal Observations and Results by W. Bell Dawson [Canada]
    Part D: Hydrography (no author listed) (never published)

Volume 11: Geology and Geography (1924)
    Part A: The Geology of the Arctic Coast of Canada, West of Kent Peninsula by *J.J. O'Neill* [Canada]
    Part B: Maps and Geographical Notes by *Kenneth G. Chipman* [Canada] and *John R. Cox* [Canada][55]

Volume 12: Life of the Copper Eskimos by *Diamond Jenness* [Canada] (1922–1923)[56]

Volume 13: Physical Characteristics and Technology of Western and Central Eskimos
    Part A: The Physical Characteristics of the Western and Copper Eskimos by *D. Jenness* [Canada] (1923)
    Part B: The Ostelogy of the Western and Central Eskimos by John Cameron [Canada][57] (1923)
    Part C: The Technology of the Copper Eskimos (no author listed) (never published)

Volume 14: Eskimo Folk-Lore and Language
    Part A: Folk-Lore, with Texts, from Alaska, the Mackenzie Delta, and Coronation Gulf by *Diamond Jenness* [Canada] (1924)
    Part B: Comparative Grammar and Vocabulary of the Eskimo Dialects of Point Barrow, the Mackenzie Delta, and Coronation Gulf by *Diamond Jenness* [Canada] (1928 and 1944)

Volume 15: Eskimo String Figures and Songs
    Part A: String Figures of the Eskimos by *Diamond Jenness* [Canada] (1924)
    Part B: Songs of the Copper Eskimos by Helen H. Roberts and *Diamond Jenness* [Canada] (1925)

Volume 16: Archaeology
    Contributions to the Archaeology of Western Arctic America (no author listed) (1946)[58]

NOTES

For their assistance in my thinking and writing here, I would like to thank Gregg Mitman, Rick Keller, Richard Staley, Bill Cronon, Lynn Nyhart, Janice Cavell, Edward Jones-Imhotep, Tina Adcock, and two anonymous reviewers. I thank Fulbright Canada and Carleton University for the institutional and financial support for archival research completed for this chapter in early 2016.

1 *Ottawa Journal*, April 29, 1926. See also Janice Cavell, "A Circumscribed Commemoration: Mrs. Rudolph Anderson and the Canadian Arctic Expedition Memorial," *Journal of the Canadian Historical Association* 23, 1 (2012): 249–82.
2 Trevor Levere, "Vilhjalmur Stefansson, the Continental Shelf, and a New Arctic Continent," *British Journal for the History of Science* 21, 2 (1988): 235–36. Virtually the only mention of the scientific reports of the Canadian Arctic Expedition highlights the simmering feud between Stefansson and Rudolph Martin Anderson. Their disagreements prevented them from finishing their narratives of the two sections of the expedition.
3 Gísli Pálsson, "Hot Bodies in Cold Zones: Arctic Exploration," *Scholar and Feminist Online* 7, 1 (2008): 7, http://sfonline.barnard.edu/ice/palsson_01.htm. See also Gísli Pálsson, *Travelling Passions: The Hidden Life of Vilhjalmur Stefansson*, trans. Keneva Kunz (Lebanon, NH: Dartmouth College Press, 2005). These are not the only ways that the Canadian Arctic Expedition has been remembered. Travelogues released by expedition members and recent trade books highlight the dangers of travel in the Arctic and the loss of eleven lives aboard the vessel *Karluk*. See Robert Bartlett, *The Last Voyage of the* Karluk (Toronto: McClelland, Goodchild and Stewart, 1916); and Jennifer Niven, *The Ice Master: The Doomed 1913 Voyage of the Karluk* (London: Pan Books, 2001).
4 Recently, environmental historian Adam Sowards examined the multiple forms of claims making that emerged during and after the Canadian Arctic Expedition with attention to scientific reports on geology and geography, on the one hand, and, on the other, the popular travelogues of Stefansson. Sowards demonstrated how environmental factors, intellectual trends, and political aspirations shaped how these writers claimed space and truth about the Arctic. He used only two of the *Report*'s volumes as a substitute for the entire *Report* and did not examine the process of producing, printing, or circulating the *Report*, thus in many ways complementing the analysis that I present here. Adam Sowards, "Claiming Spaces for Science: Scientific Exploration and the Canadian Arctic Expedition of 1913–1918," *Historical Studies in the Natural Sciences* 47, 2 (2017): 164–99.
5 Government appropriations for printing the *Report* ended in 1926, but one laggard report was published in 1946. See the appendix.
6 A.W. Greeley, *The Polar Regions in the Twentieth Century: Their Discovery and Industrial Evolution* (Boston: Little, Brown, 1928), 73.
7 Frédéric Regard, ed., *The Quest for the Northwest Passage: Knowledge, Nation, and Empire, 1576–1806* (New York: Pickering and Chatto, 2013); Frédéric Regard, ed., *Arctic Exploration in the Nineteenth Century: Discovering the Northwest Passage* (New York: Pickering and Chatto, 2013).
8 Janice Cavell, "Arctic Exploration in Canadian Print Culture, 1890–1930," *Papers of the Bibliographic Society of Canada* 44, 2 (2006): 7–42.

9 Janice Cavell and Jeffrey David Noakes, *Acts of Occupation: Canada and Arctic Sovereignty, 1918–1925* (Vancouver: UBC Press, 2010); Janice Cavell, "The Second Frontier: The North in English-Canadian Historical Writing," *Canadian Historical Review* 83, 3 (2002): 364–89.
10 Investigating the "sociology of translation" in material ways was the original intent of Bruno Latour's work on the "circulating reference." Bruno Latour, "Circulating Reference: Sampling the Soil in the Amazon Forest," in *Pandora's Hope: Essays on the Reality of Science Studies* (Cambridge, MA: Harvard University Press, 1999), 24–79.
11 Lissa Roberts, "Situating Science in Global History: Local Exchanges and Networks of Circulation," *Itinerario* 33, 1 (2009): 9–30.
12 On science in the Atlantic world during the early modern era, with reference to British North America, see Joyce Chaplin, *Subject Matter: Technology, the Body, and Science on the Anglo-American Frontier, 1500–1676* (Cambridge, MA: Harvard University Press, 2001). On the Caribbean and Pacific worlds of the twentieth century, see Camilo Quintero, "Trading in Birds: Imperial Power, National Pride, and the Place of Nature in US-Colombia Relations," *Isis* 102, 3 (2011): 1–25; and Gregory T. Cushman, *Guano and the Opening of the Pacific World: A Global Ecological History* (Cambridge, UK: Cambridge University Press, 2013). I have considered the fieldwork of the Canadian Arctic Expedition from these vantage points elsewhere. See Andrew Stuhl, *Unfreezing the Arctic: Science, Colonialism, and the Transformation of Inuit Lands* (Chicago: University of Chicago Press, 2016).
13 Robert E. Kohler, "Finders, Keepers: Collecting Sciences and Collecting Practice," *History of Science* 45, 4 (2007): 428–54.
14 G.J. Desbarats and R. McConnell to R.M. Anderson, January 4, 1917, file 23b, box 53, 1996–077 Series A – R.M. Anderson, Canadian Museum of Nature Archives (hereafter this series is referred to as CMNA).
15 Minutes of the Meeting of the Arctic Biological Committee, May 20, 1920, file 23b, box 53, CMNA; R.M. Anderson to R.E. Lyons, June 16, 1924, file 14, box 57, CMNA. When Hewitt and Macoun died in 1920, their positions were filled by M.O. Malte, director of the National Herbarium, and A.G. Huntsman, director of the Atlantic Biological Station. At nearly the same time, the geologists, geographers, and anthropologists of the Canadian Arctic Expedition had returned from service in the First World War and sought to expand the agenda for the *Report* beyond biology. To reflect this change, the Arctic Biological Committee added Dr. Edward Sapir, chief of the Division of Anthropology for the Geological Survey, and changed its name to the Arctic Publications Committee and later to the Canadian Arctic Expedition Committee. On presentation before the Subcommittee of Council on Government Printing, see Fred Cook, F.C.T. O'Hara, and F.C.C. Lynch, "Report No. 14, Ottawa, February 15, 1918," file 23a, box 53, CMNA. On the "new standard," see J.M. Macoun to C.G. Hewitt, January 24, 1917, file 23a, box 53, CMNA.
16 J.M. Macoun to E.E. Prince, March 24, 1917, file 23a, box 53, CMNA.
17 Anderson blamed the "exigencies of war" on the delay in lining up specialists to help work on the results of the Canadian Arctic Expedition. "Report of the Canadian Arctic Expedition, 1913–1918," vol. 10, file 13, MG30-40, Library and Archives

Canada (hereafter this record group is referred to as LAC). Hewitt and Prince told the Subcommittee of Council on Government Printing that "the Smithsonian Institution at Washington is well supplied with Alaskan Arctic materials in some groups, and the British Museum with material from various Arctic expeditions, while the Greenland region is best represented by Danish and Norwegian collections, consequently a number of groups of specimens have been forwarded to some of these countries for determination." See Cook, O'Hara, and Lynch, "Report No. 14." The majority of specialists brought on board represented the biological sciences – particularly entomology, marine biology, and botany. The expedition's anthropologist (Diamond Jenness), geographers (John Cox and Kenneth Chipman), and geologist (J.J. O'Neill) worked directly with their field notes and collections to write up their analyses. I did not include in the tables the Canadian scientists who supervised the creation of reports but who themselves did not author reports.

18 Rudolph Martin Anderson, "Field Study of Life-Histories of Canadian Mammals," *Canadian Field-Naturalist* 33 (1919): 87.

19 Sophie Lemercier-Goddard and Frédéric Regard, "Introduction: The Northwest Passage and the Imperial Project: History, Ideology, Myth," in Regard, *Quest for the Northwest Passage*, 7.

20 Cook, O'Hara, and Lynch, "Report No. 14." See also Minutes of the Meeting of the Arctic Biological Committee, May 31, 1918, file 23a, box 53, CMNA. Some specialists also received specimens not identified as "unique" or "type" specimens by the committee and thus considered "surplus."

21 Lynn K. Nyhart, "Voyaging and the Scientific Expedition Report 1800–1940," in *Science in Print: Essays on the History of Science and the Culture of Print*, ed. Rima D. Apple, Gregory J. Downey, and Stephen L. Vaughn (Madison: University of Wisconsin Press, 2012), 65–85.

22 Canada, Parliament, *House of Commons Debates*, 14th Parl., vol. 60, no. 63 (May 8, 1925).

23 Marianne Gosztonyi Ainley, "Rowan vs Tory: Conflicting Views of Scientific Research in Canada, 1920–1935," *Scientia Canadensis: Canadian Journal of the History of Science, Technology, and Medicine* 12, 1 (1988): 3–21. See also Tina Adcock's chapter in this volume.

24 The *Report* did not circulate by request alone. The Arctic Biological Committee drafted "distribution lists" starting in 1922 to send separate reports and bound volumes to "distinguished scientists" as well as "essential" libraries and institutions across the fields of biology, anthropology, geology, and geography. The list for the anthropological volumes alone included between seven hundred and eight hundred addresses. On distribution lists, see Minutes of the Meeting of the Canadian Arctic Expedition Committee, February 6, 1922, file 23b, box 53, CMNA.

25 Foster H. Benjamin to Director, National Museum of Canada, December 27, 1929, file 29a, box 49, CMNA; Francis Pospisil to R.M. Anderson, January 14, 1930, file 29a, box 49, CMNA. Because the full list of volumes was printed with each separate report, scientists knew that they lacked certain documents, especially the volumes that were never published.

26 R.M. Anderson to Mr. F. McVeigh, October 30, 1928, file 29a, box 49, CMNA.

27 These rules stated that, when a species was transferred from the original genus or the specific name was combined with any generic name other than that with which it was originally published, the name of the author of the specific name was retained in the notation but placed in parentheses.
28 Zoologist, Biological Division, Geological Survey, to E.E. Prince, June 12, 1919, file 24, box 49, CMNA.
29 R.M. Anderson to E.E. Prince, November 24, 1922, file 24, box 49, CMNA.
30 The exact wording of the title became a point of contention between the Arctic Biological Committee and Vilhjalmur Stefansson. Initially, the committee favoured a title that included only the years 1913–16. Any other date range, C. Gordon Hewitt noted, would draw "undeserved criticism for the amount of work carried out" by the southern party. Because his northern party did not return to Ottawa until 1918, Stefansson objected and lobbied Deputy Minister of the Department of Naval Service G.J. Desbarats to use 1913–18. Although they bristled at Stefansson's power play, members of the committee agreed to the full range of dates for the cover of bound volumes while adding *Southern Party 1913–1916* to the title page of any separate report reflecting the work of that group. See C. Gordon Hewitt to R.G. McConnell, February 27, 1919, file 23a, box 53, CMNA.
31 Vladimir Walters, "The Fishes Collected in the Canadian Arctic Expedition, 1913–1918, with Additional Notes on the Icthyofauna of Western Arctic Canada," *National Museum of Canada Bulletin* 128 (1953): 257–74; Kamal Khidas, "The Canadian Arctic Expedition 1913–1918 and Early Advances in Arctic Vertebrate Zoology," *Arctic* 68, 3 (2015): 283–92.
32 As a sample, requests came from the University of Adelaide (Australia), Presbyterian Collegiate School (Canada), Northwest Territories and Yukon Branch (Canada), Moravian Museum (Czechoslovakia), Liverpool Public Libraries (England), National Library of Saxony (Germany), Christensen's Whaling Museum (Norway), Bureau of Science (Philippines), Pacific Scientific Fishery Research Station (Siberia), Royal University Library (Sweden), and Alaska Game Commission (United States). Requests for the years 1927–29 are found in file 29a, box 49, CMNA.
33 Adrian Howkins, *Frozen Empires: An Environmental History of the Antarctic Peninsula* (Oxford: Oxford University Press, 2016).
34 Peter J. Bowler, *The Fontana History of the Environmental Sciences: Geography, Geology, Oceanography, Meteorology, Natural History, Paleontology, Evolution Theory, Ecology* (London: Fontana, 1992), 391; David N. Livingstone, "Geography," in *Companion to the History of Modern Science*, ed. R.C. Olby, G.N. Cantor, J.R.R. Christie, and M.J.S. Hodge (New York: Routledge, 1990), 743–58; Janet Browne, *The Secular Ark: Studies in the History of Biogeography* (New Haven, CT: Yale University Press, 1983), 117–37.
35 Bowler, *The Fontana History of the Environmental Sciences*, 380.
36 C. Gordon Hewitt, "Introduction and List of New Genera and Species Collected by the Expedition," in *Report of the Canadian Arctic Expedition, 1913–1918*, vol. III, *Insects* (Ottawa: Thomas Mulvey, Printer to the King's Most Excellent Majesty, December 10, 1920), v; Albert Mann, "Part F: Marine Diatoms," in *Report of the Canadian Arctic Expedition, 1913–1918*, vol. IV, *Botany* (Ottawa: Thomas Mulvey,

Printer to the King's Most Excellent Majesty, November 12, 1922), 4F; Harrison G. Dyar, "Part C: Mosquitoes," in *Report of the Canadian Arctic Expedition, 1913–1918*, vol. III, *Insects* (Ottawa: Thomas Mulvey, Printer to the King's Most Excellent Majesty, July 14, 1919), 31C.

37 C. Hart Merriam, *Life Zones and Crop Zones of the United States*, Bulletin 10 (Washington, DC: US Department of Agriculture, 1898), 19. See also Rexford F. Daubenmire, "Merriam's Life Zones of North America," *Quarterly Review of Biology* 13, 3 (1938): 327–32.

38 Lars Brundin, "Transantarctic Relationships and Their Significance, as Evidenced by Chironomid Midges," in *Foundations of Biogeography: Classic Papers with Commentaries, Parts 5–8*, ed. Mark V. Lomolino, Dov F. Sax, and James H. Brown (Chicago: University of Chicago Press, 2004), 658–67.

39 Rudolph Martin Anderson, "Recent Zoological Explorations in the Western Arctic," paper read by M.W. Lyon Jr. before the Biological Society of Washington, April 5, 1919, reprinted in *Journal of the Washington Academy of Sciences* 9, 11 (1919): 312–15. The "staple thesis" – that Canada's politics, culture, and economy have been shaped by the exploitation of fur, fish, wood, wheat, metals, and fossil fuels – was developed by political economist Harold Innis at about the same time. For a review of the work of Innis, see W.T. Easterbrook and M.H. Watkins, "Introduction" and "Part 1: The Staple Approach," in *Approaches to Canadian Economic History* (Ottawa: Carleton University Press, 1984), ix–xviii, 1–15. Rudolph Martin Anderson, "Field Studies of the Life Histories of Canadian Mammals," *Canadian Field-Naturalist* 33, 5 (1919): 86–89.

40 Sowards, "Claiming Spaces for Science."

41 Rudolph Anderson to Chas E. May, November 1, 1916, MG 30–40, vol. 2, file "Aug–Dec 1916," LAC.

42 Stefansson's ideas about the desirability of the northern country and the effects of the whaling industry on Inuit life are found in his books. See Vilhjalmur Stefansson, *My Life with the Eskimo* (New York: Macmillan, 1913); Vilhjalmur Stefansson, *The Friendly Arctic: The Story of Five Years in Polar Regions* (New York: G.P. Putnam's Sons, 1921); and Vilhjalmur Stefansson, *The Northward Course of Empire* (New York: Harcourt, Brace, 1922).

43 Rudolph Martin Anderson to J.E. Anderson, November 14, 1919, vol. 3, file "Correspondence, 1919," LAC. The toxic relationship between Stefansson and Anderson has been the focus of much analysis by Stefansson's biographers. See Gísli Pálsson, *Writing on Ice: The Ethnographic Notebooks of Vilhjalmur Stefansson* (Hanover, NH: University Press of New England, 2001), 1–13.

44 Rudolph Martin Anderson to J.E. Anderson, November 14, 1919, vol. 3, file "Correspondence, 1919," LAC. The quotation is from "Denmark Conducting Unique Experiment," *Ottawa Citizen*, May 3, 1924, found in vol. 26, file 3, "Clippings, 1921–1923," LAC. For more on Anderson's views of northern development in the 1920s, see Rudolph Anderson, "Canada's Arctic Regions," vol. 24, file 2, "Lecture Notes, 1923, 1929, n.d.," LAC. See also "Zoological Work in the Arctic Regions," *Natural Regions of Canada* 7, 10 (1928), found in RG 45, vol. 67, file 4079, Department of the Interior, LAC.

45 Anderson, "Field Study of Life-Histories of Canadian Mammals," 87.
46 "Report of the Canadian Arctic Expedition, 1913–1918," vol. 10, file 13, LAC.
47 "Stefansson Bobs Up Again," *Montreal Standard*, January 21, 1922. Anderson might have been responsible for an editorial in this newspaper claiming that Stefansson had been behaving "more like an exploiter than an explorer." Anderson did not hesitate to point out that he held a PhD, whereas Stefansson did not.
48 Anderson thus engaged in a kind of political and epistemic boundary-work that Adcock documents in this volume. On his involvement in conservation, see John Sandlos, *Hunters at the Margin: Native People and Wildlife Conservation in the Northwest Territories* (Vancouver: UBC Press, 2007).
49 Cavell and Noakes, *Acts of Occupation*; Sandlos, *Hunters at the Margin*.
50 There is now considerable interest in the circulation of Arctic science after the Second World War among a new generation of historians. Stephen Bocking, "Cold Science: Arctic Science in North America during the Cold War, 1945–1991 – A Workshop," http://environmental-history-science.blogspot.ca/2016/04/cold-science-arctic-science-in-north.html. On the persistence of the North as a frontier, see Andrew Stuhl, "The Politics of the 'New North': Putting History and Geography at Stake in Arctic Futures," *Polar Journal* 3, 1 (2013): 94–119.
51 The Canadian Museum of Civilization (now the Canadian Museum of History) featured an exhibit on the Canadian Arctic Expedition in January 2011. The exhibition spawned several educational websites on the expedition. Canadian Museum of History, "Canadian Arctic Expedition," http://www.historymuseum.ca/cmc/exhibitions/tresors/ethno/etp0200e.shtml; Canadian Museum of History, "Northern People, Northern Knowledge: The Story of the Canadian Arctic Expedition 1913–1918," http://www.historymuseum.ca/cmc/exhibitions/hist/cae/splashe.shtml; Greyhound Information and Mountain Studios, "Canadian Arctic Expedition 1913–1918," http://canadianarcticexpedition.com/2013-expedition; Stuart E. Jenness, *Stefansson, Dr. Anderson, and the Canadian Arctic Expedition, 1913–1918* (Gatineau, QC: Canadian Museum of Civilization, 2011); Anne Watson, "Canada's Unsung Expedition," *Canadian Geographic* 133, 1 (2013): 21–22.
52 Adriana Craciun, "The Franklin Mystery," *Literary Review of Canada*, May 2012, http://reviewcanada.ca/magazine/2012/05/the-franklin-mystery/.
53 Eventually authored by A. Dendy [England] and L.M. Frederick [England].
54 Eventually authored by J.H. Ashworth [Scotland].
55 Published as *Geographical Notes on the Arctic Coast of Canada*.
56 Volumes 12–16 were reorganized and retitled between 1922 and 1946. Some reports were moved to other volumes. *Songs of the Copper Eskimos* became its own volume (14). The titles here reflect the plan for reports as of December 1920.
57 Eventually authored by John Cameron [Canada], S.G. Ritchie [Canada], and J. Stanley Bagnall [Canada].
58 Eventually authored by *Diamond Jenness* [Canada].

12

North Stars and Sun Destinations
Time, Space, and Nation at Trans Canada Air Lines/Air Canada, 1947–70

BLAIR STEIN

> *I thought to myself – what an age – what would the Fathers of Confederation say. Here they are, the two symbols of this age of speed – the babies of the twentieth century – up in the clouds the Airways of TCA – below the Airwaves of CBC – both working like giant needles, knitting this country together ... closing the gaps which worried the Fathers.*
>
> – John Fisher[1]

In the late 1940s, CBC personality John Fisher took several publicity flights on Trans Canada Air Lines' newest aircraft, the Canadair DC-4M2 North Star. His characteristically patriotic reports on the flights – he called his scripts "pride builders" – focused on the role of technology in building the nation, pairing two Crown corporations as the "babies of the twentieth century" responsible for transcending Canada's vast distances. Communications scholar Maurice Charland has coined the term "technological nationalism" for this phenomenon, unique in its fervour to English-speaking Canada, "which ascribes to technology the capacity to create a nation by enhancing communication."[2] Charland is largely interested in the Canadian National Railway, but recent works by Robert MacDougall, Liza Piper, and Caroline Desbiens show how other systems such as power generation, freighting, and telephony not only allowed literal access to distant regions but also helped to develop a transnational connectedness in the Canadian imagination.[3] Air travel has been largely neglected in this analytic despite the ease with which it can be folded into modern Canadian technonational paradigms.[4] Aviation has especially resonant consequences in the Canadian context because of the centrality of geography and climate to Canadian national identity and because of aviation's potential to disrupt space and time. It also fits Charland's dual argumentative interests in Canadian technological rhetoric and state management of technological systems; Trans Canada Air Lines (TCA) was a state actor until Air Canada was decentralized in the 1980s and was therefore the most conspicuous source of Canadian air travel storytelling. Even if relatively few Canadians used TCA's services, the airline's discourse of environment, technology, and nation was widely circulated in newspaper and magazine advertisements, in direct-mail campaigns, and even in patriotic media reports such as Fisher's.[5] TCA's promotional material between the 1940s and the 1970s shows how air travel was a site of negotiation for the process of technological modernity since its shifting treatment of the place- and time-based aspects of Canadian national identity reflects larger concerns with how to express nationalism as technological systems appeared to be smoothing over the nuances of place.

Technologies that mediate Canadians' interactions with their vast and wintry surroundings sit at the nexus of cultural, technological, and environmental co-constructions and reflect Canadian experiences of modernity through the twentieth century, especially as they relate to place, space, and time. As Bernhard Rieger has suggested in his work on technology and modernity in Europe, the complexity of large turn-of-the-century technologies meant that their manufacturing processes were black-boxed, making them appear to "burst into the present from nowhere."[6] This lack of a past contributed to the spatiotemporal disruption associated with the experience of modernity, as did the perceived collapse of distance through new communications and transportation systems. By mid-century, the compression of time and space was joined by what David Harvey has called the "speed-up," associated with faster economic cycles and connected to new human interactions with mundane, small-scale technologies of mass production and consumption.[7] Aviation began the twentieth century as what Rieger might call a "modern wonder" and ended it as part of regular habits of consumption, making it critical to the relationship between Canadians and their technological perceptions of the geographic and climatic nation.

Canada's geography and climate have long been seen as central to the construction of its national identity. Carl Berger has shown that Canada's harsh winters, at least compared with those of Britain and the United States, were the basis for Confederation-era boosterism.[8] Building upon the imperial climatic sciences justifying European expansion and control over non-European places, politicians and natural historians argued that Canada's northern conditions awakened a long-lost Norseness in its citizens rendered dormant by temperate British climates.[9] Geography's role in the interrelated "staples" and "Laurentian" theses of Canadian history shows how *scale* has been seen as a major determinant of Canadian economic development.[10] Although these themes have been nuanced and decentred in recent literature, they persist in both academic and popular discourse.[11] Historian Gillian Poulter has argued that still-popular winter pastimes such as snowshoeing and winter carnivals emerged in the nineteenth century as conspicuous national climate performance, and cultural scholars Patricia Cormack and James Cosgrave have analyzed the theme of "the sheer occupation of empty space" in Canadian popular music.[12] The contributions of Stephen Bocking and Daniel Macfarlane to this volume (Chapters 10 and 13, respectively) also show the intransigence of geography to Canadian technoscientific culture. Canadian self-fashioning as a seemingly inexhaustible space with seemingly inexhaustible resources inhabited by cold-weather people has been a relative constant through the past 150 years of nationalism.

Aviation holds the paradoxical status of both celebrating and obfuscating Canadian geography and climate. Interwar aerial surveying by surplus First World War fighters and eventually purpose-built "bush planes," for instance, was a key method by which state actors made the Canadian Arctic legible, but it also homogenized the North in nationalist mythmaking.[13] Furthermore, describing Canada as an insurmountably large and harsh country made the transcendence of distance and weather by air all the more exceptional but also essentially erased the discursive power of that size and that harshness. Passenger aviation, more than military flying or aerial surveying, had the potential to navigate this paradox because it opened this experience to everyday Canadians; as a Crown corporation with a virtual monopoly on domestic air routes, TCA used promotional strategies for engaging with geography and climate that were something of a state-sanctioned representation of the nation to itself.

In this chapter, I outline three of these strategies to illustrate the changing relationships between place, time, technology, and nationalism in modern Canada. First, promotional material from TCA's tenth anniversary in 1947 echoed a prewar triumphalist narrative in which Canadian technologies were especially suited to transcend the unique barriers to mobility inherent in the nation's geography. The airline positioned itself as the heir to historical travel traditions, from early modern ships of exploration to interwar Canadian bush pilots. The perceived uniqueness of Canadian geography was seen as a benefit to TCA since the "shrinking" of distance made possible by aviation was even more impressive when the distances were so large. Second, I discuss the introduction and promotion of the Canadair DC-4M North Star, TCA's first new postwar airliner. Unveiled as part of the decennial and implemented on most of TCA's regular routes by 1950, the North Star incorporated a number of celebrated wartime diffusion technologies to enhance passenger comfort. However, the airline was suffering financially through the late 1940s, which officials attributed to seasonal traffic fluctuations and anxieties about Canadian winter mobility. This situation inspired a series of advertising campaigns highlighting the "all-weather" capabilities of the North Star. Third, this response coincided with the introduction of "sun destination" routes, including Bermuda, Jamaica, Trinidad, and Florida. The airline took the contradictory position of selling winter mobility while suggesting that passengers leave their identity-forming winters behind. During the 1940s and 1950s, while the North Star was the main means by which Canadians accessed sun destinations by air, TCA mitigated these contradictions by suggesting that visiting these destinations during the winter was an act of time travel. By exchanging January for June, instead of one type of January for another, passengers could retain the special status of their Canadian Januaries. As jet-powered airliners were introduced at Air Canada, it became increasingly difficult to keep Canadian winters discursively intact. Advertising for sun destinations from the 1960s and 1970s portrays Canadian winters as foils to the much more pleasant climates elsewhere. This was not simply a result of jet travel's status as mass transit, but also a reflection of the destabilization of space and time inherent in mid-century modernity, since only twenty years earlier TCA saw Canada's size and climate as necessary to its success. These three anecdotes from TCA/Air Canada's first three postwar decades show how air travel was a site of negotiated technological modernity. In a nation long defined by its size and climate, the potential of air travel to negate that size and make that climate easy to escape called into question the legitimacy of place-based nationalism in the modern world. TCA/Air Canada's interrelated strategies for selling itself, its machines, and its new routes are evidence that everyday experiences of space and time had large-scale impacts on the development of nationalisms through the twentieth century.

Selling Itself

Trans Canada Air Lines, now Air Canada, made its inaugural flight in the fall of 1937 as the air arm of the Canadian National Railway (CNR) and the official operator of the recently constructed Trans-Canada Airway. Much like the CNR, the airway and the airline operating on it were trumpeted as symbols of unity; air routes finally allowed Canada's sprawling population centres to be physically linked.[14] In its role as a Crown corporation, the airline was derailed by the Second World War almost immediately, chauffeuring personnel and supplies across the North Atlantic using repurposed Avro Lancaster bombers known as "Lancastrians."[15] Therefore, its decennial celebrations, which continued through 1948, represented a number of turning points for the airline. The decennial was used to promote an airline relatively free of its wartime obligations and able to open its services to increasingly large groups of Canadians.[16] The North Star, with its wartime diffusion technologies such as cabin pressurization and long-range navigation, and the inauguration of transatlantic routes represented this emergent modern infrastructure. The celebration of new aircraft and new routes, promoted under the umbrella of TCA's tenth anniversary, was a space where the airline could articulate its place, and the place of aviation more generally, inside Canadian history and geography.

Introducing transatlantic routes allowed TCA to situate itself and the North Star on a grand historical technological timeline that began with the "little dragon ships" of the Norse, followed by Cabot's and Cartier's ships of exploration, and finally the first Atlantic crossing by air in 1919. "After a thousand years," one promotional brochure claimed, "the wide ocean has been reduced to a narrow pool."[17] The transatlantic teleology was echoed in the promotion of TCA's renewed transcontinental routes, comparing mid-nineteenth-century travel – "an arduous trip by canoe, ox-cart, and on horseback" – with the speed and comfort of flying.[18] Blending technological boosterism with the discursive manipulation of geography and history was a common tactic that allowed anniversary promotions to look back at national technological development and forward to a modern technoscientific future. One TCA pamphlet devoted four of its eighteen pages to Canadian aviation history, explaining how "Canada has been air-minded since ... the Wright brothers were unfledged youngsters." It mentioned the experimental kites and aerodromes of Alexander Graham Bell, who watched Canada's first controlled powered flight in February 1909 "with his long beard blowing in the winter wind," and New Brunswick–born inventor of the variable pitch propeller W.R. Turnbull. However, the most important moment in Canadian aviation history, TCA promotional materials suggested, was the interwar use of airplanes to access, map, and develop the nation, since "Canadians began to realize the value of wings in reaching the outposts of their vast northern wilderness."[19]

Canada's "bush flying" heritage has been seen by both actors and analysts as a key technological myth, especially because of the difficulties inherent in state mapping of territories as large and as uninhabited (by settlers) as the Canadian North. Marionne Cronin has argued that unique grid-mapping techniques, as well as aircraft designs, were responses to the specific conditions of Canada's geography, suggesting that technological systems can be geographically and socially constructed simultaneously.[20] Furthermore, those constructions were multidirectional; just as northern conditions shaped bush flying, so too did the surveying and transporting undertaken by bush planes shape the mining communities of the North. Bush planes and their pilots also influenced cultural perceptions of Canadian northernness as pilots were imagined as national heroes, especially able to transcend those evergreen obstacles of geography and climate. Interwar media reports claimed that bush flying represented "the romance of transportation ... and another real epoch of human pluck and endurance" as well as an especially Canadian narrative of technology, geography, and nation.[21]

Bush flying served several discursive purposes in postwar TCA material, but they all emphasized the airline's connection to its role in national development, both in physical terms and in the national imagination. This was technically true since the state-funded Trans-Canada Airway was originally built to consolidate the ad hoc assemblage of private and public routes created by various bush flying outfits, and TCA was then established to *use* that airway. Even though TCA was a literal outgrowth of the bush flying tradition, promotional materials worked especially hard to tie the regularity of air travel service to the adventure and romance of bush flying. This often took the form of emphasizing the "northland experience" of TCA personnel, as exemplified in a "personalities" advertising campaign through the anniversary years. It profiled a series of "typical personnel who contribute to TCA's record of comfortable, reliable scheduled flying."[22] One of the advertisements featured Winnipeg hangar supervisor Frank Kelly, who "for six years flew throughout the north (to the Great Bear Lake area; from Fort McMurray to Aklavik, etc.)" before joining TCA.[23] The facts that this advertisement was specific about *where* in the North Kelly flew and that it was reissued several times in 1949 and 1950 suggest how valuable bush flying was to the Canadian aviation imaginary and how TCA positioned itself as the heir to the interwar bush flying tradition.

Focusing on the value of aviation to both transatlantic travel and Canadian history in decennial publicity maintained the relatively popular status quo in which Canadian communications technologies such as airplanes were seen as removing distance-related obstacles to mobility. Tenth anniversary promotions reflected on how "air travel has conquered time and space for the vacationist," especially since "Canada has often been referred to as a land of magnificent distances. That was before the coming of Trans-Canada Air Lines." The airplane, and TCA in particular, were assigned a massive amount of power: "Canada ten years ago was five days wide. Today, one can leave Halifax in mid-morning and watch tomorrow's sunrise in Vancouver. So effective has been the attack of TCA upon Canadian distances that already they have lost much of their old significance ... A new sense of nationhood is being fostered by Canada's new accessibility to its citizens."[24] It was with great violence, apparently, that the North Star eliminated distance in Canada. Not only had Canadian distances been "attacked" by TCA, but also, as a 1947 newsletter suggested, "the arrival of the North Stars ... will scatter our old concepts of distance and travel times to the four winds ... [and] will demonstrate their superior power ... by slashing" flying times.[25]

All of this attacking and slashing was dramatically different in tone from how decennial-related promotions advertised the actual *experience* of travelling by air. Postwar air travel was substantially more comfortable than it had been at TCA's founding, largely thanks to the wartime diffusion technologies discussed below, but brochures and advertisements from the late 1940s placed the experience of viewing Canada from above ahead of the "drawing room luxury" that the "modern aircraft" could offer passengers. One informational pamphlet claimed that "there's no boredom in air travel [because of] forest and farm land, wide prairies whose winding rivers are cut in the soil, rolling foothills and the majesty of the Canadian Rockies ... The landscape is always changing."[26] Fisher was perhaps the greatest proponent of TCA's aerial views. Flying over southern Ontario in 1948, he described cars "crawling like ants" on Hamilton's "payrollish" streets, the "patchwork quilt" of farms "blessed by geography," and the "aqua blue of Lake Ontario," without which "Canada would not hold the world position she does today."[27] Passengers also seemed to be fixated on the view from above, writing to the airline requesting guidebooks indicating landmarks below and asking that the wings be "painted a drab black" to reduce glare.²⁸ One passenger claimed in 1950 that only a "poet-scientist" could fully describe the feeling of flying, since it

> gives one a new and unusual sense of the one-ness of Canada. As you watch the provinces slip beneath your eyes in all their colorful beauty – the breath-taking magnificence of the Rockies, rich-chequered Prairies, small lonely farms, brilliant welcoming cities – you discover with a freshness and impact never achieved by history books or geographies, that this is *one* country, our own.²⁹

This deliberate selling of the aerial view reinforced and supported a place-based Canadian national identity even as aviation was erasing it. Promoting the views afforded by air travel was not unique to Canada, as David Courtwright and Chandra Bhimull have shown for the United States and Britain, but Canada's seemingly overwhelming scale, and the already-established role of aviation in reducing that scale, made the manipulation of geography by airplane especially evocative.³⁰ TCA's decennial was an opportunity for the airline to position itself as both a continuation of Canada's unique aviation past and a modern expression of space, time, and travel. The airline could call on Canadian geography and climate to support the role of aviation in making the nation legible and at the same time "slash" and "attack" it with its modern technological systems. Besides, as one 1949 booklet reminded passengers, "the bleak winter landscape passes by a lot more quickly in a [North Star] Skyliner."³¹

Selling Machines

The North Star itself was something of a technological hybrid. Built in Cartierville, Quebec, and used almost exclusively by Canadian operators, it featured the basic fuselage design of the American Douglas DC-4 transport paired with British Rolls-Royce "Merlin" engines, which had proven their mettle on some of the most celebrated fighters of the Second World War.³² More importantly here, the North Star incorporated for the first time a number of all-weather flying technologies developed during the war, mobilized to combat what TCA officials believed to be the biggest obstacle facing potential passengers: a fear of wintertime travel bolstered by a century of climatic folklore. These technologies acted as stand-ins for the inherent Canadian ability to overcome perceived seasonal barriers to mobility, but marketing the North Star as an all-weather or all-season aircraft slowly decentred climate from self-constructions of Canada, just as decennial promotions highlighting TCA's new routes decentred geography.

At the war's end, there was a great deal of anticipation that aviation technologies developed for military use, such as long-range navigation, autopilot mechanisms, cabin pressurization, and electric deicing, might make postwar air travel much more accessible, reliable, and comfortable. The North Star was the first airliner in Canada to incorporate all of these diffusion systems, and they were a focus of the popular press and airline promotions, especially cabin pressurization.[33] The plane showcased both "parlour car" comfort features, which included upholstered chairs and soothing wall colours, and "operational" features that had less direct impact on the passenger experience and instead affected operation of the aircraft itself, such as radio navigation. North Star promotional materials alternated between treating cabin pressure as a "luxury" and an "operations" feature, but the focus was always on how pressurized cabins changed the feeling of travel by air. Design historian Gregory Votolato argues that cabin pressurization was "possibly the single greatest technical innovation that changed the nature of passenger air travel" because it provided passengers with "a level of flying comfort previously unknown."[34] Everyday Canadian passengers were primed for an experience unlike any possible before the war, and invoking the North Star, with its celestially evocative name, was an ideal way to herald these changes to a modern public.

However, passengers did not seem to be interested. TCA operated in peacetime at a constant deficit, met with varying degrees of alarm by executives, employees, the government, and the public. TCA's operating deficits reached a head in terms of both size and political interest in 1949. It was a federal election year, and Conservatives concerned with large government expenditures targeted Crown corporations, especially those with large operational deficits.[35] There were a number of potential culprits, such as the expensive creation of an "Atlantic" branch of TCA to deal with overseas travel and the costs associated with designing, manufacturing, and maintaining the North Stars, but TCA officials focused on seasonality: the overwhelming passenger preference for summer over winter travel.[36] TCA carried 65,000 passengers in the summer of 1948 and 45,000 in the off-season, which President of the TCA Gordon McGregor called "a serious matter" in his memoirs.[37] McGregor was tasked with explaining the airline's struggles and justifying its choices in interviews and public speeches through 1949 and 1950, and he consistently blamed a "widely-held opinion that if it is snowing outside, it is no day for flying, [and] that airline performance is more irregular in the winter" for the traffic imbalances limiting TCA's growth. Of course, there was "no logical justification" for this belief, which was "100% wrong" thanks to the North Star's new technologies such as cabin pressurization, radio communication, and weather equipment.[38] Connecting the all-weather flying suite of technologies to cold-weather performance made seasonality an alarming problem, but one that taxpayers troubled by the airline's losses could understand because it fit into already-established paradigms of Canadian geographic and climatic identity.

It also appeared to be easily solvable through advertising and educational campaigns, which fell within McGregor's pet interest in passenger relations.[39] TCA's discrete Advertising Department began operations in early 1949, and almost immediately it engaged Montreal-based firm Cockfield, Brown and Company, which had worked as a state advertiser as early as the 1930s and was favoured by the Liberal Party of Canada.[40] TCA's Advertising Department was described as having been purpose-built to address seasonality. Since "one of Advertising's biggest concerns is the 'ironing out' of" seasonal imbalances, claimed Director of Advertising D.C. Bythell in 1949, the majority of its funding was allocated to the winter months because, despite the North Star's all-weather technologies, "the general public still take a quick look at the wintery blasts before flying." "Consequently," Bythell suggested, "TCA Advertising has as one of its major tasks the selling of the idea that 'anytime is flying time.'"[41] As the airline's most advanced piece of equipment, the North Star played a role in these efforts and represented a convergence of environment, technology, and identity that TCA, as a state actor, could articulate to everyday Canadians.

Bythell used the Advertising Department's first institutional campaign – which featured a group of tobogganing children claiming that the North Star was "the real way to get around in winter!" – as a frequent example (Figure 12.1). The copy trumpeted that TCA had carried over three-quarters of a million passengers "in *wintertime*" thanks to the higher operational ceilings made possible by cabin pressurization.[42] Bythell saw this advertisement as a smashing success; it appeared in at least a dozen newspapers, as a direct-mail flyer, and as a two-colour magazine ad through 1950. More importantly, it was "designed to help dispel the widely held belief that winter flying is more hazardous, less reliable than travel at other seasons."[43] The key word here was *belief*, making the fear keeping Canadians from flying appear to be completely imagined and a consequence of the historical construction of Canadian climates. The selective use of capitalization in the advertisement's copy even implied that "winter" might be different from "Canadian Winter." Another campaign, with the tagline that the easiest way to escape the weather was to "go over it, or go around it," using a pressurized North Star, showed how the Canadian climatic imaginary could be both centred and destabilized by air travel (Figure 12.2). Although Advertising Manager Donald S. McLauchlin claimed that this campaign was an attempt at "selling the Canadian public on air travel all year round," its "cheerful approach" and "cartoon technique" would also "combat an all too frequent attitude on the part of the public that winter increases the hazards of air transportation."[44] The ads in this campaign were not set in any particular season, and "winter" appeared only once: "Winter or summer, it makes no difference to your travel when you fly TCA." However, the bullet points highlighting the "smoothest routes," "constant comfort," and "veteran pilots" were stylized snowflakes.[45] Whether or not they were explicitly about winter, such advertisements needed to both elevate and dismiss the place that winter held in the Canadian technological consciousness.

North Stars were repeatedly depicted as rising above uniquely Canadian geographic and seasonal barriers to mobility, but that uniqueness still had to remain in order for the advertising to be effective. In the eyes of TCA advertising officials, the lore of Canadian climates had the unintended consequence of making Canadians approach flying in general, and TCA in particular, with trepidation. North Star promotional materials painted seasonality as a Canadian problem that could be solved through new technological systems, advertising, and education. Canadians, because of their geographic and climatic realities, were still portrayed as having a specific connection to aviation, but those realities were increasingly displaced by air travel's modern technological amenities.

Selling Places

All-weather technologies helped air travel to become quicker, easier, and cheaper through the postwar decades and allowed increased access to TCA's services, including flying to "sun destinations." While on his seasonality speech tour in 1950, McGregor detailed two defences against the "seasonal fluctuation evil": advertising TCA's all-weather performance and introducing air routes to destinations with climates opposite to those of Canada's "winter doldrums."[46] TCA inaugurated flights to Bermuda in 1948, followed by Nassau, Jamaica, Trinidad, and select Florida cities by the end of the decade. These routes originally served a number of purposes, including extending national economic influence to current and former British

Left to right:

Figure 12.1 This 1949 institutional advertisement was designed to foster all-weather confidence in air travel but still foregrounded Canadian winters. | Air Canada Collection, Canada Aviation and Space Museum.

Figure 12.2 1951 TCA advertisement suggesting that the easiest way to escape the weather was to "go over it, or go around it." | Air Canada Collection, Canada Aviation and Space Museum.

colonies through the construction of airfields and an injection of tourism dollars, but these underlying concerns were less publicly visible than the discourse of sun and sand. The eventual popularity and ubiquity of sun destination travel made advertisements for these destinations a staging ground for emerging anxieties about how best to preserve identities based upon geography and climate once modern technological systems had destabilized them. In the 1940s and 1950s, when air travel to tropical areas was still a novelty, TCA sold its southern routes as what might more accurately be called "summer destinations," claiming that passengers could exchange Canadian winters for conditions that felt like Canadian summers. The arrival of turbine-powered airliners saw a relative dismissal of this strategy in favour of a wholesale disparaging of Canadian winter weather; passengers appeared to be abandoning their winters entirely rather than simply shortening them. This manipulation of time, and the ease with which TCA/Air Canada subverted the discourse of Canadian seasonality, show some of the ways in which modern technological systems were considered responsible for the separation of place and time from national identity.

The root of sun destination discourse at TCA was climatic contrast. The airplane's ability to transcend climates was extended by showing that planes could take passengers to *other* climates. TCA's employee newsletter introduced flights to Bermuda by explaining that travellers could board "a big North Star at Dorval, leave the snow-covered Laurentians far behind and look down on Bermuda's green hills and blue inlets" four hours later.[47] Coverage of early flights to the Caribbean described going "from pines to palms" and "leaving galoshes and mufflers in Montreal or Toronto" in favour of "silver beaches."[48] Even when a parliamentarian could not attend a publicity trip in November 1948 and asked for a "rain-check," McGregor kindly suggested that he use "some other name, in the case of the Caribbean."[49] This was ground on which to tread carefully since the sun destination imaginary at TCA could negate the place that climate held in Canadian national self-fashioning by foregrounding other climates and making them appear more appealing.

At first, the airline developed some complex, albeit unintentional, discursive techniques to reconcile Canadian climate nationalism with air travel's modern disruption of space and time. These techniques generally framed visiting sun destinations as an act of time travel. Passengers could, for instance, buy "your ticket to Summer" or "take a TCA North Star Skyliner south where ... it's always 'June in January.'"[50] This was not an entirely new strategy. Historian Mimi Sheller traces this phenomenon as far back as the nineteenth century, when popular travel narratives described "the wonderful transition from winter to late spring" after a week at sea, transforming February into "a moist June morning."[51] These tropes took on a new meaning in the postwar context, as suggested by the treatment of the North Star as a "magic carpet" (one *Vancouver Sun* reporter on a publicity flight literally called the plane a "black box"), a marked departure from all-weather flying ads.[52] Selectively demystifying the North Star shows how the modern technologies of air travel could be mobilized to perform different duties. It was still January in Bermuda, as it was in Toronto or Montreal, but framing sun destination travel as movement through time implied that passengers were replacing their special winters not with another type of winter but with summer.

These strategies kept Canadian climates discursively intact, but once jet-powered airliners became popular at TCA through the 1960s, advertisements encouraging passengers to "climb aboard in the middle of winter and climb off ... in the middle of summer" were gradually replaced by those highlighting the power of the jet to remove Canadians from the miseries of their winters.[53] This was partially because of air travel's emerging status as a form of mass transit thanks to the jet engine, but also because of the relative ease with which it, along with other mid-century technologies of modernity, could minimize place in national discourse. This technological transformation allowed everyday Canadians to divorce the material experience of their geography and climate from the cultural aspects that in turn helped to reconcile the apparent contradiction of being a "true north" people while, for example, watching hockey games in heated indoor arenas. As historian Godefroy Desrosiers-Lauzon has suggested, "Canadian winter, despite its length and its intensity, despite its identity-making function, [and] despite its integration inside popular culture, *is no longer what it used to be.*"[54] This created anxieties about the roles that geography and climate could still play in Canadian nationalisms.

In the jet age, sun destinations were portrayed as far more exhilarating than Canadian ones. Instead of propping up Canadian climates, as advertisements had delicately done in the previous decade, Air Canada promotional materials saying that sun destinations were a "great way to warm up to winter" were destabilizing the Canadian place-based imaginary.[55] One late 1960s campaign, for instance, asked, "How could a numb-fingered, bone-chilled, slush-splattered puddle-hopper become a winter-lover in just a few hours?" It juxtaposed images of a woman in a winter coat, the same woman in a bikini, and the jet that would whisk her away from winter (Figure 12.3).[56]

Figure 12.3 Advertisements through the 1960s made sun destination winters appear to be increasingly more appealing than Canada's winters. | Air Canada Collection, Canada Aviation and Space Museum.

Jet-powered aircraft could remove every geographic and climatic barrier to mobility and replace that geography and that climate with apparently more appealing tropical ones. And once their allure was established in promotional materials, they began to appear insidious. One 1970 brochure encouraged travellers to join Club Calypso, the airline's fictionalized frequent-southern-fliers club, by calling it "Air Canada's secret society of sun-worshippers." Accompanying the list of Club Calypso destinations was a "warning: each is out to win your heart and soul – to become your perennial 'place in the sun.' So take heed, hedonists!"[57] Once they drew Canadians in, it appeared, warm-weather destinations might not let them leave.

This rhetoric reflects an implicit change in the ways that space and time were perceived and experienced in the modern world, suggesting that technological systems such as air travel were causing place to become peripheral to national identity and running counter to a century of nationalist folklore of harsh conditions and character-building winters. The Canadian intelligentsia grappled with how to reconcile winter identity with air travel culture, especially once jets opened access to tropical destinations. This struggle was represented in the mid-1970s by concerns about a "travel deficit": Canadians spent more money travelling abroad than foreigners spent travelling in Canada. For example, MP John Crosbie suggested in a 1977 session of Parliament that "all of us have heard of the 'ban the bomb' campaign. What we need is a 'ban the tan' campaign. Every Canadian who dares to go south next winter and returns to Canada with a tan should be ostracized." He graciously added that, "if hon. Members get a tan from the lights in this Chamber, we can explain that."[58] Although ostensibly he was talking about tourism dollars, Crosbie articulated the jet age concern about how to maintain a national identity based upon geography and climate when they were so easy to abandon. Desrosiers-Lauzon has connected these state efforts to mitigate the "travel deficit" to the "technological conquest," and subsequent cultural devaluation, of Canadian winter, since by the 1970s the prevailing opinion was "that a distaste for winter and a taste for Florida sunshine were evidence of a declining commitment to Canada."[59] As this analysis of TCA/Air Canada promotional materials shows, however, it was not simply about commitment to Canada but also about commitment to geography and climate in the modern Canadian national imaginary.

Conclusion

By the time MPs were suggesting that Canada "ban the tan," the shifting connections between space and time so prevalent in twentieth-century modernity were giving way to the acceleration and volatility of production and consumption associated with postmodernity. David Harvey has linked the experience of postmodern acceleration to the disorienting effect of communication and transportation technologies a century earlier. The emerging ephemerality of space and time encouraged a "search for more secure moorings and longer-lasting values in a shifting world."[60] Because Canadians could travel anywhere in the world in hours by airplane, the explicit call to place-based identity in the 1970s represented a grasping return to what remained uniquely Canadian about Canada: its geography and climate. And the connections between aviation and Canadian place-based nationalism have not entirely disappeared; in a 2014 editorial, the CEO of Air Canada pointed out that it had been best suited to fly during 2013's "polar vortex" since "safety will always be a top priority at Air Canada just as surely as winter is a part of the Canadian reality."[61]

In his "pride builder" report on TCA's tenth anniversary, John Fisher told his listeners to celebrate "the men and women who have carried the Maple Leaf high and far ... [and who] have helped give Canada a feeling of nationhood."[62] Indeed, Canadian airplanes such as the North Star and their associated technologies have been folded into the Canadian nation-building pantheon, celebrated as agents of unity, paired with and compared to hardy Canadian people, and blamed for the dilution of geographic and climatic nationalism. As a state actor, TCA was responsible for crafting two complementary narratives: the everyday concrete interactions between Canadians and modern aviation technologies and a self-reflexive narrative of technology, history, identity, and geography. The changing geographic and climatic rhetoric in TCA's advertising and promotional materials from the 1940s through the 1970s reveals the power of modern technological systems to distance users from the lived experience of place and its implications for the maintenance of Canadian national identity.

NOTES

I would like to thank Hunter Heyck, Marcia Mordfield, Sylvie Bertrand, and the rest of the team at the Canada Aviation and Space Museum library and archives, the anonymous reviewers, and the editors of this volume for their comments, support, and assistance that contributed to this chapter.

1 "John Fisher Reports: Up and Down" (script), May 2, 1948, Air Canada fonds, RG 70, vol. 254, Library and Archives Canada (hereafter LAC).

2 Maurice Charland, "Technological Nationalism," *Canadian Journal of Political and Social Theory* 10, 2 (1986): 197. See also Marco Adria, *Technology and Nationalism* (Montreal/Kingston: McGill-Queen's University Press, 2009), especially 33–71.

3 Robert MacDougall, "The All-Red Dream: Technological Nationalism and the Trans-Canada Telephone System," in *Canadas of the Mind: The Making and Unmaking of Canadian Nationalisms in the Twentieth Century*, ed. Norman Hillmer and Adam Chapnick (Montreal/Kingston: McGill-Queen's University Press, 2007), 46–62; Liza Piper, *The Industrial Transformation of Subarctic Canada* (Vancouver: UBC Press, 2010); Caroline Desbiens, *Power from the North: Territory, Identity, and the Culture of Hydroelectricity in Quebec* (Vancouver: UBC Press, 2013). For more on railways, see A.A. Den Otter, *The Philosophy of Railways: The Transcontinental Railway Idea in British North America* (Toronto: University of Toronto Press, 1997); and R. Douglas Francis, *The Technological Imperative in Canada: An Intellectual History* (Vancouver: UBC Press, 2009).

4 In the introduction to a recent environmental history of mobility in Canada, the editors lamented that aviation remains "a topic that needs more attention." Ben

Bradley, Jay Young, and Colin M. Coates, eds., *Moving Natures: Mobility and the Environment in Canadian History* (Calgary: University of Calgary Press, 2016), 16.

5 Passenger numbers climbed steadily through this period: TCA carried approximately 185,000 passengers in 1945, 540,000 in 1948, and over 1.1 million in 1951.

6 Bernhard Rieger, *Technology and the Culture of Modernity in Britain and Germany, 1890–1945* (Cambridge, UK: Cambridge University Press, 2005), 34. See also Wolfgang Schivelbusch, *The Railway Journey: The Industrialization of Time and Space in the Nineteenth Century* (Berkeley: University of California Press, 1986).

7 David Harvey, *The Condition of Postmodernity: An Inquiry into the Origins of Cultural Change* (Oxford: Blackwell, 1989), 265–66. "Everyday" technological interactions appear in recent postcolonial studies of modernity such as David Arnold, *Everyday Technology: Machines and the Making of India's Modernity* (Chicago: University of Chicago Press, 2013); and Rudolf Mrázek, *Engineers of Happyland: Technology and Nationalism in a Colony* (Princeton, NJ: Princeton University Press, 2002). For more on aviation, see Chandra Bhimull, "Empire in the Air: Speed, Perception, and Airline Travel in the Atlantic World" (PhD diss., University of Michigan, 2007).

8 Carl Berger, "The True North Strong and Free," in *Canadian Culture: An Introductory Reader*, ed. Elspeth Cameron (Toronto: Canadian Scholars Press, 1997), 83–103.

9 Robert Grant Haliburton, "The Men of the North and Their Place in History," *Ottawa Journal*, March 20, 1869.

10 For the staples thesis, see Harold Innis, *The Fur Trade in Canada* (New Haven, CT: Yale University Press, 1930); for the Laurentian thesis, see Donald Creighton, *The Commercial Empire of the St. Lawrence, 1760–1850* (Toronto: Ryerson Press, 1937).

11 On the Laurentian thesis, see, for instance, Daniel Macfarlane, *Negotiating a River: Canada, the US, and the Creation of the St. Lawrence Seaway* (Vancouver: UBC Press, 2014), especially 13–15. On the staples thesis, see Eric W. Sager, "Wind Power: Sails, Mills, Pumps, and Turbines," in *Powering Up Canada: A History of Power, Fuel, and Energy from 1600*, ed. R.W. Sandwell (Montreal/Kingston: McGill-Queen's University Press, 2016), 162–85.

12 Gillian Poulter, *Becoming Native in a Foreign Land: Sport, Visual Culture, and Identity in Montreal, 1840–85* (Vancouver: UBC Press, 2009); Patricia Cormack and James Cosgrave, *Desiring Canada: CBC Contests, Hockey Violence, and Other Stately Pleasures* (Toronto: University of Toronto Press, 2013), 52.

13 For Canada's "northern myth" and the Arctic, see Sherrill Grace, *Canada and the Idea of North* (Montreal/Kingston: McGill-Queen's University Press, 2007); and Renée Hulan, *Northern Experience and the Myths of Canadian Culture* (Montreal/Kingston: McGill-Queen's University Press, 2002).

14 Jonathan Vance, *High Flight: Aviation and the Canadian Imagination* (Toronto: Penguin Canada, 2002), 179.

15 TCA was officially assigned to the Canadian Government Trans-Atlantic Air Service in early 1943, but the airline's personnel and fleet had unofficially been mobilized years before.

16 Then President Gordon McGregor claimed in his memoirs that 1948 was the first year that the airline could operate free of wartime influence. Gordon McGregor, *Adolescence of an Airline* (Montreal: Air Canada, 1970), 6.
17 *North Star over the Atlantic* (brochure), 1947, Air Canada Collection, Canada Aviation and Space Museum (hereafter CASM).
18 *Horizons Unlimited* (booklet), 1949, Air Canada Collection, CASM.
19 *Canada's National Air Service* (brochure), c. 1946, 5–6, Air Canada Collection, CASM.
20 Marionne Cronin, "Flying the Northern Frontier: The Mackenzie River District and the Emergence of the Canadian Bush Plane, 1929–1937" (PhD diss., University of Toronto, 2006), 17–22. See also Marionne Cronin, "Shaped by the Land: An Environmental History of a Canadian Bush Plane," in *Ice Blink: Navigating Northern Environmental History*, ed. Stephen Bocking and Brad Martin (Calgary: University of Calgary Press, 2017), 103–30.
21 "Air Service from Waterways Links Simpson to Steel," *Edmonton Journal*, January 8, 1929.
22 D.C. Bythell, "Institutional Advertising – Personalities" (memorandum), October 28, 1950, Air Canada Collection, CASM.
23 The others were a pilot, an electrical engineer, a flight attendant, and a purser, but only Kelly's and the engineer's profiles were reissued. "Check ... Check ... Check ..." (advertisement), 1949, Air Canada Collection, CASM.
24 "Doing a Job for Canada," *Between Ourselves*, March 1947, 12.
25 "North Stars across the Continent," *Between Ourselves*, December 1947, 16–17.
26 "Air Lines Map of Trans Canada Air Lines," c. 1947, Air Canada Collection, CASM.
27 "John Fisher Reports: Up and Down" (script), May 2, 1948, Air Canada fonds, RG 70, vol. 254, LAC.
28 "What Others Think of Us – 'Crow's Wing,'" *Between Ourselves*, January 1953, 3.
29 "What Others Think of Us," *Between Ourselves*, September 1950, 15.
30 Bhimull, "Empire in the Air," 67–122; David Courtwright, *Sky as Frontier: Adventure, Aviation, and Empire* (College Station: Texas A&M University Press, 2005); David Courtwright, "The Routine Stuff: How Flying Became a Form of Mass Transportation," in *Reconsidering a Century of Flight*, ed. Roger Launius and Janet Daly Bednarek (Chapel Hill: University of North Carolina Press, 2003), 209–22.
31 *Horizons Unlimited* (booklet), 1949, Air Canada Collection, CASM.
32 North Stars were operated by TCA, Canadian Pacific Airways, the Royal Canadian Air Force, and the British Overseas Airways Corporation, where they were known as "Argonauts."
33 The first North Stars in service at TCA in April 1947 (dubbed the DC-4M1) were borrowed from the Royal Canadian Air Force and not pressurized since the project had to be rushed; by June 1948, they had been replaced by pressurized DC-4M2s. Larry Milberry, *The Canadair North Star* (Toronto: CanAv Books, 1982), 42–46. For more on the North Star's deicing systems, see Rénald Fortier, *Propellers* (Ottawa: National Aviation Museum, 1996), especially 22–24. For more on the history of deicing technologies, see William Leary, "A Perennial Challenge to Aviation Safety:

Battling the Menace of Ice," in Launius and Bednarek, *Reconsidering a Century of Flight*, 132–50.
34 Gregory Votolato, *Transport Design: A Travel History* (London: Reaktion Books, 2007), 198.
35 See, for instance, Warren Baldwin, "Drew Wants More Data on North Stars Deal," *Globe and Mail*, April 9, 1949, 3; and James Hornick and William P. Snead, "Secret TCA Plans Bared to Replace North Star Engines," *Globe and Mail*, May 16, 1949, 1.
36 Seasonal imbalances in passenger traffic were not unique to TCA, nor were generalized air travel anxieties. See Courtwright, *Sky as Frontier*, 140–45; and Richard Popp, "Commercial Pacification: Airline Advertising, Fear of Flight, and the Shaping of Popular Emotion," *Journal of Consumer Culture* 16, 1 (2016): 61–79.
37 He was president from early 1948 until 1968. McGregor, *Adolescence of an Airline*, 8.
38 Address by Gordon McGregor, February 14, 1950, MG 30 E283, vol. 13, Gordon R. McGregor fonds, LAC.
39 McGregor was traffic manager before his promotion to president, and he maintained an interest in traffic, which included advertising and public relations, during his tenure.
40 For more on Cockfield, Brown, see Daniel J. Robinson, "Cockfield, Brown & Company," in *The Advertising Age Encyclopedia of Advertising*, ed. John McDonough and Karen Egolf (Chicago: Fitzroy Dearborn, 2003), 341–43. For government advertising in Canadian history, see Jonathan Rose, "Government Advertising and the Creation of National Myths: The Canadian Case," *International Journal of Nonprofit and Voluntary Sector Marketing* 8, 2 (2003): 153–65; and Jonathan Rose, *Making "Pictures in Our Heads": Government Advertising in Canada* (Westport, CT: Praeger, 2000).
41 D.C. Bythell, "30,000,000 'Fly TCA's,'" *Between Ourselves*, December 1949, 5–6.
42 "TCA Tops the Weather" (advertisement), 1949, Air Canada Collection, CASM.
43 Memorandum by D.C. Bythell, December 19, 1949, Air Canada Collection, CASM.
44 Memorandum by D.S. McLauchlin, "Institutional Advertising – All-Weather Flying," November 26, 1951, Air Canada Collection, CASM.
45 "Why Doesn't Somebody Do Something about the Weather?" (advertisement), 1951, Air Canada Collection, CASM.
46 Address by Gordon McGregor, March 27, 1950, MG 30 E283, vol. 13, Gordon R. McGregor fonds, LAC.
47 "Bermuda Bound," *Between Ourselves*, April 1948, 8.
48 "TCA Goes to the Caribbean," *Between Ourselves*, January 1949, 4.
49 Gordon McGregor to C.D. Howe, November 10, 1948, RG 70, vol. 254, Air Canada fonds, LAC.
50 "Now There's Time for Everything" and "Your Ticket to Summer" (advertisements), 1950, Air Canada Collection, CASM.
51 Mimi Sheller, *Consuming the Caribbean: From Arawaks to Zombies* (London: Routledge, 2003); Charles Stoddard, *Cruising among the Caribbees* (New York: Scribner, 1903), 14–19.
52 Jack Scott, "Our Town," *Vancouver Sun*, May 1948, RG 70, vol. 254, Air Canada fonds, LAC.

53 "We're the Only Airline that Flies Non-Stop to Florida" (advertisement), February 1967, Air Canada Collection, CASM.
54 Godefroy Desrosiers-Lauzon, "Nordicité et identities québécoise et canadienne en Floride," *Globe revue international d'études québécoises* 9, 2 (2006): 143; my translation (emphasis in original). See also Sophie-Laurence Lamontagne, *L'hiver dans la culture québécoise* (Québec: Institut de recherche sur la culture québécoise, 1983).
55 "How You Feel about Winter Often Depends on Where You Sit" (advertisement), September 1967, Air Canada Collection, CASM.
56 "It's Easy to Be a Good Sport about Winter" (advertisement), October 1967, Air Canada Collection, CASM.
57 *Island Vacations Winter 1970/1971* (brochure), RG 70, vol. 117–15.4, Air Canada fonds, LAC.
58 Canada, *House of Commons Debates*, October 25, 1977 (John Crosbie), http://parl.canadiana.ca/view/oop.debates_HOC3003_01/254?r=0&s=1.
59 Godefroy Desrosiers-Lauzon, *Florida's Snowbirds: Spectacle, Mobility, and Community since 1945* (Montreal/Kingston: McGill-Queen's University Press, 2011), 214, 219.
60 Harvey, *The Condition of Postmodernity*, 292.
61 Calin Rovinescu, "Let It Snow," *EnRoute Magazine*, December 2014, 29.
62 "John Fisher Reports" (script), April 6, 1947, MG 27, vol. 94, C.D. Howe fonds, LAC.

13 Negotiating High Modernism
The St. Lawrence Seaway and Power Project

DANIEL MACFARLANE

The St. Lawrence Seaway and Power Project was one of the major river modification ventures of the twentieth century. Built cooperatively by Canada and the United States between 1954 and 1959, it is both a deep-draft canal system (the Seaway) and a hydroelectric endeavour (the Power Project). The St. Lawrence project is an extremely revealing episode in terms of how government planners, engineers, and the public understood the interplay among progress, technology, nationalism, and water during the early Cold War. Building upon my previous work on the St. Lawrence Seaway and Power Project, I use this megaproject to explore the concept of "high modernism."[1]

In the words of James Scott in his 1998 book, high modernism

> is best conceived as a strong, one might even say muscle-bound, version of the beliefs in scientific and technical progress ... At its core was a supreme self-confidence about continued linear progress, the development of scientific and technical knowledge, the expansion of production, the rational design of social order, the growing satisfaction of human needs, and, not least, an increasing control over nature.[2]

High modernist plans are predicated on a synoptic view informed by bureaucratic and technocratic expertise at the expense of local knowledge and without recognition of the limitations of a top-down approach. They involve large-scale attempts to make "legible" social and natural environments through simplification, standardization, and ordering so as to control them and prescribe utilitarian plans for their betterment.³ Scott identifies nineteenth-century modernist antecedents, arguing that German mobilization in the First World War was the first instance of high modernism, with the period from the 1930s to the 1960s being the height of the high modernist élan.

The simple elegance of a beefed-up modernist outlook that transcends borders and political ideologies makes a great deal of intuitive sense for many commentators. As a result, high modernism has become a sort of shorthand for any sufficiently large twentieth-century project, usually organized by a government. I aim to demonstrate that high modernism is an appropriate concept for characterizing and understanding large-scale state projects in mid-twentieth-century liberal democratic settings, provided that the concept is recalibrated and nuanced to take into account specific locations and cultures. More specifically, I intend to show that the St. Lawrence Seaway and Power Project should be considered a fully high modernist project. A high modernist approach was certainly apparent in the organizing logic and imperatives that drove plans for the St. Lawrence project, in terms of both the river and the people who lived near it. The project was a nation-building exercise controlled by centralized bureaucracies with the aim of regimenting the natural environment for the sake of progress and in turn attempting to organize, regulate, and improve Canadian society. But it was also attuned to liberal, capitalist, and democratic principles refracted through the prism of Cold War imperatives and modalities as well as Canadian St. Lawrence nationalism and American imperialism. As I explain further in the following pages, this leads me to identify a *negotiated* form of high modernism.

Creating a Seaway

A consideration of the role that rivers and water have played in fostering Canadian nationalism and mythology makes the significance of the St. Lawrence River readily apparent. Canadians "consider water part of their natural identity."⁴ Moreover, "rivers are Canadian cultural icons; they have consistently communicated the idea of Canada, its meta-narrative of nation-building and collective identity."⁵ Donald Creighton's "Laurentian thesis" is the most famous example of framing the St. Lawrence River as a central organizing metaphor for Canadian history.⁶ The seaway offered nation-building parallels to the transcontinental railways, promoting and facilitating Canadian identity, national unity, progress, and prosperity while linking the country in an east-west orientation, in contrast to the north-south pull of the United States. Moreover, a St. Lawrence deep waterway continued the Canadian state's historical tendency to deal with the country's often hostile spatial setting by heavily subsidizing or building large transportation networks portrayed as benefiting general society even if they would most directly line the pockets of private business and industry.

The history of canal building in the St. Lawrence River goes back several centuries. Initial discussions about a joint Canadian-American deep waterway in the river date to the 1890s, and in the following decades the idea of pairing it with hydroelectric development became entrenched. Before flowing through Quebec, the St. Lawrence forms the border between the province of Ontario and the state of New York, meaning that permission from both countries is necessary to change the river's water levels.[7] A modernist ethos had underpinned the engineering plans for the St. Lawrence since the late nineteenth century – during the Second Industrial Revolution, Canada was a world leader in hydroelectric technology, as James Hull points out in this volume.[8] However, in the decades after the First World War and with the concomitant change in scale, technology, and ambition, the St. Lawrence Seaway and Power Project planning became high modernist. Bilateral negotiations and transnational engineering studies led to formal agreements in 1932 and 1941 to develop the river, but in both cases the US Congress nixed the accords, largely because of the opposition of sectional interests. As demands for the St. Lawrence project increased because of the exigencies of the early Cold War, the seaway grew into a major bilateral problem. Tired of American congressional inaction, and fed by various forms of nationalism that framed the St. Lawrence as an exclusively Canadian river, the St. Laurent government attempted to go ahead with an "all-Canadian" seaway entirely in Canadian territory on the north shore. After an Ontario–New York power development was approved, American pressure induced Canada to acquiesce to a joint seaway in a 1954 agreement.

In his study of the "technological imperative" in Canada, historian R. Douglas Francis suggests that "technological nationalism has characterized the Canadian state's rhetoric concerning identity."[9] According to Francis, in the early twentieth century technology was seen as enabling the United States to dominate Canada; however, technology was a "double-edged sword" because Canadian adoption of modern technology later in the century offered Canada the potential to reduce its dependence on the United States. For many Canadians in the 1950s, the seaway represented this type of emancipatory technology. Grounded in an expansive post–Second World War national confidence, support for an all-Canadian seaway blended various expressions of Canadian nationalism (e.g., geographic, environmental, political, economic) with a technological nationalism that coalesced around the waters of the St. Lawrence, forming a type of "hydraulic nationalism."

The river was remade into a new type of hybrid envirotechnical system.[10] The St. Lawrence Seaway and Power Project features the largest navigable inland waterway in the world (292 kilometres), running from Montreal to the terminus of the Welland Canal at Lake Erie. At the time of its completion, it was the largest transborder power project in the world, and the major power station, the Moses-Saunders power dam, was the second-largest hydro dam in North America. The international border cut right through the middle of the structure. The total cost for the entire project was more than US$1 billion. The seaway features a continual minimum depth of twenty-seven feet and fifteen locks with a depth of thirty feet (see Figure 13.1). The larger Great Lakes–St. Lawrence water route system provides a network of deep canals, dredged channels, and locks that stretch some 3,700 kilometres from the heart of the continent to the Atlantic Ocean. The power reservoir, formed by two new control dams that worked in conjunction with the Moses-Saunders power dam, raised the water level and made twenty-seven-foot navigation feasible.

Hydroelectric development has been called "the most prominent manifestation of the high modernist impulse,"[11] and eminent Canadian historical geographer Graeme Wynn has pointed to the St. Lawrence Seaway and Power Project as the epitome of a Canadian high modernist enterprise.[12] The relocation of people, houses, and infrastructure, necessary because of the new reservoir, was the largest rehabilitation project in Canadian history. A number of Canadian communities were affected by the raised water levels, which flooded approximately 20,000 acres. About 6,500 people were affected in Ontario. Most had lived in the submerged towns and hamlets – called the "Lost Villages" – along the north shore of the international section of the St. Lawrence. About 18,000 acres were inundated, and 1,100 people moved, on the less populated New York side (see Figure 13.2).

As William Becker and Robert Passfield have shown, the project involved a range of engineering, scientific, and technological advances.[13] Canadian engineers were at the forefront of applying the latest techniques when it came to the science of soil mechanics for the design of dikes and embankments; the use of jet piercer technology; the "air-cushion method" for underwater blasts close to canal walls; and new ways of placing concrete, particularly in winter.[14] The largest bridge-raising operation anywhere in the

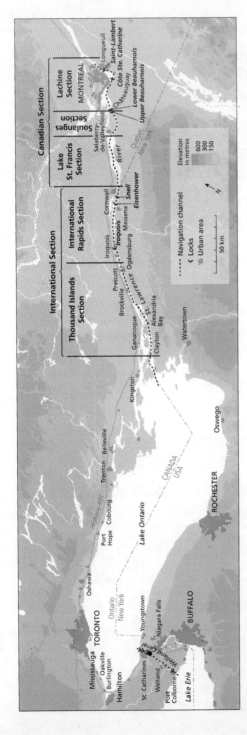

Figure 13.1 Map of the St. Lawrence Seaway | Cartography by Eric Leinberger. Reprinted from Daniel Macfarlane, *Negotiating a River: Canada, the US, and the Creation of the St. Lawrence Seaway* (Vancouver: UBC Press, 2014), 4–5.

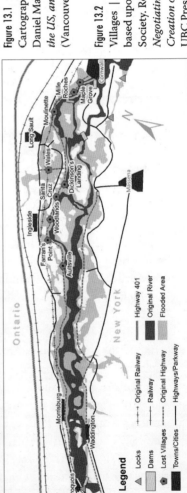

Figure 13.2 Map of Lake St. Lawrence and Lost Villages | Cartography by Daniel Macfarlane, based upon a map from the Lost Villages Historical Society. Reprinted from Daniel Macfarlane, *Negotiating a River: Canada, the US, and the Creation of the St. Lawrence Seaway* (Vancouver: UBC Press, 2014), 140.

world to that date was undertaken in Montreal as part of the seaway construction. Another major engineering advance was the extensive use of high-precision scale hydraulic models that replicated long stretches of the river in minute detail: the topography, the shoreline, the river channels, the contours of the river bottom, and the turbulence and velocity of the currents in all areas of the river in its natural state at both high- and low-water levels.[15] The first computer used in Canada was employed to calculate backwater flows for an all-Canadian seaway in the early 1950s.[16] There were other technological and planning advances to reduce and control ice formation. For relocation of the Lost Villages, discussed in more detail below, mechanical house-moving machines were employed in Canada for the first time, and the Canadian National Research Council used nine houses from a Lost Villages community to test their resistance to fire. The results were used to revise the Canadian Fire Code and reputedly led to the adoption of smoke rather than heat detectors.[17] In 2000, the American Public Works Association included the St. Lawrence project in its list of the ten most important public works projects of the past century.

Despite all of these achievements, the major engineering advances of the St. Lawrence project arguably lie more in the scale on which technologies and techniques were used: that is, it was organized on a larger scale than any other previous project. The results were impressive considering the complexity and diversity. From engineering and administrative perspectives, the project was an organizational triumph, particularly in the development of a project management system that in many ways was the forerunner of critical path methods. Indeed, the fact that the entire project was finished on time was remarkable given its scope and size.

I list these accomplishments chiefly to establish that this was a key case of both Canadian nation building and Canadian "big engineering." The St. Lawrence project represented progress, and the Canadian and Ontario governments went to great lengths to show off the power of the state. Tour buses and viewing stations were provided, and over 1 million people ultimately came to watch construction, including many dignitaries, political leaders, and engineers from abroad.[18] The seaway was reputed to be the largest construction project in the world at the time, and Canadians took great pride in their technological ability to remake such a historic river system.

Big science can be defined as state-sponsored projects of a sufficient physical scale that require massive technological and financial inputs. In his chapter in this volume, David Theodore asserts that "normal science in Canada is small science." However, I would contend that Canada has been prone

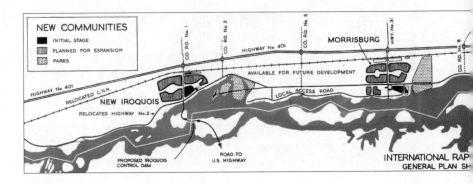

to big science in certain fields of applied science (e.g., transportation and hydroelectric engineering), usually as a result of challenges and opportunities stemming from Canada's geographic and spatial setting. Canada is a vast country, making large transportation infrastructure such as canals and transcontinental railways attractive. It is also a country with abundant water power sites (and relatively little coal); as a result, the nation played a leading global role in developing hydroelectric facilities. I propose that this selective technological adoption be labelled "functional" big science. I draw the term "functionalism" from the realm of Canadian foreign policy, in which it is a classic principle holding that Canada should choose which areas of international affairs to actively engage on an issue-by-issue basis depending on its abilities and capacities (e.g., serving at the United Nations). Granted, functionalism as a guiding principle of Canadian foreign policy has been embellished since it was only selectively invoked and in many ways was essentially a cover for simply acting in self-interest.[19] Nonetheless, the field of Canadian engineering reveals its own historical version of the functional principle: in those special cases in which Canada was uniquely equipped, or in which necessity required it, the country engaged in big science. The St. Lawrence Seaway and Power Project was one of those cases.

Lost River, Lost Villages

Nothing could stand in the way of progress, not even whole communities. Mass displacement in the St. Lawrence Valley was a small price to pay for the production of electricity and the increased accessibility of iron ore deposits. Flooding out thousands of people in the Lost Villages and surrounding rural areas (including Mohawk reserves) was justified in the name of progress and for the benefit of the wider nation. The reorganization and resettlement of those affected by the power development would be for their

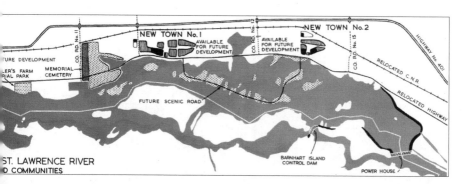

Figure 13.3   Ontario Hydro schematic outline of Lake St. Lawrence and rehabilitated area, including Lost Villages. | © Ontario Power Generation.

own benefit since they would be placed in consolidated new towns – instead of scattered about in inefficient villages, hamlets, and farms – with modern living standards and services (see Figure 13.3). The Hydro-Electric Power Commission of Ontario (Ontario Hydro) was responsible for the Canadian hydroelectric aspect of the dual St. Lawrence project and for the majority of the rehabilitation on the Ontario shore. The three new towns, constructed from scratch in farmers' fields to absorb the displaced residents, provided a sort of tabula rasa on which planners could utilize the latest in planning principles. Occupants could move their old residences or get new homes with basements, modern sewers, water and hydro facilities, and sewage treatment plants. Whereas the previous towns had spread along the waterfront in a long and narrow grid, the new communities used curved streets and crescents.[20] Major services and amenities were grouped strategically together in centralized plazas and new strip malls, with schools, churches, and parks placed to facilitate easier and safer access (see Figure 13.4).

As high modernist projects are wont to do, history was literally erased in favour of a heroic future. The submerged area included communities originally settled by United Empire Loyalists as well as the location of the Battle of Crysler's Farm in the War of 1812. The battle site was replaced by a hilltop monument beside Upper Canada Village, a replica pioneer village built to commemorate life at the time of Confederation as part of a wider effort to create tourist and recreation amenities in the area. Locals countered by later forming the Lost Villages Historical Society and creating their own living history museum that better represented their seaway experience.

The transportation and mobility networks along and across the river were reconfigured and upgraded – and not just on a local or regional level – as a

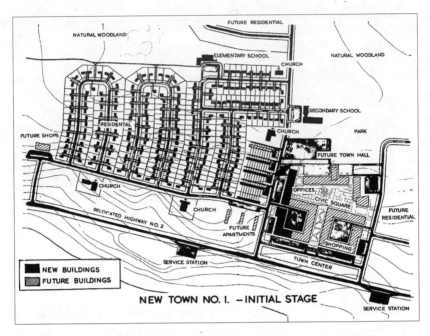

Figure 13.4 Ontario Hydro blueprint of plans for New Town No. 1 (Ingleside). | © Ontario Power Generation.

deep-draught waterway fundamentally changed national, continental, and international scalar transportation relationships in much the same way as railways and airplanes (on the latter, see Blair Stein's chapter in this volume).[21] Such organizational and settlement plans imposed state-defined political, economic, and social values and enabled the various levels of government to control how these communities fit into the emerging postwar order.

The state used the St. Lawrence project as a spectacle to demonstrate its power and prove its legitimacy to its citizens. Sampling, polling, surveying, testing, and modelling were used extensively, for as fundamental techniques of a high modernist approach they allowed the state to control information, set the terms of debate, and "manufacture consent": if people knew the facts, then the rationality of the project would inevitably compel them to accept its logic.[22] Ontario Hydro created observation platforms, and millions of people came to watch the construction. Efforts to educate and reassure those affected by new technologies are also apparent in Jan Hadlaw's study in this volume of Bell Canada's switch to dial telephony. Indeed, other chapters in this volume uncover various ways that the Canadian public understood and engaged with modern science and technology. The residents of the Lost Villages were repeatedly promised that they would benefit materially since the seaway would turn the upper St. Lawrence region into Canada's leading industrial area. Many residents of the area secured employment on the project, though the boom times of the construction period proved to be short-lived.

Ontario Hydro repeatedly went door to door and held public and town hall meetings.[23] At these meetings, it compromised on some aspects of the relocation – the most prominent example was the concession to use house movers so that people could keep their original homes. However, Ontario Hydro was willing to do this mainly because moving houses was cheaper than building new ones.[24] At the insistence of the provincial government, the amount of compensation for forceful taking was increased, and a commission for appeals was established. However, this commission generally ruled in favour of Ontario Hydro.[25]

A societal deference to experts and the government was apparent. For the governments involved, as well as for the general public, the idea that it was all a sacrifice worth making was pervasive. Certainly, some resisted in different ways, but for many the project carried an aura of inevitability. Moreover, those dislocated by the power pool generally expected that the St. Lawrence project would bring with it great prosperity, and they bought into the general logic of progress.

Ontario Hydro was more responsive to the preferences of those whom it dislocated compared with its American counterpart and relative to other previous megaprojects, but the power utility was ultimately unable to see beyond the momentum and high modernist framework of the St. Lawrence Seaway and Power Project. The Power Authority of the State of New York (PASNY) was responsible for hydro development and rehabilitation on the American side (the two federal governments were in charge of the navigation aspect). Robert Moses, who headed PASNY, is identified in *Seeing like a State* as a high modernist luminary in reference to his long career remaking the landscape of New York City.[26] Moses, every bit the populist high modernist prophet, was brought in to PASNY precisely because of his proven ability to move people out of the way and to get big projects done quickly.

As an aside, though I am referring to and making generalizations about "the state," it is important to recognize the various ways that a state's viewpoint and authority are constantly in flux. E.A. Heaman rightly points out that the multifarious Canadian state is "a process more than an institution," and the differing, often even contradictory, viewpoints and objectives within and among various levels of government authority reflect what has been called the "internal pluralism of the state."²⁷ At the same time, what is so striking about the creation of the seaway is how governments at various levels, as well as those whom they governed, tended to share similar views of the nationalist, technological, and state-building implications of the St. Lawrence project: that is, a high modernist vision permeated Canadian government and society.

Engineered River

Engineers and experts viewed nature as something to be controlled and ordered through technology with little to no consideration of the wider environmental impacts. Because of the cultural prestige attached to the engineering field, this view extended to the state and society. The rhetoric used by experts and governments focused on defeating, dominating, and exploiting the river. A megaproject ethos is also revealed by the type of language not used: namely, acknowledgment of the limits and repercussions inherent in a project on the scale of the St. Lawrence Seaway and Power Project.

As was the case for so many other modern Canadian technological and scientific projects addressed in this volume, the remaking of the St. Lawrence was characterized by extreme optimism. But in many ways the term "optimism" does not fully capture the extent to which planners believed that they were bettering society and nature. The engineering and government planning exhibited a hubris regarding the ability to control and replumb the St. Lawrence, the second largest river on the continent. The International Joint Commission was responsible for developing a river profile so that construction could proceed, and engineers spent the duration of the construction phase trying to establish a satisfactory range of minimum and maximum water levels for both the St. Lawrence River and Lake Ontario. This involved ascertaining the "natural" levels of the river and lake, essentially a self-defeating endeavour since "natural" was too difficult to determine because of a lack of information and previous modifications to the Great Lakes–St. Lawrence basin. The engineers kept producing methods of regulation to prescribe the water levels, but because of faulty approaches and a lack of full knowledge they were repeatedly forced to revise them over the span of a few years. As a result, they strove to attain levels "as nearly as may be."²⁸ Soon after the project was complete, however, water levels in the Great Lakes–St. Lawrence system approached lows that had not been factored into planning, and in the winter ice created "hanging dams" that interfered with power generation. The engineering process revealed many errors, assumptions, guesses, and partisan opinions; in public, however, engineers projected an aura of precision and confidence.

To be fair, the planners were in many ways products of their training and societal ideals, and they were subject to dominant national and transnational ideas that promoted the collaboration of industrial capital and the state as necessary to maximize the development of natural resources in the name of economic and social progress. They believed that they were wisely maximizing natural resources. There was great societal and occupational pressure on the "experts" to provide answers and to do so confidently: in addition to employment and economic factors, national and organizational pride, and the role of technology and expertise in East-West Cold War tensions, personal and professional stature was at stake. Whereas scientists studying pollution issues in the Great Lakes–St. Lawrence basin a decade later could openly admit their uncertainty, public disclosure of doubt was unthinkable for the St. Lawrence engineers.[29]

They labelled anything beyond their control as an "Act of God," suggesting that, if it was unknowable by their scientific techniques, it was beyond comprehension.[30] It was not that the governments and planners involved could not comprehend the complexity of their task but that they chose to ignore or mentally bracket the uncontrollable aspects of the St. Lawrence environment in order to persist in their belief that they had perfect conceptual understanding. I suggest that the high modernist wave in North America crested in the 1950s and 1960s and then began to recede, precisely because, in the wake of an undertaking such as the St. Lawrence Seaway and Power Project, experts were forced to become cognizant of the limits to their knowledge and ability to control nature and people.

Part of the problem stemmed from the faith that the engineers placed in their models (see Figure 13.5). The planning authorities, including the International Joint Commission, were enamoured of these models and took every opportunity to show them off to the public. The enormous models, some taking up whole warehouses, were believed to be indispensable for determining the future fluvial geomorphology, and thus they were central to the engineering plans. But models were often found to be wrong. Sometimes the error stemmed from incorrect knowledge, such as faulty gauge data, taken from the river.[31] Because of the scale of a model, a slight error would be distorted out of proportion when applied to actual excavations or structures in the river. Such distortions also occurred in attempts to simulate the turbidity of the river by increasing the model roughness factor.[32] Sometimes

Figure 13.5 Ontario Hydro model in warehouse at Islington, Ontario. | © Ontario Power Generation.

partisan and national rivalries were at play as engineers from the various nations and agencies sought to prove that their methods and experts were the best. This was particularly apparent when it came to model results, and the resolution of such disputes was often based upon political rather than engineering considerations, indicating yet another way that the process was continually negotiated.

However, they were able to overcome these engineering miscalculations, and we should not overlook the fact that in the end, broadly speaking, the construction and operation of the St. Lawrence project went according to plan. Compared with preproject conditions, over the long term, St. Lawrence and Lake Ontario water levels were more predictable and controllable, and the range of water levels was compressed (i.e., extreme highs lower, extreme lows higher). From the perspective of the experts, there were relatively few unintended environmental consequences, since the more pernicious (e.g., invasive species) could likely have been prevented with appropriate enforcement, while planners considered other negative repercussions to be tolerable by-products. This is not to excuse the damage wrought by the

Figure 13.6 Moses-Saunders power dam under construction | Photograph by Eleanor L. (Sis) Dumas. © Dumas Seaway Photograph Collection, Mss. Coll. 124, Special Collections, St. Lawrence University Libraries.

St. Lawrence project but to highlight that environmental ramifications were assumed and accepted.

The shape of the St. Lawrence Seaway and Power Project was a consequence of its Cold War context, for the Soviet Union and China were simultaneously undertaking large-scale hydroelectric and river basin industrialization projects.[33] It was also the product of a global engineering fraternity and the transnational spread of engineering techniques, ideologies, and planners. Uhl, Hall and Rich, the lead American engineering firm, had participated in major hydraulic projects at home, such as the Tennessee Valley Authority (TVA) dams, and around the globe.[34] The TVA was of course initiated by President Franklin Roosevelt, who had been deeply influenced by the Ontario Hydro model when as governor of New York he had created PASNY. Several engineering consultants were involved with all of these public power organizations. St. Lawrence engineers and tradespeople had worked on many other dam-building projects throughout the continent, often following them like a moving caravan. Officials from other countries travelled to view the seaway while it was under construction. Because of its size and complexity, the St. Lawrence undertaking served as a sort of graduate school for hydraulic engineers. After its completion, St. Lawrence engineers were sought for their expertise on other megaprojects in North America and virtually every other continent. While the St. Lawrence project was under way, the United States and Canada were finishing another transborder water control megaproject, upstream at Niagara Falls, and about to begin another on the Columbia River.[35]

Negotiating High Modernism

Scholars often apply high modernism without recognition of the parameters that James Scott puts on the concept. In particular, he insists that fully high modernist projects – what he calls "authoritarian" or "ultra" high modernism – can occur only in authoritarian states and are bound to fail because of their intrinsic contradictions. Scott lists three elements of what he calls "the most tragic episodes of state development in the late nineteenth and twentieth centuries": in addition to the actual high modernist ideology, he points to the unrestrained use of the immense powers wielded by modern states and an incapacitated or prostrate civil society.[36] Scott further avers that truly high modernist projects cannot even get fully off the ground in liberal political economies because of three major barriers: private sphere of activity, private sector of the economy (i.e., free market), and effective democratic institutions.[37]

To illustrate, Scott suggests that high modernism took varied forms, or had staggered developments, across eras and locations. He ascribes a "softer" form of authoritarian high modernism to Tanzanian village making in the 1970s but labels the Brazilian creation of a new capital city from scratch as "ultra-high modernism."[38] However, Scott does not provide much detail in *Seeing like a State* about what separates or defines these high modernist variations, though in a separate study he does address how the high modernist aspirations of the TVA were derailed by its liberal democratic setting.[39]

Several scholars have attempted to modify aspects of high modernism for North American circumstances. For example, in his study of state-led reform in the context of US agriculture, Jess Gilbert identifies a "low modernism" stemming from the participatory democratic and citizen betterment goals of New Deal authorities.[40] In the Canadian context, Tina Loo and Meg Stanley have shown that there was actually an intimate engagement with place in Canadian postwar dam-building efforts, a "high modernist local knowledge" defined by detailed awareness of specific environmental conditions.[41] Loo has furthered this analysis elsewhere, also taking issue with the declensionist narrative intrinsic to Scott's original formulation of high modernism as well as his critique of the synoptic vision on which high modernism relies.[42] On that score, David Pietz's analysis of Chinese dam building demonstrates that, though privileging synoptic expertise carries potential dangers, the inverse can also be true. Place-based local knowledge and labour ("walking on two legs," as Pietz labels it) were substituted for technological expertise (though this was partially the result of a lack of expertise, capital, and equipment) in northern Chinese water projects, which led to a number of problems and ecological disturbances, with more than two thousand dams built during the Mao period eventually failing.[43]

The St. Lawrence River was abstracted and reduced to simplified schematics in order to make the riverine environment legible, but on-the-ground conditions were not ignored. In fact, the two power entities, the various governments, and the transnational agencies had spent decades closely scouring, studying, and analyzing the St. Lawrence bioregion to ascertain specific place information. These microlevel details were then translated and projected into macrolevel plans. St. Lawrence engineers repeatedly modified and adapted their plans based upon what they faced on the ground; however, even though they were repeatedly made aware of their fallibility and limits, they continued to ignore the underlying flaws in their methods and maintained full trust in their models and technological expertise. In the end, it was in the belief among the engineers that they could actually know and master every square inch of the environment, rather than ignorance of the river basin that they were manipulating, that the hubris of the St. Lawrence undertaking was most apparent.

The first two elements of Scott's tripartite definition – a high modernist ideology and the unrestrained use of immense powers – are so readily apparent in the St. Lawrence project that, even though it was undertaken by liberal democracies rather than authoritarian states with weak civil societies, it should still be considered fully high modernist. In the context of the seaway, the argument that private sectors make the economic reins too complicated to handle does not hold up, nor does the idea that democratic spheres and institutions invariably thwart high modernist schemes. True, consent for high modernist schemes had to be manufactured continuously. However, when consent was achieved, through town hall meetings, surveys, and elections, it gave high modernist plans greater popular legitimacy. People thought that they had been consulted. At least in the case of the St. Lawrence project, one must conclude that democratic institutions and the free market only served to complicate and partially modify, rather than block, high modernist interventions.⁴⁴ Moreover, despite the engineering mistakes and compromises, the St. Lawrence project largely worked as intended, thus undermining the claim that high modernist projects inevitably fail.⁴⁵

The Canadian and American governments in the 1950s would not be considered by most definitions as authoritarian, but simply calling the St. Lawrence case a "softer" form of ultra high modernism does not adequately capture its unique qualities and variations. We see what I have termed *"negotiated* high modernism": high modernism in conception, negotiated politically in practice. Although the St. Lawrence project was fundamentally about control, the governments still needed continually to mediate and reassert their authority, accomplished by involving people in the process as workers, admirers, and relocatees. Lacking the centralized and autocratic authority simply to impose schemes without some measure of approval from civil society, the various levels of the involved states repeatedly had to adapt, negotiate, and legitimize themselves and their high modernist St. Lawrence vision to both the specific natural environments and the societies that they aimed to control.

The "negotiated" character of the St. Lawrence project can be further distinguished by the range of diplomatic discussions that took place not only within and among the various authorities and bureaucracies in one nation but also between the governments of two countries.⁴⁶ In the final tally, the Canadian government's approach to the St. Lawrence Seaway and Power Project was manifestly a product of the uniquely Canadian cultural conceptions of the St. Lawrence River; national links among identity, technology, and the natural environment; Canada's relationship with the United States; the very existence of Canada as a negotiated state (i.e., a compromise among various ethnicities and cultures); and the distinctive Canadian relationship between the state and civil society.

NOTES

I would like to thank Shirley Tillotson, Tina Loo, James Scott, Henry Trim, H.V. Nelles, the anonymous reviewers, and the editors of this volume for comments and thoughts that contributed to this chapter.

1 Daniel Macfarlane, *Negotiating a River: Canada, the US, and the Creation of the St. Lawrence Seaway* (Vancouver: UBC Press, 2014).
2 James C. Scott, *Seeing like a State: How Certain Schemes to Improve the Human Condition Have Failed* (New Haven, CT: Yale University Press, 1998), 89.

3   Scott, ibid., 90, asserts that in premodern and/or early modern Europe, the state used the forerunners of high modernist techniques to amass "descriptive" information (e.g., maps), whereas high modernist states are "prescriptive." In this volume, Stephen Bocking and Andrew Stuhl also discuss the issue of "legibility."
4   Carolyn Johns, "Introduction," in *Canadian Water Politics: Conflicts and Institutions*, ed. Mark Sproule-Jones, Carolyn Johns, and B. Timothy Heinmiller (Montreal/Kingston: McGill-Queen's University Press, 2008), 4. See also Environment Canada, "Water, Art, and the Canadian Identity," n.d., accessed February 12, 2013, http://www.ec.gc.ca/eau-water/default.asp?lang=En&n=5593BDE0-1.
5   Jean Manore, "Rivers as Text: From Pre-Modern to Post-Modern Understandings of Development, Technology, and the Environment in Canada and Abroad," in *A History of Water, Series 1*, vol. 3, *The World of Water*, ed. Terje Tvedt and Eva Jakobsson (London: I.B. Tauris, 2006), 229.
6   "Laurentian Thesis," *Canadian Encyclopedia*, http://www.thecanadianencyclopedia.com/en/index.cfm?PgNm=TCE&Params=A1ARTA0004556; Donald Creighton, *The Empire of the St. Lawrence* (Toronto: Macmillan, 1956), originally published in 1937 as *The Commercial Empire of the St. Lawrence, 1760–1850*; Donald Creighton, *Dominion of the North: A History of Canada* (Toronto: Houghton Mifflin, 1944). For an excellent discussion of the power of the river narrative in Canadian history, see Christopher Armstrong, Matthew Evenden, and H.V. Nelles, *The River Returns: An Environmental History of the Bow* (Montreal/Kingston: McGill-Queen's University Press, 2009), 10–20.
7   Generally, this was done through the International Joint Commission (IJC), though binational agreements could be made outside the IJC. On the history of the IJC and Great Lakes–St. Lawrence basin water levels, see Murray Clamen and Daniel Macfarlane, "The International Joint Commission, Water Levels, and Transboundary Governance in the Great Lakes," *Review of Policy Research* 32, 1 (2015): 40–59.
8   In addition to his chapter in this volume, see James Hull, "A Gigantic Engineering Organization: Ontario Hydro and Technical Standards for Canadian Industry, 1917–1958," *Ontario History* 93, 2 (2001): 179–200; and James Hull, "Technical Standards and the Integration of the U.S. and Canadian Economies," *American Review of Canadian Studies* 32, 1 (2002): 123–42. On electricity and the technology sublime, one must of course consult the work of David Nye in *American Technological Sublime* (Cambridge, MA: MIT Press, 1994) and *Electrifying America: Social Meanings of a New Technology, 1880–1940* (Cambridge, MA: MIT Press, 1990).
9   R. Douglas Francis, *The Technological Imperative in Canada: An Intellectual History* (Vancouver: UBC Press, 2009). See also Macfarlane, *Negotiating a River*, 78–80; Marco Adria, *Technology and Nationalism* (Montreal/Kingston: McGill-Queen's University Press, 2010); and Cole Harris, "The Myth of the Land in Canadian Nationalism," in *Nationalism in Canada*, ed. Peter Russell (Toronto: McGraw-Hill, 1966), 27–43.
10 Richard White, Joel Tarr, Martin Melosi, Mark Fiege, and Jeffrey Stine, in their works from the 1990s, can be considered as the vanguard of scholars who explore "hybrid" natures. Scholars of water and environmental history, often using approaches from

"envirotech" history, have been employing "hybridity" as a critical concept in the past decade. There are several edited collections that explore the envirotech field; see, for example, Martin Reuss and Stephen H. Cutcliffe, eds., *The Illusory Boundary: Environment and Technology in History* (Charlottesville: University of Virginia Press, 2010); and Dolly Jørgensen, Finn Arne Jørgensen, and Sara B. Pritchard, eds., *New Natures: Joining Environmental History with Science and Technology Studies* (Pittsburgh: University of Pittsburgh Press, 2013). In addition to these collections, several pieces have probed the methodology of hybridity and envirotech history; see Edmund Russell, James Allison, Thomas Finger, John K. Brown, Brian Balogh, and W. Bernard Carlson, "The Nature of Power: Synthesizing the History of Technology and Environmental History," *Technology and Culture* 52, 2 (2011): 246–59; Sara B. Pritchard, "Toward an Environmental History of Technology," in *The Oxford Handbook of Environmental History*, ed. Andrew C. Isenberg (New York: Oxford University Press, 2014), 227–58; and Paul S. Sutter, "The World with Us: The State of American Environmental History," *Journal of American History* 100, 1 (2013): 94–119. Recent envirotech monographs include Joy Parr, *Sensing Changes: Technologies, Environments, and the Everyday* (Vancouver: UBC Press, 2010); Sara B. Pritchard, *Confluence: The Nature of Technology and the Remaking of the Rhône* (Cambridge, MA: Harvard University Press, 2011); Finn Arne Jørgensen, *Making a Green Machine: The Infrastructure of Beverage Container Recycling* (New Brunswick, NJ: Rutgers University Press, 2011); Edmund Russell, *Evolutionary History: Uniting History and Biology to Understand Life on Earth* (New York: Cambridge University Press, 2011); Christopher Jones, *Routes of Power: Energy and Modern America* (Cambridge, MA: Harvard University Press, 2014); and Ashley Carse, *Beyond the Big Ditch: Politics, Ecology, and Infrastructure in the Panama Canal* (Cambridge, MA: MIT Press, 2014).

11 Tina Loo, "People in the Way: Modernity, Environment, and Society on the Arrow Lakes," *BC Studies* 142–43 (2004): 165. Other works that have addressed high modernism in Canada include Matthew Farish and P. Whitney Lackenbauer, "High Modernism in the Arctic: Planning Frobisher Bay and Inuvik," *Journal of Historical Geography* 35, 3 (2009): 517–54; James L. Kenny and Andrew G. Secord, "Engineering Modernity: Hydroelectric Development in New Brunswick, 1945–1970," *Acadiensis* 39, 1 (2010): 3–26; Tina Loo and Meg Stanley, "An Environmental History of Progress: Damming the Peace and Columbia Rivers," *Canadian Historical Review* 92, 3 (2011): 399–427; Benjamin Forest and Patrick Forest, "Engineering the North American Waterscape: The High Modernist Mapping of Continental Water Projects," *Political Geography* 31, 3 (2012): 167–83; and Tina Loo, "High Modernism, Conflict, and the Nature of Change in Canada: A Look at *Seeing like a State*," *Canadian Historical Review* 97, 1 (2016): 34–58. Recent books on Canadian hydroelectricity include Armstrong, Evenden, and Nelles, *The River Returns;* Thibault Martin and Steven M. Hoffman, eds., *Power Struggles: Hydro-Electric Development and First Nations in Manitoba and Quebec* (Winnipeg: University of Manitoba Press, 2009); Meg Stanley, *Voices from Two Rivers: Harnessing the Power of the Peace and Columbia* (Vancouver: Douglas and McIntyre, 2011); David Massell, *Quebec Hydropolitics: The Peribonka*

Concessions of the Second World War (Montreal/Kingston: McGill-Queen's University Press, 2011); Christopher Armstrong and H.V. Nelles, *Wilderness and Waterpower: How Banff National Park Became a Hydro-Electric Storage Reservoir* (Calgary: University of Calgary Press, 2013); Caroline Desbiens, *Power from the North: Territory, Identity, and the Culture of Hydroelectricity in Quebec* (Vancouver: UBC Press, 2013); Macfarlane, *Negotiating a River*; and Matthew Evenden, *Allied Power: Mobilizing Hydro-Electricity during Canada's Second World War* (Toronto: University of Toronto Press, 2015).

12 Graeme Wynn, *Canada and Arctic North America: An Environmental History* (Santa Barbara, CA: ABC-CLIO, 2006), 284.

13 William H. Becker, *From the Atlantic to the Great Lakes: A History of the U.S. Army Corps of Engineers and the St. Lawrence Seaway* (Washington, DC: Government Printing Office, 1984); Robert W. Passfield, "The Construction of the St. Lawrence Seaway," *Canal History and Technology Proceedings* 22 (2003): 1–55. Passfield has authored a range of publications on hydraulic engineering history in Canada, including the Soo, Trent, and Rideau Canals. See his chapter titled "Waterways" in *Building Canada: A History of Public Works*, ed. Norman R. Ball (Toronto: University of Toronto Press, 1988), 113–42. On the broader history of engineering in Canada, consult the rest of *Building Canada* as well as Norman R. Ball, *"Mind, Heart, and Vision": Professional Engineering in Canada 1887 to 1987* (Ottawa: National Museum of Science and Technology, 1988).

14 The Canadians kept pouring concrete in the winter while the Americans stopped.

15 IJC, Canadian Section, 64-4-4-2:2, Preservation & Enhancement of Niagara Falls, International Niagara Falls Engineering Board, March 1, 1953.

16 Scott M. Campbell, "Backwater Calculations for the St. Lawrence Seaway with the First Computer in Canada," *Canadian Journal of Civil Engineering* 36, 7 (2009): 1164–69.

17 Ronald Stagg, *The Golden Dream: A History of the St. Lawrence Seaway at Fifty* (Toronto: Dundurn Press, 2010), 216.

18 Daniel Macfarlane, "Fluid Meanings: Hydro Tourism and the St. Lawrence and Niagara Megaprojects," *Histoire sociale/Social History* 49, 99 (2016): 327–46.

19 Adam Chapnick, "Principle for Profit: The Functional Principle and the Development of Canadian Foreign Policy, 1943–1947," *Journal of Canadian Studies* 37, 2 (2002): 68–85.

20 Writing as the St. Lawrence project was completed, Peter Stokes, critical of many elements of the rehabilitation, contended that "the improvement of the loop streets [is] unappreciated since the previous towns weren't large enough to appreciate the traffic hazards of the old grid system." Peter Stokes, "St. Lawrence, a Criticism," *Canadian Architect*, February 1958, 43–48. See also Sarah Bowser, "The Planner's Part," *Canadian Architect*, February 1958, 38–40.

21 For a further exploration of issues connected to transportation and mobility, see Daniel Macfarlane, "Creating the Seaway: Mobility and a Modern Megaproject," in *Moving Natures: Environments and Mobility in Canadian History*, ed. Ben Bradley, Colin M. Coates, and Jay Young (Calgary: University of Calgary Press, 2016), 127–50.

22 Hydro-Electric Power Commission of Ontario (hereafter HEPCO), SPP series, "Memorandum to Lamport: House to House Survey – Village of Farran's Point," February 10, 1955; HEPCO, SPP series, "Memorandum to Carrick: Property Transactions – St. Lawrence Seaway," July 12, 1954. See also Loo, "People in the Way"; and Loo and Stanley, "An Environmental History of Progress."

23 For example, HEPCO, SPP series, "St. Lawrence Rehabilitation: Meeting at Osnabruck," November 23, 1954.

24 HEPCO, SPP series, "Report on the Acquisition of Lands and Related Matters for the St. Lawrence Power Project (by Property Office)," 1955–56; HEPCO, SPP series, "The Acquisition of Lands and Related Matters for the St. Lawrence Power Project," supplementary report to James S. Duncan, chairman, and HEPCO commissioners, January 2, 1957.

25 Government of Ontario, RG 19–61–1 – Municipal Affairs, Research Branch – Special Studies, St. Lawrence Seaway Study, box 21, file 14.1.5 – Minutes of Meetings – St. Lawrence Seaway #1, "Memorandum of Meeting Re: Iroquois," December 21, 1954; HEPCO, SPP series, "Report on the Acquisition of Lands and Related Matters for the St. Lawrence Power Project (by Property Office)," 1955–56.

26 Scott, *Seeing like a State*, 88.

27 E.A. Heaman, *A Short History of the State in Canada* (Toronto: University of Toronto Press, 2015), 222; Armstrong and Nelles, *Wilderness and Waterpower*, 215.

28 Macfarlane, *Negotiating a River*, 118–24.

29 In addition to Stephen Bocking's contribution to this volume, there is a large body of literature on scientific uncertainty and policy making concerning the Great Lakes environment, such as Terence Kehoe, *Cleaning Up the Great Lakes: From Cooperation to Confrontation* (Dekalb: Northern Illinois University Press, 1997).

30 The quotation is from IJC, Canadian Section, docket 68–2-5:1–9–St. Lawrence Power Application. See also Executive Session 1957/04 and 1957/10, IJC, St. Lawrence Power Development, Semiannual Meeting, Washington, DC, April 9, 1957.

31 IJC, Canadian Section, St. Lawrence Power Application Model Studies – vol. 1, "The Importance to Canada of the Construction of a Hydraulic Model for the Determination of the Effects of the Gut Dam and Channel Improvements in the Galops Rapids Section of the St. Lawrence River," December 9, 1953. On the US Army Corps of Engineers and hydraulic models in the twentieth century, in addition to the enormous body of literature on the Army Corps of Engineers' various districts and projects, see the work of Martin Reuss, particularly "The Art of Scientific Precision: River Research in the United States Army Corps of Engineers to 1945," *Technology and Culture* 40, 2 (1999): 292–323.

32 IJC, Canadian Section, docket 68–8-6:3–St. Lawrence Power Application; Federal Power Commission in the United States Court of Appeals 1953–54, St. Lawrence Power Application, Model Studies – vol. 1, "Associate Committee of the National Research Council on St. Lawrence River Models" (draft), October 15, 1953.

33 Paul Josephson, *Industrialized Nature: Brute Force Technology and the Transformation of the Natural World* (Washington, DC: Island Press/Shearwater, 2002); Richard P. Tucker, "Containing Communism by Impounding Rivers: American

Strategic Interests and the Global Spread of High Dams in the Early Cold War," in *Environmental Histories of the Cold War*, ed. J.R. McNeill and Corinna R. Unger (Cambridge, UK: Cambridge University Press, 2013), 139–63; Dorothy Zeisler-Vralsted, *Rivers, Memory, and Nation-Building: A History of the Volga and Mississippi Rivers* (New York: Berghan Books, 2015); David Pietz, *The Yellow River: The Problem of Water in Modern China* (Cambridge, MA: Harvard University Press, 2015).

34 Linda Nash, "Traveling Technology? American Water Engineers in the Columbia Basin and the Helmand Valley," in *Where Minds and Matters Meet: Technology in California and the West*, ed. Volker Janssen (Oakland: University of California Press, 2012), 135–58.

35 On Niagara Falls, see Daniel Macfarlane, "Creating a Cataract: The Transnational Manipulation of Niagara Falls to the 1950s," in *Urban Explorations: Environmental Histories of the Toronto Region*, ed. L. Anders Sandberg, Stephen Bocking, Colin Coates, and Ken Cruikshank (Hamilton: L.R. Wilson Institute for Canadian History, McMaster University, 2013), 251–67; and Daniel Macfarlane, "'A Completely Man-Made and Artificial Cataract': The Transnational Manipulation of Niagara Falls," *Environmental History* 18, 4 (2013): 759–84.

36 Scott, *Seeing like a State*, 88.

37 Ibid., 101–2.

38 Scott, ibid., 224, identifies three elements: improvement, bureaucratic management, and aesthetic dimension.

39 Scott notes in the introduction to *Seeing like a State* that he had initially included the TVA as an American example of a high modernist project in an earlier draft of his book. This research was later published as "High Modernist Social Engineering: The Case of the Tennessee Valley Authority," in *Experiencing the State*, ed. Lloyd I. Rudolph and John Kurt Jacobsen (Toronto: Oxford University Press, 2006), 1–21.

40 Jess Gilbert, "Low Modernism and the Agrarian New Deal: A Different Kind of State," in *Fighting for the Farm: Rural America Transformed*, ed. Jane Adams (Philadelphia: University of Pennsylvania Press, 2003), 129–46; Jess Gilbert, *Planning Democracy: Agrarian Intellectuals and the Intended New Deal* (New Haven, CT: Yale University Press, 2015).

41 Loo and Stanley, "An Environmental History of Progress."

42 Loo, "High Modernism, Conflict, and the Nature of Change in Canada."

43 Pietz, *The Yellow River*, 256–57.

44 From a contemporary perspective, some might argue that North American political institutions in the 1950s were not adequately representative or democratic.

45 Granted, the seaway was a failure in the sense that it never carried the traffic that had been predicted and thus failed to be self-amortizing.

46 For a detailed discussion of the political negotiations, see Macfarlane, *Negotiating a River*.

Epilogue
Canadian Modernity as an Icon of the Anthropocene

DOLLY JØRGENSEN

Perhaps nothing is more modern than money. Although humans have been using coins and shells as media of exchange for millennia, the paper money carried around in wallets is a modern phenomenon.[1] Unlike a coin, which has inherent value in its materials, paper currency, also known as a banknote, is exchanged for goods and services not because of any inherent value in the paper but because it has symbolic value. The parties in the exchange accept it because they believe that it is worth something, often because someone in power has guaranteed the value. From the start of the twentieth century, paper money has been issued almost exclusively by nation-states, changing a practice that had included independent banks issuing banknotes prior to that time. As James Scott pointed out, modern state building has often relied on making nature as well as urban life legible and controllable.[2] Nation-states attempt to standardize and organize many aspects of society, the environment, and technology. Official state paper money is one way to create order and control. Issuing currency is one of the fundamental expressions of modern sovereign nations.

Unlike coins made from durable materials that last decades, paper money requires constant replacement from wear and tear: the average bill is in circulation less than two years. This gives states the opportunity to refashion their paper money through new designs, often with the intent of reducing the number of counterfeit bills in circulation. Most states redesign their paper money every one to two decades, meaning that when the bills are issued they tend to represent recent thinking. The "significance, universality, selectivity, and regular updating" make banknotes particularly useful as indicators of identity content.[3] Who or what stands behind and stands for the nation on the face of its currency is a deliberate choice. The paper (or now in some cases plastic) that the note is printed on becomes representative of the issuer of the money presumed to stand behind it. Users of the currency need to have faith in the medium of exchange, and this happens on a symbolic level. The images appearing on banknotes are always deliberately selected to contribute to the acceptance of the paper as a financial medium.[4] Designers must create a representation of the economic viability of the state both for the people who will see the bill on a daily basis and for foreigners who will encounter the paper money as outsiders. What appears on its money says something fundamental about a nation's self-image.

So, when I was asked to write a closing reflection on this volume, I immediately thought of paper money as a symbol of modernity and one specific currency issue in particular: a Canadian $5 bill from 1935 (see Figure 14.1). This banknote was part of the first series of banknotes issued by the Bank of Canada. The notes were designed by security printers in consultation with the federal government and issued in English and French versions. This $5 bill encapsulates the confluence of science, technology, and the modern in the form of an allegorical image of electrical power on the back of the banknote.[5]

Beginning in the nineteenth century, many of the images on currency from Western Europe and North America were allegorical, a visual choice linked to the interrelationship of culture and economics.[6] Representations of personified nature, industry, virtues, and vices were standard cultural images of the time made recognizable and legitimate through wide circulation. Allegorical images, particularly those invoking the Greek and Roman classical past, were associated with empire and authority, so they often appeared on the currencies of countries wanting to claim autonomy and significance in the eighteenth and early nineteenth centuries.[7] This particular visual genre was part and parcel of claiming the nation-state as modern while building on Western cultural heritage. On the allegorical Canadian $5 bill, bodies, technologies, and environments converge to speak to the making of modern Canada.

The first element of the allegory is the body. The muscular young male labourer sits in the foreground atop a powerplant with his foot resting on a large gear. He represents Electricity. Although mechanical parts do the immediate work of converting water into energy once a dam is constructed, it

Figure 14.1 Canadian $5 bill issued in 1935. | National Currency Collection, Bank of Canada Museum.

is human labour that constructed it and human ingenuity that designed it. The body is the conqueror of untamed nature through skills and knowledge, not unlike the multipurpose human bodies of the Putnam Eastern Arctic Expeditions in Tina Adcock's chapter. Considering the beneficial role that electricity was believed to play in the body, as discussed by Dorotea Gucciardo in her chapter, it is perhaps no accident that the figure is muscular and in perfect form. As seen in Beth Robertson's chapter, modern energy sources were equated with modernity and transformative power.

Of course, the image is not of just any body – it is the white male European body in prime form that becomes the prime mover of modernity. Indigenous bodies, which had long worked the land, are not present. Europeans and settler Canadians considered them inferior and in need of civilization, as recorded in Richard King's ethnographic encounters detailed in Efram Sera-Shriar's chapter. Animal bodies, which became the boundary objects of the Putnam Eastern Arctic Expeditions, are likewise erased from the scene. The carcasses piling up as fish perished in their futile attempt to migrate upstream beyond the new dam are invisible.[8] The hypermasculinity of the body, with extreme musculature and bare chest, defines labour as the realm of manly men, ignoring the central role of women in building up Canada through their own labour.[9] Although this banknote hails the male form as the type that built the nation, in reality it was only one of many bodies. The choice for this banknote tells us something about the vision of Canadian modernity held by central government actors: Canada would be built by and on the bodies of white men, and harnessing electric power could make those bodies even stronger. As Jane Nicholas has demonstrated with her study of

Epilogue 351

the Diamond Jubilee of 1927, men came to embody the Canadian national vision of progress in the interwar period. The working man was supposed to take on the role of "literal builder of the country" with pride.[10] The visual choice of the male body on the $5 bill becomes an even more obvious ideological statement when we realize that the figure of Electricity need not have been a man at all – the Mexican 500 peso bill from 1925 features Electricity as a woman, and the US $2 bill from 1896 represents Electricity along with Science, Steam, Commerce, and Manufacture as women.

The second element of the allegory is the technological artifact. The modern concrete dam as shown in the engraving is typical of early-twentieth-century dams in Canada. It is similar to the design shown in early photographs of the Calgary Power facilities at Horseshoe Falls, opened in 1911, and Kananaskis Falls, opened in 1913.[11] Canada was an early adopter of electricity-producing hydropower plants, with its first generating station near Quebec City opening in 1885. The mid-1920s to early 1930s witnessed a growth in hydroelectric development for the aluminum industry and interprovincial transmission of power, a story well represented in the literature on twentieth-century Canada.[12] Electricity through hydropower was making Canada into a modern industrial nation, an image stressed on the banknote.

The industrial aspect of the technology is noteworthy. In the twenty-first century, the term "modern" (or "postmodern") often invokes computing and information; but in the early twentieth century, "modern" meant gears and factories. Hydropower was useful because of the gears that it could turn. Moving parts put people to work. Thus, the figure sits atop the powerhouse with gear underfoot, a clear sign that this is a vision of the Second Industrial Revolution. As James Hull notes in his chapter, the factory of this revolution was a new rational one and part of a larger trend of social rationalization beyond the factory walls.

The electrical technology shown on the $5 banknote of 1935 was set within a constellation of technologies envisioned as part of Canada's "modernization." The series of new notes issued in 1935 used allegories of technology to define modern Canada.[13] Agriculture featured prominently: a female Agriculture allegory appears on the $1 bill and $20 bill, and a seated Harvest is engulfed in agricultural bounty on the $10 bill (the $500 bill has a figure that is also Harvest or Fertility). As demonstrated by Eda Kranakis in her chapter on Monsanto and genetically modified crops, agriculture should never be ignored in discussions of modernity; science and technology are part and parcel of agricultural development. The $2 bill showed Mercury to represent speed flanked by trains (both steam and electric) and ships. These transportation modes, like the later airplanes discussed by Blair Stein in her chapter, served to compress both space and time. The highest-value notes, and thus the ones least likely to be in general circulation, also highlighted invention: an allegory of Modern Innovation (albeit one looking very ancient Greek) graces the $50 bill, and an allegory of Commerce shows a steamship to a young companion in front of a boat dock on the $100 bill. Taken together these images paint a picture of a Canada built upon agriculture, transportation, and new energy technologies, even if changes in those industries tended to be slow and nonrevolutionary in Canadian history.[14] Of course, these visions of modern technology as large and imposing obscure the reality that small-scale science and technology likely drove many modern advances, as elucidated by David Theodore in his study of a one-man lab.

The $5 bill image focuses on the production of electricity but hides the changes to work processes that accompanied this newfound electric power. It also masks the ways in which electricity modified social habits, such as the rotary telephone development elucidated in Jan Hadlaw's chapter. How to use a rotary phone instead of a human operator was not necessarily obvious to new users. New technologies are not instantly assimilated into daily life – they are integrated through negotiation and adjustment of both the technology and the user, a process termed "domestication."[15] The banknote image also obscures the connection of Canadian technology to global networks. As Edward Jones-Imhotep shows in the case of Bull's cannon launchers, and as Andrew Stuhl explores in the Canadian Arctic Expedition of 1913–18, Canadian innovation and science were often transnational. Even hydropower in Canada was transnational – the dams built on the Canadian side of Niagara Falls exported two-thirds of their power to US industrial customers by 1910.[16]

The final element of the allegory is the rugged wild landscape that would be turned into work and power by the dam. The raging river with dangerous falls on the left has been transformed by the figure in the middle into a modern dam on the right. Labour and technology have harnessed and converted the landscape. As Daniel Macfarlane indicates for the case of the St. Lawrence Seaway and accompanying hydropower dam, the waterscapes of Canada were heavily remade through engineering and high modernist ideals. This transformation of the landscape was a key element in attempts to make Canadian water resources legible for the nation-state and convertible into wealth and power.

In the background of the image, snow-topped peaks rise behind a pine forest. Although no chapter in this collection focuses on forests, we should remember that concerns about forest depletion and management spurred on Canada's early conservation movement at the end of the nineteenth century.[17] Unlike its southerly neighbour, Canada was concerned not to protect wilderness but to rationalize forestry production, a thoroughly modernist philosophy in which efficiency and objectivity were seen as ideals. Claiming resources and transforming them put them to work. But modernity also involved documenting those resources; they had to be categorized, systematized, and comprehended before claiming and transforming could proceed. These, then, are the kinds of landscapes that the scientists discussed by Stephen Bocking and Andrew Stuhl attempted to capture and understand in order to turn them into resources for the fledgling state.

Although this book stresses the role of pure and applied science in the making of modern Canada, science does not appear explicitly in the series of banknotes from 1935. It could have. Other global currencies – including the US $100 bill of 1864 and $2 bill of 1896, France's 20-franc bill of 1939–42, and Australia's £10 bill of 1954–59 – all feature an allegorical Science, so its inclusion would not have been out of place.[18] What the Canadian choice reveals is that, while historians can recognize the role of science in the creation of modernity, practical physical technologies of agriculture, transport, and power more obviously symbolized progress to Canadians at the time. That is partly what makes this volume important – to expose those parts of the modernization project that lie below, beyond, or beside the obvious story. Science infused understandings of and modifications to bodies, technologies, and environments.

Looking again at the overall visual impression of the 1935 $5 banknote, we see that the human dominates the modern scene. Even though the man allegorically represents Electricity, his human form matters to the interpretation of the image. Electricity is man, and man is Electricity, in the modern age. All things both technological and natural are under this man's control in this representation of progress. Domination is a central element of modernist progressivist thinking, as the volume's introduction demonstrates. The upshot of this domination is the Anthropocene, the age of human technological transformation of the planet's fundamental geology and biology.[19] The legacy of hydropower projects, nuclear science, large-scale hunting, and genetically modified crops is a radically altered Earth. At the same time, the conglomerate effects on the planet of smaller individual actions that generate carbon dioxide, such as flying south to get away from winter weather, cannot be underestimated as geological forces.

There have been many debates about when the Anthropocene began, but the general trend has been to focus either on the Industrial Revolution starting in the eighteenth century or on the immediate post–Second World War era as its start.[20] Regardless of which of those two dates is chosen, Canada is an Anthropocenic nation. The year 2017 marked the 150th anniversary of the founding of Canada as a federation, which puts the nation's primary development within the timeline of the Anthropocene.[21]

Dipesh Chakrabarty proposed that the idea of the Anthropocene should force us to rethink histories of modernity because "the mansion of modern freedoms stands on an ever-expanding base of fossil-fuel use."[22] In this vein, I would like to propose that the attempt of this book to define the role of science and technology in the making of modern Canada could be reframed as an inquiry into the role of science and technology in the making of *a localized Anthropocene*. Canada at its core is about the radical modification of land and water to serve human needs and to build up the new nation-state, both locally in Canada and globally through worldwide networks.[23] Although Canada has made heavy investments in hydropower, the tar sands and other oil and gas extraction in the North fit Chakrabarty's description of an Anthropocene defined by increasing fossil fuel use. The nation has been built within the capitalistic economic system, which has also been heavily implicated as a source of the Anthropocene.[24] Canada's modernity cannot be separated in time or space from the Anthropocene in which it occurred.

Canada's particular timeline as a product of the Anthropocene stands in contrast to European nations that led the charge in creating the new epoch. The countries of Western Europe make their claims to modernity in other ways. Their banknotes rely primarily on images of cultural production (literary figures, musicians, artists), history, and political leaders. Although Canadian modernity is often deliberately compared with European modernity, as the editors of this volume do in the introduction, I wonder if a more fruitful comparison would be to look at the modernization of other nations founded within the Anthropocene. That would mean looking south and east at nations that made deliberate claims of being created as modern nations contemporaneously with Canada. These places share with Canada the aspirational modernity discussed in the introduction.

The image of the man-made hydropower dam taming nature took on iconic status on the currencies of many nations in the mid-to-late twentieth century. From Egypt and India in the 1940s to Saudi Arabia and El Salvador

Epilogue 355

in the 1970s, postcolonial nations put hydropower images on their banknotes as a claim of modernity. The only nation from Western Europe to deploy a similar iconography was Austria, whose 1,000-schilling bill from 1961 included a hydropower dam; from Eastern Europe, only Romania in 1952 and Czechoslovakia in 1960 did the same. These particular nations were rebuilding infrastructure after the Second World War and wanted to show how far they had come in doing so. The hydropower dam became an ultimate symbol of energy freedom, economic independence, and nationhood. The images on these banknotes show human control of technology, the environment, and bodies in order to claim the status of modernity.

The Canadian deployment of this image, as well as the other technological images in the 1935 series of banknotes, tells us that the Canadian claim to modernity is more like those of "developing" nations than we might have considered before. What similarities exist between Canada in 1935 and India, Egypt, and other such countries in 1940? Why is this? Perhaps similarities exist because all of these countries came into their own, solidifying as nation-states, in the Anthropocene. Perhaps it is because they all have colonial legacies, so maybe Canada needs to be thought of as a postcolonial space of development. Perhaps they make the same kinds of claims to modernity because of the prominent role of engineers or bureaucrats in their governments. Perhaps they all share a desire for the legibility of local environments and peoples in order to compete in the global marketplace. These are suggestions that would need to be tested.

So, as a final provocation to close this volume, I want to propose that looking at the contestations over science and technology in the making of these younger nation-states would be a relevant comparison for understanding the making of modern Canada. Their histories would be instructive because nations really *of* the Anthropocene are infused with a mentality of complete human domination of environments, technologies, and bodies. An Anthropocenic nation, then, is both a way of being and a way of understanding the world. The nations that came into being in the Anthropocene never *became* modern – they were *born* modern – and never knew a premodern state. Canada is one of these nations.

NOTES
1 Paper money actually appeared first in twelfth-century Tang China, but it was not replicated elsewhere until much later. The first banknotes in Europe were issued in 1661 by Sweden. Until the mid-1800s, most banknotes were issued by private independent banks (thus the term "banknotes") rather than by the nation-state.

2 James C. Scott, *Seeing like a State: How Certain Schemes to Improve the Human Condition Have Failed* (New Haven, CT: Yale University Press, 1998).
3 Jacques E.C. Hymans, "International Patterns in National Identity Content: The Case of Japanese Banknote Iconography," *Journal of East Asian Studies* 5, 2 (2005): 315–46.
4 Emily Gilbert, "'Ornamenting the Façade of Hell': Iconographies of 19th-Century Canadian Paper Money," *Environment and Planning D: Society and Space* 16, 1 (1998): 57–80.
5 The image on the front of the note is a portrait of Edward, Prince of Wales, the first son of King George V and Queen Mary, in a colonel's uniform. Both front and back engravings were made by master engraver Harry P. Dawson.
6 Gilbert, "'Ornamenting the Façade of Hell.'"
7 Ibid. A Canadian example of this turn to antiquity and allegory can be seen in a print issued by the Robert Simpson Company in conjunction with the Diamond Jubilee celebrations of 1927. In the image, allegories of abundance and fertility in classical garb give their gifts to a personified Canada. The image is reproduced in Jane Nicholas, "Gendering the Jubilee: Gender and Modernity in the Diamond Jubilee of Confederation Celebrations, 1927," *Canadian Historical Review* 90, 2 (2009): 268.
8 Matthew D. Evenden explores the conflict between hydropower development and fish bodies in *Fish versus Power: An Environmental History of the Fraser River* (Cambridge, UK: Cambridge University Press, 2004).
9 For more on the labour of Canadian women, see, for example, R.W. Sandwell, *Canada's Rural Majority: Households, Environments, and Economies, 1870–1940* (Toronto: University of Toronto Press, 2016); and Joy Parr, "What Makes Washday Less Blue? Gender, Nation, and Technology Choice in Postwar Canada," *Technology and Culture* 38, 1 (1997): 153–86.
10 Nicholas, "Gendering the Jubilee," 266.
11 Photographs of those two dams are available from the Glenbow Museum's photographic archive, http://ww2.glenbow.org/search/archivesPhotosSearch.aspx. For example, see photographs PD-365-1-93 and PD-365-1-95.
12 See the notes in the chapters by James Hull and Daniel Macfarlane in this volume for references to this copious literature.
13 Images of the entire 1935 series are available from the Bank of Canada Museum, https://www.bankofcanadamuseum.ca/complete-bank-note-series/1935-first-series/.
14 The chapters in R.W. Sandwell, ed., *Powering Up Canada: The History of Power, Fuel, and Energy from 1600* (Montreal/Kingston: McGill-Queen's University Press, 2016), show that Canadians tended to use many different energy sources even after "modern" electricity and fossil fuels became available. In this context, the images of an older steam train and a new electric train as both representing "modern" Canada on the currency series make sense.
15 See Nelly Oudshoorn and Trevor Pinch, eds., *How Users Matter: The Co-Construction of Users and Technology* (Cambridge, MA: MIT Press, 2003); and Thomas Berker, Maren Hartmann, Yves Punie, and Katie J. Ward, *Domestication of Media and Technology* (Maidenhead, UK: Open University Press, 2006).

16 Karl Froschauer, *White Gold: Hydroelectric Power in Canada* (Vancouver: UBC Press, 1999), 57.
17 Laurel Sefton MacDowell, *An Environmental History of Canada* (Vancouver: UBC Press, 2012), Chapter 4.
18 I have used the most-referenced catalogues for paper money enthusiasts to perform searches for currency featuring this kind of imagery: *Standard Catalog of World Paper Money, General Issues 1368–1960*, 12th ed., ed. George S. Cuhaj (Iola, WI: Krause, 2008); and *Standard Catalog of World Paper Money, Modern Issues 1961–Present*, 14th ed., ed. George S. Cuhaj (Iola, WI: Krause, 2008).
19 Although originally proposed as a geological era by Paul J. Crutzen and Eugene F. Stoermer in "The Anthropocene," *Global Change Newsletter* 41 (2000): 17–18, and in Paul J. Crutzen, "Geology of Mankind: The Anthropocene," *Nature* 415 (2002): 23, the term "Anthropocene" has come to represent the human ability to modify the Earth's ecosystem on large scales, including biologically. See, for example, Will Steffen, Jacques Grinevald, Paul Crutzen, and John McNeill, "The Anthropocene: Conceptual and Historical Perspectives," *Philosophical Transactions of the Royal Society A* 369 (2011): 842–67; and Rudolfo Dirzo, Hillary S. Young, Mauro Galetti, Gerardo Ceballos, Nick J. B. Isaac, and Ben Collen, "Defaunation in the Anthropocene," *Science* 345 (2014): 401–6.
20 For an overview of the dating options, see Simon L. Lewis and Mark A. Maslin, "Defining the Anthropocene," *Nature* 519 (2015): 171–80. The initial recommendation of the Working Group on the Anthropocene of the International Commission on Stratigraphy was to focus geological inquiry defining the Anthropocene on the nuclear age, which began with nuclear detonation and testing at the end of the Second World War.
21 We should remember that, though Canada celebrated its 150th anniversary in 2017, Europeans have been settling in its territories since the early sixteenth century, and Indigenous populations have lived in the land now known as Canada for thousands of years. The anniversary thus marked the founding of a "modern" nation, making it appropriate as a reference point for this book on modern Canada.
22 Dipesh Chakrabarty, "The Climate of History: Four Theses," *Critical Inquiry* 35, 2 (2009): 208.
23 Pierre Bélanger, ed., *Extraction Empire: Sourcing the Scales, Systems, and States of Canada's Global Resource Empire* (Cambridge, MA: MIT Press, 2017).
24 Chakrabarty discusses the issue of rethinking capitalist globalization as an Anthropocenic phenomenon. Some scholars have even advocated calling the period "Capitalocene." See Donna Haraway, "Anthropocene, Capitalocene, Plantationocene, Chthulucene: Making Kin," *Environmental Humanities* 6 (2015): 159–65; and Jason Moore, "The Capitalocene, Part I: On the Nature and Origins of Our Ecological Crisis," *Journal of Peasant Studies* 44, 3 (2017): 594–630.

Contributors

Tina Adcock is a cultural and environmental historian of modern Canada and an assistant professor of history at Simon Fraser University. Her research considers the relationship between colonialism, modernity, and the production of knowledge, with a special focus on the twentieth-century North. She has published work in Swedish, Norwegian, Canadian, and American scholarly journals and volumes and is currently completing a project on the cultural history of northern Canadian exploration between 1920 and 1980. She is an associate (2017–20) of the L.R. Wilson Institute for Canadian History at McMaster University.

Stephen Bocking is a professor of environmental policy and history in the School of the Environment at Trent University. He teaches courses on science and politics, environmental history, and global political ecology. His research examines the evolution of environmental knowledge and politics in a variety of contexts, including northern Canada, the salmon aquaculture industry, and biodiversity conservation. His publications include *Ice Blink: Navigating Northern Environmental History* (co-edited with Brad Martin); *Nature's Experts: Science, Politics, and the Environment*; *Biodiversity in Canada: Ecology, Ideas, and Action*; and *Ecologists and Environmental Politics: A History of Contemporary Ecology*.

Dorotea Gucciardo is the president of the Canadian Science and Technology Historical Association. She works in the Academic Dean's Office and teaches history at King's University College in London, Ontario. She has guest-edited themed issues for *Scientia Canadensis* and the *Bulletin of Science, Technology, and Society*, and she has published in *ICON* and *A Companion to Women's Military History*. She is currently working on a new manuscript, "Modernizing Canada: Electricity and Everyday Life."

Jan Hadlaw is a historian of technology, media, and design whose publications focus on the material and visual culture of twentieth-century technologies. She is the co-editor of *Theories of the Mobile Internet: Materialities and Imaginairies*, and her work has been published in *Space and Culture*, *Material Culture Review*, *Média et Information*, and *Design Issues*. Her current research project is a cultural and business history of the modern telephone in the United States.

James Hull is a member of the Department of History and Sociology at the Okanagan Campus of the University of British Columbia. A former editor-in-chief of *Scientia Canadensis*, he has published dozens of articles and book chapters on topics ranging from the history of industrial research to Canada-US economic integration. He is currently working on a study of British Columbians' responses to the Canadian Manufacturers Association's "Made-in-Canada" campaign.

Edward Jones-Imhotep is a cultural historian of science and technology and an associate professor of history at York University. His work explores the intertwined histories of technology, trust, and social order in modern Europe and North America. His first book, *The Unreliable Nation: Hostile Nature and Technological Failure in the Cold War*, won the Sidney Edelstein Prize in the history of technology. His next book, *Reliable Humans, Trustworthy Machines*, explores how people throughout the modern period experienced machine failures as a problem of the self.

Dolly Jørgensen is a professor of history at the University of Stavanger, Norway. She is an environmental historian broadly interested in exploring the intersections of animals, technologies, and humans. Her publications include *New Natures: Joining Environmental History with Science and*

Technology Studies (co-edited with Finn Arne Jørgensen and Sara B. Pritchard), *Northscapes: History, Technology, and the Making of Northern Environments* (co-edited with Sverker Sörlin), and *Visions of North in Premodern Europe* (co-edited with Virginia Langum).

Eda Kranakis is a professor in the Department of History at the University of Ottawa and a past president of the Canadian Science and Technology Historical Association (CSTHA). Trained in history of science and technology, she has published on topics in history of intellectual property, aviation history, and history of engineering. She was awarded the 2015 Jarrell Prize by the CSTHA and the 1985 Usher Prize by the Society for the History of Technology. Her book *Constructing a Bridge: An Exploration of Engineering Culture, Design, and Research in Nineteenth-Century France and America* was designated an Outstanding Academic Book of the Year by the American Library Association's *Choice Magazine*.

Daniel Macfarlane is an assistant professor in the Institute of the Environment and Sustainability at Western Michigan University. He is the author of *Negotiating a River: Canada, the US, and the Creation of the St. Lawrence Seaway* and co-editor (with Lynne Heasley) of *Border Flows: A Century of the Canadian-American Water Relationship*. He is currently writing a book on the history of engineering Niagara Falls and co-editing a collection (with Murray Clamen) on the history of the International Joint Commission.

Beth A. Robertson is a feminist historian of science, technology, and medicine. She currently teaches in the Department of History at Carleton University and is the author of *Science of the Seance: Transnational Networks and Gendered Bodies in the Study of Psychic Phenomena, 1918–40*. Robertson's awards include, among others, a Social Sciences and Humanities Research Council Doctoral Fellowship, a Mitacs Accelerate Postdoctoral Fellowship, and the Automated Computer Machinery History Fellowship.

Efram Sera-Shriar is a lecturer in modern history at Leeds Trinity University, UK. He is interested in the intersection of voyages of exploration, race, visual culture, science, medicine, and society throughout the British Empire from the late eighteenth century to the early twentieth century. He has published extensively on the history of the human sciences, including his books *The Making of British Anthropology, 1813–1871* and *Historicizing*

Humans: Deep Time, Evolution, and Race in Nineteenth-Century British Sciences. His current research examines British anthropology's engagement with modern spiritualism during the late Victorian era.

Blair Stein is a PhD candidate in history of science, technology, and medicine at the University of Oklahoma. She is especially interested in the connections between environment, technology, identity, and the uneven experiences of modernity. Her work has appeared in *Technology and Culture, Scientia Canadensis,* and the *Journal for the History of Astronomy,* and she currently manages social media for the Canadian Science and Technology Historical Association.

Andrew Stuhl is an assistant professor of environmental humanities at Bucknell University. He is the author of *Unfreezing the Arctic: Science, Colonialism, and the Transformation of Inuit Lands* and of several articles, essays, and book chapters on the intertwined colonial and environmental histories of the western Arctic. He is a receipient of the 2017 Young Scholars Prize from the International Union of History and Philosophy of Science and Technology.

David Theodore is Canada Research Chair in Architecture, Health, and Computation in the Peter Guo-hua Fu School of Architecture, McGill University. His research focuses on architectural history, hospital life, digital technology, and social theory. He has published on healthcare and architecture in *Social Science and Medicine, Technology and Culture,* and the *Canadian Medical Association Journal.* He also writes on contemporary design issues and currently serves as a regional correspondent for *Canadian Architect* and as a contributing editor at *Azure*.

Index

Note: "(i)" following a page number indicates an illustration; "(t)" indicates a table.

Adams, Wellington, 97
Adcock, Tina, 12, 18, 19, 20, 86, 254, 260, 373*n*7, 301*n*23, 304*n*48, 350
administration: environment as object of, 266–67; in relation between science and modernity, 253, 257
Advisory Board of Wild Life Protection, 67, 73–74, 76
Agnew, Harvey, 108–9
agriculture: "green revolution," 217; impact of technoscience on, 23; modernity and, 216–17, 233; seed regimes, 237, 239*n*1; seed saving, 224–25, 228, 229, 237, 247*n*108
Air Canada. *See* Trans Canada Air Lines (TCA)
air travel: Canadian identity and, 319–20; development of, 311, 315; as experience of modernity, 306, 308, 311–12; promotion of sun destinations, 317–18; in winter, 314–15
Akaitcho (chief), 45, 46, 48, 49
Alaska Boundary Dispute, 63

Albanese, Catherine, 104
Alberta: environmental concerns, 266; resource industries, 265–66
Alexander, Jennifer Karns, 132
Algonquin Provincial Park: multiple-use management, 258, 259; recreational opportunities of, 264
American Geographical Society (AGS), 69
American Museum Greenland Expedition, 66–68
American Museum of Natural History (AMNH), 66–67, 69, 74, 75, 76
American Telephone and Telegraph Company (AT&T), 146, 147, 153, 154
Anderson, Rudolph Martin, 282, 283, 284, 286–87, 288, 291–92, 293, 294, 295, 299*n*2, 300*n*17, 304*n*47
animal bodies, as boundary objects, 350
Anthropocene: concept of, 253, 272, 353–55, 357*nn*19–20
antimodernism, 8, 9–10, 30*n*33, 260

Index

Apostoli, George, 93
Arctic: boundary-work in, 62, 63–64; British expeditions to, 42; bush flying in, 307, 310; Canada's claim of sovereignty in, 63, 64, 65–66, 74, 76–77, 280–81, 290, 294; in Cold War geographies, 17; division into life zones, 290, 291, 292; environmental issues, 268; exploration of, 19, 63–64, 67–68, 77, 281; geopolitical and scientific imaginaries of, 294; as international territory, 257; natural history of, 290–91; new vision of, 24; political and economic importance of, 290, 291; RCMP posts in, 63; reference works on, 290; study of landscapes of, 24
Arctic Biological Committee, 282, 283, 286, 288, 300n15, 302n30
Arctic Publications Committee, 300n15
Armstrong, Christopher, 133, 164n36
"assured destruction," idea of, 200–1
Atlantic Missile Range, 195(i)
AT&T. *See* American Telephone and Telegraph Company
atomic power: in Canada, use of, 120n55; first nuclear test, 109; Hiroshima and Nagasaki bombings, 113; idea of "peaceful," 113; medical applications of, 113–14; optimistic view of, 113, 115; paranormal phenomena and, 105, 108–9, 113; in public imagination, 109, 116; theories of, 20
Augustus (Inuk interpreter), 47
aviation: bush flying, 310; development of, 307; innovations in, 24; technology, 313
Axel Heiberg Island, 63, 64

Back, George, 42–47
Baker, Hugh Cossart, Jr., 144
Ball, Norman, 130
Banff National Park, 263

banknotes: 1935 $5 bill, 349–50, 350(i), 353; agriculture allegory on, 351; allegorical Electricity on, 352, 353; allegorical Science on, 353; allegories of abundance and fertility on, 356n7; allegory of modern innovation on, 351–52; introduction of, 348, 355n1; nature images on, 352–53; regular redesign of, 348–49; science and technology images on, 349–50, 351, 354–55; as symbol of modernity, 349–50
Barbados: commemorative postage stamps, 198; HARP installation in, 185, 193, 196–98, 198(i), 199; McGill University activities on, 17, 212n47; radar and telemetry infrastructure, 196, 197(i); strategic location of, 195(i), 196, 197, 197(i)
Barnard, Monroe Grey, 81n45
Bartholow, Robert, 91
Bartlett, Bob, 71, 72, 81n52
BASF corporation, 127, 237
Beard, George, 87, 95, 99n15
Becker, William, 329
Beer, Jeremy de, 232, 233
Bélanger, Pierre, 15
Belisle, Donica, 134
Bell, Alexander Graham, 164n36, 309
Bell Canada Telephone Company: advertising, 21–22, 149(i), 154; communication with subscribers, 155–56; conversion to automatic system, 145–48, 150–51, 152–53, 154, 160–61, 163n16; employment policy, 148–49, 155, 163n20; expansion of, 163n10; labour costs, 146; leasing policy, 147–48; operator protocol, 151; promotional materials, 154, 156(i); public criticism of, 151; quality of equipment, 147; storefront displays, 156–58, 157(i); "subscriber problem," 148; telephone etiquette, 150; user education, 153–54

Bellairs, Carlyon, 196
Bell's subscriber education program: demonstrations of dial service, 158–59, 159(i), 160(i), 160–61, 164n36, 165n47; how-to-dial instructions, 154–55, 156(i); one-on-one training, 158, 159; overview, 161
Benjamin, Foster H., 286, 287
Benson, Clara, 134
Berger, Carl, 14, 306
Berger, Thomas, 269
Berman, Marshall, 85
Bhimull, Chandra, 312
big building, concept of, 173
big science, 22, 170–71, 173, 177, 179–80, 180n9, 331–32
Biggs, Lindy, 131
Binnema, Ted, 14
Binnie, Ian, 230, 232
biofact, notion of, 240n11
biogeography, 289
Biological Board of Canada, 258
birds, as boundary objects, 70–71, 78
Bledstein, Burton, 134
Bocking, Stephen, 10, 23, 24, 289, 307, 343n3, 346n29, 353
body(ies): of animals as boundary objects, 70–71, 78, 350; atomic energy and, 105, 111, 114, 115–16, 120n57; as battery, idea of, 86, 88; in Canadian $5 bill allegory, 349–50, 351; electrotherapeutic treatment of, 90–91, 92; environment and, 87; in ethnographic writings on Indigenous peoples, 39–41, 48–49; idea of electricity as natural life force of, 85–86, 88, 90, 96, 97; idea of rejuvenation of, 105, 111, 115; Lacey's teachings on transformation of, 105, 110, 111–12, 114, 115; modernity and, 20; notion of "etheric," 111; occultists' hypotheses about, 107–8, 111, 116; power of, 114; scholarship, 18–19, 111
Bothwell, Robert, 14

Bowker, Geoffrey, 61, 62
Bowler, Peter, 289
Bowman, Isaiah, 69
Brain, Robert, 108
Bravo, Michael, 44
British Columbia: ecology studies in, 260; resources management, 259; "sustained yield" policy, 261
Browne, Janet, 289
Brush, Stephen G., 109
Buffalo Society of Natural Sciences, 69
Bull, Gerald: as arms dealer, 23, 193, 199, 203, 207, 208n1; assassination, 187, 206–7, 209n15; autobiography, 185; background and education, 186, 188; business affairs, 203–4, 206; career, 17, 188, 191, 207; CIA reports about, 206; criticism of, 202–3; idea of cannon-launched satellites, 191, 192–93, 201–2, 202(i), 207–8; indictment of, 203, 208n1; inventions, 22, 185, 186, 207; involvement in HARP, 196, 198–99; modernist vision, 187; "Project Babylon," 204–5; reaction to Cuban Missile Crisis, 199; scientific research, 185–86, 188–89, 191, 201–2, 215n88; technical genealogy of inventions, 208n1; US citizenship, 186, 203; view of Barbados, 197–98
Bull's supergun: ambiguity of, 207–8; CIA reconstruction of, 205(i), 206; depiction of, 204(i); development of, 22, 185, 188–89; flight envelopes for, 193(i); as scientific instruments, 187
Byrd, Richard, 63, 64, 67, 68
Bythell, D.C., 314

Canada: as Anthropocenic nation, 354, 355; claim of Arctic sovereignty, 63, 64, 65–66, 74, 76–77, 280–81, 290, 294; commissioner of patents, 219, 222–23; diplomacy, 74; economic development of, 17–18, 303n39,

307; electrical era in, 15, 84–85; energy sources, 356n14; environmental history, 16, 23, 252; federal courts, 216, 217, 219, 222–23, 229, 230–31, 232–33, 235–37; geography and landscapes, 12, 24, 255; hydroelectric development, 85, 127, 329, 351; modernization of, 8–9, 11, 21, 23, 351, 357n21; national vision of progress, 13, 15, 351; nationalism, 318, 320; natural resources, 353; patent rules, 219–20; regulatory state in, 133; technological nationalism, 305, 328; transnational histories of, 17–18; travel deficit, 320

Canada Centre for Inland Waters, 265

Canada Lancet, 88, 92, 93, 99n15

Canadair DC-4M North Star: all-weather flying technologies, 312–13; first service at TCA, 323n33; introduction of, 308; promotion of, 308, 309, 315; significance for nation-building, 321; as technological hybrid, 312; as winter carrier, 314

Canadian Arctic Expedition (1913–18): Arctic imaginaries and, 290; biology and, 290; claims making during, 299n4; exhibition in Canadian Museum of Civilization, 304n51; historiography, 279, 280; importance of, 279–80; influence on national identities, 291; lasting impacts of, 295; press coverage, 279; reports and travelogues about, 280, 299n3, 299n4; scientists recruited to analyze the collection of, 283, 284(t)

Canadian Arctic Expedition Committee, 300n15

Canadian Armament Research Establishment (CARDE), 188, 191

Canadian courts: scientific illiteracy of, 230–31

Canadian Field-Naturalist (Anderson), 291

Canadian Forest Product Laboratories, 131, 261

Canadian identity: climate and, 24–25; construction of, 306–7; geography and, 25, 320; modern technology and, 321; place and, 319–20; water and, 327

Canadian Manufacturers' Association, 131, 133

Canadian science and technology: historiography, 12–13, 16–17; impact of the First World War on, 128–29; inventions, 23–24; marginality of, 18; in relation to modernity, 6, 7, 20–21, 24, 355; in social and cultural development of Canada, role of, 14, 25–26; system of technical standards, 133; transnational nature of, 13, 15–16

Canadian universities, 128–29, 135, 183n40, 261, 263, 266, 283

Canadian Wildlife Service, 256–57, 261, 265

canola, 227–29, 239n4, 243n60. *See also* Roundup Ready® canola

Caribbean islands: military-scientific geography of, 194–95, 198–99, 212n57; vacationing in, 315, 317–18

Castle, William R., Jr., 74, 75

Castonguay, Stéphane, 128

Cavell, Janice, 280, 294

Chakrabarty, Dipesh, 5, 354

Chalk River nuclear laboratories, 120n55

Chandler, Alfred, 130

Charland, Maurice, 305, 306

Chavasse, Henry, 92

Chinese dam building, 341

Chipewyan people, 42, 44, 49, 50, 51

circulation: historical scholarship on, 281–82

Clark, Daniel, 87

Clark, E. Ann, 233

classification, as key practice of modernity, 61, 62

Cold War, 105, 107, 112, 186, 187, 195, 196–208, 255, 294, 326–28, 337, 339
College of Physicians and Surgeons, 100n33
Collins, H.M., 170
Comacchio, Cynthia, 9
Commission of Conservation, 258, 261
computers in medicine, use of, 167–68
conservation, 10, 71, 72, 73–74, 135, 258, 261, 264
Convention for the Protection of Migratory Birds, 73, 260
Cormack, Patricia, 307
Corn, Joseph J., 152, 153, 154
Cosgrave, James, 307
cosmic rays, introduction of idea of, 112
Courtwright, David, 312
Coventry, Arthur, 260
Craciun, Adriana, 295
Cree people, 42, 44, 49, 50, 51
Creighton, Donald, 327
Cronin, Marionne, 310
Crosbie, John, 25, 320
Crysler's Farm, Battle of, 333
Cuban Missile Crisis, 199

Dall, William H., 287
Dansereau, Pierre, 275n33
David Goes to Baffin Land (D.B. Putnam), 70–72, 73(i), 81n45
David Goes to Greenland (D.B. Putnam), 68
David Suzuki Foundation, 269
Defence Research Board (DRB), 191, 192, 200, 201, 255
Dene, 42, 44, 49, 50, 51, 294
Desbarats, G.J., 302n30
Desbiens, Caroline, 305
Desrosiers-Lauzon, Godefroy, 318, 320
dial telephony: advantages of, 164n23; geography of, 154; introduction of, 145, 146; vs manual telephony, 163n19, 164n28; newspaper coverage of, 155, 160, 163n11; promotion of, 144, 145, 150–51; social impact of, 143–44, 145–46, 162n6; system of phone numbers, 152, 155, 164n29. *See also* telephony
Digital Equipment Corporation (DEC) PDP-12 minicomputer, 166, 167(i), 175, 176(i)
digital imaging techniques, 176, 182n36
disruption: of environments, 268; in human-nature relations, 253–54, 268; of knowledge, 268–69; of politics of modernity, 269–70
Dixon, Doris, 231
Doel, Ronald E., 62
Doxiadis, Constantinos, 174
Drummond, Ian, 125, 135
Dudley, Charles, 127
Dummitt, Christopher, 8

Eastern Arctic Patrol, 63, 64, 65
ecology, 260, 268–69
ectoplasm, 107–9
Edison, Thomas, 96, 110
Edzo (Dogrib chief), 46
efficiency: as cultural construct, 132; as virtue, 135, 136
Einstein, Albert, 109, 119n31, 189
electric technologies, 96–97, 98
electric vibrator, 94
electricity: allegory of, 350–51; belief in mysterious force of, 87; development of hydro-, 130; early vision of power of, 84; in form of communication, 85; impact on social habits, 352; interest in medicinal qualities of, 85
electrochemistry, 130
electroencephalography (EEG), 166, 176
electrotherapy: criticism of, 97; description of practice of, 85, 89, 90; electric bath treatment, 90–92; experiments with, 86; in gynecology, use of, 92, 93–94; handbook on, 89; lack of information about, 90;

popularity of, 88–89; portable devices for, 95–96, 97; practitioners of, 97; pseudoscientific accounts of, 86; rise and fall of, 20, 97; treatment of specific afflictions and ailments, 92–93, 95; university courses on, 89–90
Elton, Charles, 260
environment: Canadians' relations with, 23, 251, 263; disruption of, 87, 268, 272; expansion of capitalism into local, 262, 263; geographies of hostile, 17; government programs and, 266–67; human health and, 87, 268; knowledge and, 268, 269, 270, 272, 276n34; in knowledge production, role of, 179; modernity and, 135, 253, 267, 272; as object of administration, 24, 253, 264, 266–67, 270; protection of, 266, 267; redefinition of, 268; scholarly studies of, 12, 23, 251–52, 253, 265, 270–71, 272n2, 299n4; science and, 23; St. Lawrence Seaway's impact on, 337, 338–39, 341, 352; state's authority over, 269; technology and, 306, 314, 342, 343n10, 349, 353, 355. See also Arctic
environmental science, 265, 266–67, 269–70
environmentalism, 264, 265, 267
Esquimaux people. See Inuit
"etheric body," notion of, 111
ethnographic guidebooks, 47, 58n39
ethnography, 40–41, 43
Ethnological Society of London (ESL), 39, 40
European Patent Office, 220
Evans, Clinton, 227
Experimental Lakes Area, 264, 265
Explorers Club, 63, 67, 77

farmers, 227, 229, 234
Federation of Ontario Naturalists, 260
Feindel, William, 166

Fernow, Bernhard, 257, 261
Ferranti-Packard company, 130
Fife, David, 254
Finnie, O. S., 65, 66, 67, 68, 69, 70, 75, 77
Fisher, John, 305, 311, 321
fisheries: impact of dams on, 277n46; management, 258
Fisheries Research Board, 258, 265
Fisk, Dorothy, 189–90, 190(i), 201
Forest Products Laboratories of Canada, 135
forests: management, 258, 261; transformation of, 257–58
Fox, Kate, 104
Fox, Leah, 104
Fox, Margaret, 104
Foxe Basin, 68
Francis, R. Douglas, 328
Franklin, John, 42, 47, 50–51, 281
Fuller, Buckminster, 174
functionalism, 332

Galilei, Galileo, 189, 190
Galison, Peter, 173, 182n33
Galvan, Jill, 104
gene transfer methods, 228
genetically modified crops, 216, 217
Geological Survey of Canada, 65, 254, 261, 283
germ theory, 87
Gieryn, Thomas, 61
Gilbert, Jess, 340
Gingras, Yves, 14, 126
Gloor, Pierre, 166
Gluck, Carol, 6
glyphosate, 216, 218, 221, 228, 230, 235
glyphosate resistance (tolerance), 219, 220, 221–22, 223, 224, 231, 235
Goodwin, W.L., 129
Gordin, Michael, 107
Gould, Laurence M., 69
Graham, Maxwell, 73
Gray, David, 295
Great Fish River Expedition, 39

Great Fish River survey, 42
Great Lakes, 264, 265, 267
Greeley, Adolphus, 280
Green, Fitzhugh, 67
Green, Venus, 163n18, 163n20
Greenland: economic development of, 292, 293
Greenwood, Jeremy, 128
Griffiths, Marie, 111
Group of Seven, 10
Gucciardo, Dorotea, 14–15, 20, 105, 145, 209n14, 350
gun-launched rockets, 192–93, 193(i)

Hadlaw, Jan, 17, 21, 209n14, 334, 352
Hamilton, T. Glen, 108
Hare, Kenneth, 196
Harkin, J.B., 66, 69, 73
HARP. *See* High Altitude Research Project (HARP)
Harper, Stephen, 269
Harvey, David, 306, 320
Heaman, E.A., 335
Hess, Victor, 112
Hewitt, C. Gordon, 283, 296, 300n15, 300n17, 302n30
Hibbert, Harold, 139n29
High Altitude Research Project (HARP): CIA reports on, 206; description of, 186–87; launch site in Barbados, 17, 185, 196–98, 197(i), 199; test facility in Antigua, 199; withdrawal of funding, 203
high modernism: agriculture and, 217–18, 226, 233–39; concept of, 10–11, 326–27, 341; destructive aspects of, 11; forms of, 340; historiography, 12; megaprojects of, 347n39; negotiated, 11, 25, 327, 340–41, 342; state and, 343n3
Hill, Theo, 196
Hjort, Johan, 261
Hodgkin, Thomas, 40, 54
Hogg, Jonathan, 109
Horsch, Robert, 231, 232

Hudson's Bay Company (HBC), 19, 42, 52–53
Hughes, Thomas, 98n6
Hull, James, 17, 21, 209n14, 258, 259, 261, 328, 351, 356n12
human-wildlife relations, 259–60
Huntsman, A.G., 300n15
Hussein, Kamil, 204
Hussein, Saddam, 17, 203–4
hybrid envirotechnical system, 329, 343n10
Hydro Quebec, 130
Hydro-Electric Power Commission of Ontario (Ontario Hydro), 130, 333, 334(i), 334–35
hysteria: causes and symptoms, 92–93; treatment, 94

Image and Logic (Galison), 173
Indigenous knowledge, 268–69
Indigenous peoples: craftsmanship skills, 51; encounters with modernization, 8; ethnographic studies of, 39, 40, 42; European perceptions of intelligence of, 50, 51; European settlements and, 40, 50, 51–53; methods of civilizing, 53–54; reliance on firearms, 52; starvation, 50; trading system, 44; visual representations of, 48–49; Western perception of, 43
Industrial Revolution, 4, 125
industrialism, 136, 183n47
Innis, Harold, 303n39
insulin, discovery of, 25
intellectual property, 18
International Joint Commission (IJC), 257, 343n7
Inuit, 42, 45–46, 47, 48, 49, 69, 71, 75, 80n37, 268, 294, 295
Inuit Land Use and Occupancy Study, 269
Iraq: Canadian relations with, 17; modernization of, 204
Ireland, 237

Iroquois people, 42, 49
islands: in history of imperialism, 193–94

Jarrell, Richard, 14, 105, 251
Jenness, Diamond, 287, 298, 300n17
Jones Sound, 68
Jones-Imhotep, Edward, 16, 17, 18, 22, 23, 352
Jordan, MacKay, 88

Kahn, Herman, 200
Karafyllis, Nicole C., 240n11
Kelly, Frank, 310
Kelly, Patrick, 221, 230
Kennedy, Frank M., 147, 155, 164n23
Kenny, Nicolas, 8
King, Richard: background and education, 41–42; civilizing mission of, 41, 49–54; on customary trading system, 44; drawings of Indigenous peoples, 48–49; ethnographic research, 39–40, 41, 44, 45, 54–55; expeditions, 19, 39, 41, 54; first meeting with Inuit, 46–47, 47–48; at Fort Reliance, 44; friendship with Akaitcho, 45; on HBC trade practices, 53; on names of geographic landmarks, 45; observation of Indigenous culture, 40, 42–43, 44–46, 47, 50–51, 53; publications, 39, 41; reputation, 43; study of races, 13, 41; travel accounts, 40, 54
King, Richard, Sr., 41
King, William Lyon Mackenzie, 74, 76
Kitchener-Waterloo technology cluster, 104
Knorr Cetina, Karin, 184n50
Kohler, Robert, 61, 282
Koolhaas, Rem, 173
Kranakis, Eda, 11, 17, 18, 23, 138n28, 351
Kuklick, Henrika, 61
Kuyek, Devlin, 239n1

LaBrosse, Edith, 188
Lacey, Thomas: on atoms and atomic energy, 113, 115–16; on development of cyclotrons, 114; on future of science, 114; as leader of spiritualist group, 105–6; lectures of, 105, 106; meetings with spirit guides, 110, 111; predictions of, 111–12; publications of, 105; reference to cosmic rays, 112; "The Room Alone Lectures" recordings, 106; on transformation of the body, 105, 110, 111–12, 114, 115; understandings of materiality, 116
Laing, Decima, 106
Laing, John M., 106
landscapes: transformation of, 11–12, 264
Large Hadron Collider, 170
Latour, Bruno, 246n95, 247n102, 300n10
Lauct, Christoph, 109
Laurier boom, 128
Lavine, Matthew, 109, 112
Lawrence, William, 43–44
Lazonick, William, 139n30
Leopold, Aldo, 260
Levere, Trevor, 14, 251, 279
Lewis, Harrison F., 70, 72, 74
Lewis, Wilfred Bennett, 120n55
"life zone" concept, 290, 291
Lily Dale spiritualist retreat, 105, 117n8
Little Science, Big Science (Price), 170
Livingstone, David, 289
Lodge, Oliver, 119n31
Logan, Robert A., 63
Loo, Tina, 10, 11–12, 340–41
Lost Villages Historical Society, 333
low modernism, 340
Lowe, Graham, 134

Macallum, A.B., 283
MacDonnell, Robert, 89
MacDougall, Robert, 305
Macfarlane, Daniel, 11, 25, 145, 178, 262, 263, 307, 352

MacGill, Elsie, 134
Mach, Ernst, 189
Mackenzie, George P., 64
Mackenzie Valley Pipeline Inquiry, 269
MacMillan, Donald Baxter, 63, 64, 67, 68
MacMillan-Byrd expedition, 62–63, 64, 67, 74, 78
Macoun, J.M., 254, 283, 300n15
Maines, Fred, 117n13
Maines, Minnie, 117n13
Maines, Rachel, 94
Malte, M.O., 300n15
Martin, Michèle, 151, 158, 163n18
Marvin, Carolyn, 144
Massey, Vincent, 75, 76
Matthews, J., 93
Mau, Bruce, 173
maximum sustained yield (MSY), 262
McBryde, W.A.E., 129
McCalla, Douglas, 128, 135
McConnell Brain Imaging Centre, 174
McCullough, Warren, 174
McGregor, Gordon, 313, 315, 323n16, 324n39
McInnis, Marvin, 126, 128, 129, 135
McLaren, Angus, 95
McLauchlin, Donald S., 315
McLuhan, Marshall, 174
McTaggart-Cowan, Ian, 260
Mead, Margaret, 174
medical electricity. *See* electrotherapy
Merriam, Clinton Hart, 290, 291, 294
micro-social experience, 62
middle class, emergence of, 134
Migratory Birds Convention Act, 66, 70, 73, 76
Millikan, Robert, 108, 112
mining, 15
missiles, 194–95. *See also* Atlantic Missile Range
Mitchinson, Wendy, 87, 92
Mitman, Gregg, 87
modernism, 10–11, 27n9. *See also* high modernism; low modernism; ultra high modernism

modernity: agricultural, 216, 217, 226, 233, 234, 237, 238; assumptions about, 226, 235; Canadian, 8, 9, 10, 12, 271–72, 354; challenge of nation-states, 272; characteristics of, 246n95, 252, 253–54, 270; classification as key practice of, 61, 62; critique of, 4–5, 6–7; in disarray, 226, 238, 239; European, 3–4, 13, 354; as global phenomenon, 7; human health crisis and, 87; meaning of the term, 5; potential for resistance within, 253; science and, 252–53, 271; social impact of, 145; technology and, 143, 145, 146, 161, 305–6; as variable phenomenon, 272
modernization, 8, 9, 10, 21, 27n9, 126. *See also* Second Industrial Revolution
molecular biology, 218, 230, 231
money, 348. *See also* banknotes
Monsanto: agrochemical innovations, 218–19, 220–21; contract system, 225–27; CP4 gene, 218, 219, 220–21, 224, 231; Grower's Agreement, 225, 226, 227; impact on political economy of agriculture, 243n62; implementation of Roundup Ready® system, 237; legal tactics, 23, 230, 246n97; plant regeneration techniques, 223–24; pMON546 biofact, 221; prevention of seed saving practice, 229; profit generation, 227, 234–35; Roundup Ready® gene patents, 216, 219–24, 226, 230, 239n5, 242n43; Technology Use Agreement (TUA), 225, 226, 227
Monsanto vs Schmeiser lawsuit: consequences of, 237–38; expert testimonies, 231–32; international impact of, 237–38; Monsanto's lawyers, 245n78; Monsanto's legal victory, 229–30; patent exhaustion situation, 232; press coverage, 234;

scientific-technical challenge of, 220, 221–22, 231, 232, 234, 235–36; Supreme Court decision, 216, 233, 236–37, 245n84, 247n103
Montreal Neurological Institute (MNI), 22, 166, 174, 175(i), 183n38
Mordell, Don, 191
Morgan, Cecilia, 30n33
Morgan, Isabel, 91
Moses, Robert, 335
Moses-Saunders power dam, 329, 339(i)
Munro, James, 260
Murphy, Charles, 196
mutually assured destruction (MAD), 201

Narrative of a Journey to the Shores of the Arctic Ocean (King), 41, 42, 43, 50
Nash, Linda, 87
National Bureau of Standards, 132
National Physical Laboratory, 132
National Research Council (NRC), 128, 133
nature: complexity and unpredictability of, 268; debates on definition of, 247n102; images on banknotes, 352–53; relations of humans and, 253–54, 268; scientific approach to, 260, 262, 336; transformation of, 258, 262–63
Nelles, H.V., 130, 133, 164n36
neurasthenia, 93, 95
Newell, Dianne, 129
Newton, Isaac, 189, 191, 210n22
Nicholas, Jane, 8, 350
Noakes, Jeff, 294
North Star. *See* Canadair DC-4M North Star
Northwest Game Act, 65, 66
Northwest Passage, 42, 281
Northwest Territories: management of natural resources, 74; non-British expeditions to, 65; regulation of exploration of, 62, 66–67

Northwest Territories Act, 60–61
Nyhart, Lynn K., 285

occult science, 107, 109–10, 111
Oldenziel, Ruth, 194
one-man research team, idea of, 169
Ontario: environmental issues, 266; forestry resources, 257–58
Ontario Fisheries Research Laboratory, 255
Ontario Hydro. *See* Hydro-Electric Power Commission of Ontario (Ontario Hydro)

Pacey, Arnold, 129
Pacific Proving Grounds, 194
Padgette, Stephen, 219
Palladino, Eusapia, 108
Pálsson, Gísli, 279
Paris Guns (German long-range guns), 185, 189, 190
Parr, Joy, 11, 12
Parry, William Edward, 46
Passfield, Robert, 329
Patent Act, 236
Patent Cooperation Treaty, 240n15
patent exhaustion, principle of, 232–33
patents, 23, 230, 241n22, 247n103
Pelton, Guy Cathcart, 84
Peña, Carolyn Thomas de la, 88, 95
Perry, Adele, 18
Pietz, David, 341
Pincock, Jenny O'Hara, 117n13
Pioneer Hi-Bred Ltd. v Canada (Commissioner of Patents), 222–23, 232
Piper, Liza, 305
Politics of Development (Nelles), 130
Polymer Corporation, 130
Pope, John A., 73(i), 80n37, 81n45
Pospisil, Francis, 287
Poulter, Gillian, 307
Power Authority of the State of New York (PASNY), 335, 339
Pratt, Mary Louise, 43
Pratte, J., 223

Price, Derek John de Solla, 170, 171
Prince, E.E., 283, 300n17
production revolution, 129–32
Project Babylon, 187, 204–5, 205(i), 206
projectiles, 192
protoplasm, 108
Putnam, David Binney: consumption of wildlife, 72–73; *David Goes to Baffin Land*, 70–72, 73(i), 81n45; as explorer, 68, 69; hunting episodes, 71–72; publicity, 73(i), 76; scientific collecting, 67, 69–70
Putnam, George Palmer: Arctic expeditions, 13, 19, 60, 62, 66, 68–70, 77; attitude to hunting, 71; boundary-work of, 67–68, 75; cooperation with Canadian officials, 67, 68–69, 75–76; correspondence with Finnie, 68, 70, 75; dispatches to the *New York Times*, 68; as elite sportsman and tourist, 19, 20, 62, 66, 71–73; fall from grace, 78; public apology of, 75–76
Putnam Baffin Island Expedition: archaeological and ethnological fieldwork, 69; consumption of wildlife during, 72–73, 81n52; diplomatic scandal around, 63, 74–75, 76; geographical explorations, 69; guns and ammunition carried by, 73(i), 80n37; hunting activities, 70–71, 73(i); medical experiments, 69; negative publicity, 76; oceanographical fieldwork, 69; publications arising from, 70, 75, 81n42; zoological fieldwork, 69–70

radioactive isotopes: discovery of, 113–14, 120n57
RAND Corporation, 200
Rasmussen, Knud, 72
rationality vs reason, 199–200, 213n61
Rawls, John, 213n61
Re Application of Abitibi Co., 219, 220
Red Fife (grain variety), 254

Report of the Canadian Arctic Expedition (1913–1918): archive of, 282; Canadian officials' views on, 285–86, 288; distribution and application of, 280, 282, 286–89, 301n24; economic implications of, 289–90; funding for publication of, 300–300n17; impact on Canadian science, 24, 292, 293; intellectual values of, 281, 285–86, 288–89, 294–95; interest of scientific community in, 282–85, 287, 288; as international project, 283–85, 293–94, 300n17; notions of the Arctic in, 294; plan for, 295–98; political implications of, 289–90, 294; preparation of, 280–84; publication standards, 285, 287–88, 302n27; publishing cost, 283; scrutiny of, 287; volume titles, 288, 289(i), 302n30
Rhine, J.B., 107
Richet, Charles, 107, 108
Richter, Andrew, 200
Rieger, Bernhard, 306
Roberts, Lissa, 281
Robertson, Beth, 12, 20, 209n14, 350
Rockwell, Alphonso D., 88, 95, 99n15
Rosebrugh, Abner Mulholland, 88, 90
Rosenblueth, Arturo, 172
Ross, John, 42
Roundup®, 218. *See also* glyphosate
Roundup Ready® canola: 1996 harvest of, 229; adoption by farmers, 227; commercialization of, 216, 224–25; comparison to conventional canola, 228–29; development of, 216, 217, 218; introduction in Canada, 217; legal case, 23; promiscuous spread, 228; resistance to glyphosate, 235; "value capture" system for, 228
Royal Commission on Industrial Training and Technical Education, 133
Rudy, Jarrett, 8
Rutherford, Ernest, 171

Saad 16 weapons complex, 206
Sandlos, John, 10
Sapir, Edward, 300n15
Saulteaux people, 42, 43, 49, 57n20
Schmeiser, David, 265
Schmeiser, Percy: infringement of the Roundup Ready® gene patent, 220; as modernist farmer, 238; negative publicity about, 238; as symbol of resistance to modernity, 23, 238–39; use of saved seeds by, 224. *See also* Monsanto vs Schmeiser lawsuit
Schultz, Garnet, 106
science: architecture and, 173–74; biology as noncollaborative, 184n50; boundary-work in, 61; Canadian courts and, 230–31, 232, 234–36; characteristics of Canadian, 16, 22, 251, 286; definition of, 5, 264, 272n2; development of, 107, 109–10, 265, 269, 270; environment in history of, 251–52, 270–71, 272; epistemic boundaries of, 66; historical roles of, 251–52; increasing specialization of, 183n42; interdisciplinary, 171, 172, 265; international connections in, 261–62; legitimate and illegitimate practices in, 20, 86; limits and dangers of, 237; modernist agriculture and, 226, 227, 233–36; modernity and, 252–53, 262, 263, 266; occult movement and, 20, 107; political boundaries and, 255, 256–57; professional-amateur boundary in, 62; sport vs, 19, 71, 77–78; transformation of nature and, 258. *See also* big science; Canadian science and technology; small science
Scientists and Explorers Ordinance, 19, 60–61, 62, 64–65, 65–66, 68, 77
Sconce, Jeffrey, 104
Scott, James, 10, 11, 233, 326, 327, 340, 341, 343n3, 347n39, 348

Second Industrial Revolution: in Canadian context, 125–26, 132, 135; consequences of, 86, 351; definition of, 127; economic modernization and, 126, 131; historiography, 125–26, 127, 136n6; impact on workforce, 131–32, 133, 134; origin of, 129–30; periodization of, 126, 127–28; state functions in, 132–33; technological innovations and, 127–28, 258
Seeing like a State (Scott), 335
Segal, Ariel, 109
Senior, John, 90
Sera-Shriar, Efram, 19, 24, 281, 350
Sheldon, E.W., 106
Sheller, Mimi, 317
Shelley, Percy Bysshe, 18
Sherwood, G.W.H., 66
Siegel, Andre, 131
Sinclair, Bruce, 25, 129
Skelton, O.D., 74, 75, 76
Slaton, Amy, 133
small science: architectural and spatial metaphors, 173–74; vs big science, 22, 170–71, 179–80, 331–32; as Canadian phenomenon, 177–78; concept of, 168–69, 176; obscurity of, 179; as practice specific to medical research, 178; profile of scientists involved in, 179; research program of, 176; small spaces and, 174, 178–79; use of small equipment in, 175; value of, 177
Smith, Lapthorn, 90, 93
Smith, Otto G., 106
S,M,L,XL (Mau), 173
Solis, Jeanne Cady, 95
Sowards, Adam, 291, 299n4
Space Research Corporation, 205
spiritualism, 104–5, 107, 109–10, 111
spiritualist group in Kitchener-Waterloo area, 106, 107, 110, 112, 116
St. Lawrence River, 25, 328, 341
St. Lawrence Seaway and Power Project: Canadian nation building and,

327–29, 331; Cold War context of, 339–40; communities affected by, 329, 330(i), 331, 332–33(i), 332–36, 345n20; construction of, 326, 327, 328; cost of, 329; dislocation and resettlement of residents, 335–36; education of residents about, 334–35; engineering process, 329, 331, 336–37, 338; environmental impact, 337, 338–39, 352; as high modernist project, 25, 326, 327, 328, 333, 340–42, 341–42; as hybrid envirotechnical system, 329; map of, 330(i), 332–33(i); Moses-Saunders power dam, 339(i); "negotiated" character of, 342; Ontario Hydro model, 338(i); optimism about, 336; planning of, 333, 337–38, 338(i); publicity, 331, 334; rehabilitated area, 332–33(i); resistance to, 335; significance of, 331; transportation and mobility networks, 333–34
Stanley, Meg, 10, 340
Star, Susan Leigh, 61, 62
Starnes, Cortlandt, 66
State and Enterprise, The (Traves), 133
Stearns, Peter, 125, 132
Stefansson, Vilhjalmur, 280, 292, 294, 295, 299n2, 299n4, 302n30, 304n47
Stein, Blair, 12, 18, 24, 25, 145, 334, 352
Stewart, Keith, 130
Stoermer, Eugene F., 357n19
Stokes, Peter, 345n20
Stowe, Emily, 89
Strange, Carolyn, 8
Streeter, Daniel W., 67
Strowger, Almon B., 146
Stuhl, Andrew, 12, 18, 24, 25, 65, 135, 254, 255, 257, 343n3, 352, 353
Subrahmanyam, Sanjay, 7
Supreme Court of Canada, 216, 222–24, 230–31, 232–33, 235–37
Sutherland, R.J., 201, 213n71
Suzuki, David, 265

Sweeny, Robert, 127
Swift, Jamie, 130
Szostak, Rick, 127

Taverner, Percy, 260, 295
Taylorist scientific management, 131, 132
TCA. *See* Trans Canada Air Lines (TCA)
technology: allegories of, 328; automatic telephone switching, 146; ballistics, 17, 188–90; biological, 216, 218, 219–20, 227, 233, 238, 243n62; Canadian history and, 20–21, 24, 25, 126, 305, 328; Cold War and, 208, 326, 337; communications, 143, 145, 305, 311; definition, 5–6, 131; education and, 146, 153, 158; electrical, 84, 96, 143, 351; emancipatory, 328; firearms, 52, 189–90; high, 169; human control of, 355; hydroelectric, 328; labour and, 4, 84, 93, 146, 352, 359; medical, 18, 20; modern, 256, 258–59, 328, 352; nationalism and, 305, 307, 310, 326; nature controlled through, 336; patented, 227, 232; Second Industrial Revolution and, 127–28, 258; social impact of new, 144–46, 158; transfer, 170. *See also* Canadian science and technology
Tefft, Amelia, 89
Tektronix 4002 terminal, 168(i)
telephone directories, 155
telephone operators: cartoons depicting, 150(i); complaints about, 149–50; as dial service demonstrators, 158, 159(i), 160(i); interactions with telephone users, 148–50, 163n19; public view of, 151; women as, 148–50, 163n18, 164n21
telephony: change from manual to automatic, 143, 146; feeling of distant-intimacy, 151; limitations of traditional, 147; as "personalized service," idea of, 149; shared

experience of, 148; subscription growth, 147. *See also* dial telephony
Telephony (trade journal), 156, 163*n*11
Temin, Peter, 125
Tennessee Valley Authority (TVA), 339, 347*n*39
territory: authority over, 255–57; in history of modernity, 253, 254; issues of provincial and federal, 255–56; mapping of, 254; notion of vertical, 255; relation between knowledge and, 254–55; scientific view of, 254
"textualizing technics," 153, 154
Theodore, David, 16, 18, 22, 209*n*14, 331, 352
Thomas, Lynn, 28*n*14
Thompson, Christopher: career, 166, 167–68, 174, 177, 179; education and experience, 167; expertise in computing, 175–76; research projects, 168, 169, 172, 176
Tomkowicz, Robert, 232, 233
Tomlinson, George, 139*n*29
trained acquaintance, idea of, 169, 171, 172, 173, 174, 176, 177
Trans Canada Air Lines (TCA): advertisements, 310, 314–15, 316(i), 317–18, 319(i); educational campaigns, 314–15; inaugural flight, 309; jet-powered aircraft, 318–19; operating deficits, 313; postwar operation, 322*n*15, 323*n*16; promotional materials, 306, 307–8, 309–12; publicity flights, 305; seasonal imbalances in passenger traffic, 313–14, 324*n*36; selling of the aerial view, 311–12; as state actor, 306; sun destinations, 24, 308, 315–20; as symbol of unity, 309; transcontinental routes, 309
transformation: concept of, 258; in history of modernity, 253
transnational modernity, 280, 293
Traves, Tom, 133

Trent Watershed survey, 257–58
Tropical Research Laboratory, 196
Trout, Jenny, 89
Trudeau, Pierre, 269
Truman, Harry, 195
Turnbull, W.R., 310
Two Worlds (Lacey), 105

ultra high modernism, 340, 342. *See also* high modernism; modernism
United States: acquisition of island territories, 194; diplomatic relations with Canada, 19, 62–63, 64, 67–68, 74–77, 342; exploration of the Arctic, 63–64; nuclear testing, 194
uranium: in Canadian Subarctic, discovery of, 25–26
urban landscapes, 21, 259, 262
user manuals, emergence of, 153

vanillin, production of, 130
Vardalas, John, 130
Vaver, David, 231
Velikovosky, Immanuel, 107
Verne, Jules, 193, 201
Vernon, James, 145, 162*n*6
Vernoy, S., 89
Votolato, Gregory, 313
Vries, Jan de, 134

Walden, Keith, 8
Wallace, Hugh, 41
Wallerstein, Immanuel, 5
weapons, as experimental objects, 189
Webber, Frank G., 147
Weinberg, Alvin M., 171, 180*n*9
White, Lynn, 136
"White Eagle" (spirit guide), 111
Wiener, Norbert, 169, 171, 172–73, 174, 177
wildlife, 260–61
Wilmot, Samuel, 258
Winder, Gordon, 129
Wohlstetter, Albert, 200

women: in building up Canada, role of, 350; as consumers, 134; health and, 89, 92, 93, 94–95; professions of, 134, 158
World Wildlife Fund Canada, 269
Wråkberg, Urban, 62
Wrangell Island, 292
Wright, Sidney, 117n13
Wynn, Graeme, 329

Yuma Proving Grounds in Arizona, 196

Zakreski, Terry, 231, 232
Zeller, Suzanne, 14, 41, 62
Zeman, Scott, 113